Student Solution

Thomas Engel & Philip Reid
University of Washington

PHYSICAL CHEMISTRY

THIRD EDITION

Thomas Engel
Philip Reid

PEARSON

Boston   Columbus   Indianapolis   New York   San Francisco   Upper Saddle River
Amsterdam   Cape Town   Dubai   London   Madrid   Milan   Munich   Paris   Montréal   Toronto
Delhi   Mexico City   São Paulo   Sydney   Hong Kong   Seoul   Singapore   Taipei   Tokyo

Editor in Chief: Adam Jaworski
Executive Editor: Jeanne Zalesky
Senior Marketing Manager: Jonathan Cottrell
Associate Editor: Jessica Neumann
Assistant Editor: Coleen McDonald
Managing Editor, Chemistry and Geosciences: Gina M. Cheselka
Production Project Manager and Manufacturing Buyer: Dorothy Cox
Production Management and Composition: PreMediaGlobal
Cover Design: Seventeenth Street Studios
Artwork courtesy of Kim Kopp; photo credit: Frank Huster

1 2 3 4 5 6 7 8 9 10—EB— 16 15 14 13 12

www.pearsonhighered.com   ISBN-10: 0-321-76668-7; ISBN-13: 978-0-321-76668-7

Contents

1 Fundamental Concepts of Thermodynamics

Numerical Problems

P1.1   Approximately how many oxygen molecules arrive each second at the mitochondrion of an active person with a mass of 84 kg? The following data are available: Oxygen consumption is about 40. mL of O_2 per minute per kilogram of body weight, measured at $T = 300.$ K and $P = 1.00.$ atm. In an adult there are about 1.6×10^{10} cells per kg body mass. Each cell contains about 800. mitochondria.

We first calculate the number of moles of O_2 consumed per unit time and convert this quantity into molecules per unit time using the Avogadro number.

$$n_{O_2} = \frac{PV}{RT} = \frac{1\,\text{atm} \times 40. \times 10^{-3}\,\text{L min}^{-1}\,\text{kg}^{-1}}{8.206 \times 10^{-2}\,\text{L atm mol}^{-1}\,\text{K}^{-1} \times 300.\,\text{K}} \times \frac{1\,\text{min}}{60\,\text{s}} \times 84\,\text{kg}$$

$$= 2.27 \times 10^{-3}\,\text{mol s}^{-1}$$

$$\text{Molecules } O_2 \text{ per mitochronian} = \frac{N_A \times n_{O_2}}{\text{number of cells/kg} \times \text{number of mitochondria/cell} \times \text{body weight}}$$

$$= \frac{2.27 \times 10^{-3}\,\text{mol s}^{-1} \times 6.022 \times 10^{23}\,\text{mol}^{-1}}{1.6 \times 10^{10}\,\text{cells kg}^{-1} \times 800.\,\text{cells}^{-1} \times 84\,\text{kg}} = 1.27 \times 10^{6}\,\text{molecules s}^{-1}$$

P1.2   A compressed cylinder of gas contains 2.74×10^3 g of N_2 gas at a pressure of 3.75×10^7 Pa and a temperature of 18.7°C. What volume of gas has been released into the atmosphere if the final pressure in the cylinder is 1.80×10^5 Pa? Assume ideal behavior and that the gas temperature is unchanged.

Let n_i and n_f be the initial and final number of mols of N_2 in the cylinder.

$$\frac{n_i RT}{P_i} = \frac{n_f RT}{P_f}$$

$$n_f = n_i \frac{P_f}{P_i} = \frac{2.74 \times 10^3\,\text{g}}{28.01\,\text{g mol}^{-1}} \times \frac{1.80 \times 10^5\,\text{Pa}}{3.75 \times 10^7\,\text{Pa}} = 0.470\,\text{mol}$$

$$n_i = \frac{2.74 \times 10^3\,\text{g}}{28.01\,\text{g mol}^{-1}} = 97.8\,\text{mol}$$

The volume of gas released into the atmosphere is given by

$$V = \frac{(n_f - n_i)RT}{P} = \frac{(97.8 - 0.470)\,\text{mol} \times 8.2057 \times 10^{-2}\,\text{L atm mol}^{-1}\text{K}^{-1} \times (273.15 + 18.7)\,\text{K}}{1\,\text{atm}} = 2.33 \times 10^{3}\,\text{L}$$

P1.5   A gas sample is known to be a mixture of ethane and butane. A bulb having a 230.0 cm³ capacity is filled with the gas to a pressure of 97.5×10^3 Pa at 23.1°C. If the mass of the gas in the bulb is 0.3554 g, what is the mole percent of butane in the mixture?

$$n_1 = \text{moles of ethane} \quad n_2 = \text{moles of butane}$$

$$n_1 + n_2 = \frac{PV}{RT} = \frac{97.5 \times 10^3 \text{ Pa} \times 0.230 \times 10^{-3} \text{ m}^3}{8.314 \text{ J mol}^{-1} \text{ K}^{-1} \times (273.15 + 23.1) \text{ K}} = 9.10 \times 10^{-3} \text{ mol}$$

The total mass is $n_1 M_1 + n_2 M_2 = 0.3554$ g.

Dividing this equation by $n_1 + n_2$,

$$\frac{n_1 M_1}{n_1 + n_2} + \frac{n_2 M_2}{n_1 + n_2} = \frac{0.3554 \text{ g}}{9.10 \times 10^{-3} \text{ mol}} = 39.0 \text{ g mol}^{-1}$$

$$x_1 M_1 + x_2 M_2 = (1 - x_2)M_1 + x_2 M_2 = 39.0 \text{ g mol}^{-1}$$

$$x_2 = \frac{37.0 \text{ g mol}^{-1} - M_1}{M_2 - M_1} = \frac{39.0 \text{ g mol}^{-1} - 30.069 \text{ g mol}^{-1}}{58.123 \text{ g mol}^{-1} - 30.069 \text{ g mol}^{-1}} = 0.320 \text{ mole } \% = 32.0\%$$

P1.6   One liter of fully oxygenated blood can carry 0.18 liters of O_2 measured at $T = 298$ K and $P = 1.00$ atm. Calculate the number of moles of O_2 carried per liter of blood. Hemoglobin, the oxygen transport protein in blood, has four oxygen binding sites. How many hemoglobin molecules are required to transport the O_2 in 1.0 L of fully oxygenated blood?

$$n_{O_2} = \frac{PV}{RT} = \frac{1 \text{ atm} \times 0.18 \text{ L}}{8.206 \times 10^{-2} \text{ L atm mol}^{-1} \text{ K}^{-1} \times 298 \text{ K}}$$

$$= 7.36 \times 10^{-3} \text{ mol}$$

$$N_{hemoglobin} = \frac{n_{O_2} \times N_A}{4} = \frac{7.36 \times 10^{-3} \text{ mol} \times 6.022 \times 10^{23}}{4}$$

$$= 1.11 \times 10^{21} \text{ molecules}$$

P1.12   A rigid vessel of volume 0.400 m³ containing H_2 at 21.25°C and a pressure of 715×10^3 Pa is connected to a second rigid vessel of volume 0.750 m³ containing Ar at 30.15°C at a pressure of 203×10^3 Pa. A valve separating the two vessels is opened and both are cooled to a temperature of 12.2°C. What is the final pressure in the vessels?

$$n_{H_2} = \frac{PV}{RT} = \frac{715 \times 10^3 \text{ Pa} \times 0.400 \text{ m}^3}{8.314 \text{ J mol}^{-1} \text{ K}^{-1} \times (273.15 + 21.25) \text{ K}} = 116.8 \text{ mol}$$

$$n_{Ar} = \frac{PV}{RT} = \frac{203 \times 10^3 \text{ Pa} \times 0.750 \text{ m}^3}{8.314 \text{ J mol}^{-1} \text{ K}^{-1} \times (273.15 + 30.15) \text{ K}} = 60.38 \text{ mol}$$

$$P = \frac{nRT}{V} = \frac{(116.8 + 60.38) \text{ mol} \times 8.314 \text{ J mol}^{-1} \text{ K}^{-1} \times (273.15 + 12.2)\text{K}}{(0.400 + 0.750) \text{ m}^3} = 3.66 \times 10^5 \text{ Pa}$$

P1.15   Devise a temperature scale, abbreviated G, for which the magnitude of the ideal gas constant is 5.52 J G⁻¹ mol⁻¹.

Let T and T' represent the Kelvin and G scales, and R and R' represent the gas constant in each of these scales. Then

$$PV = nRT = nR'T'$$

$$T' = \frac{R}{R'}T = \frac{8.314}{5.52}T = 1.51T$$

The temperature on the G scale is the value in K multiplied by 1.51.

P1.16   Aerobic cells metabolize glucose in the respiratory system. This reaction proceeds according to the overall reaction

$$6O_2(g) + C_6H_{12}O_6(s) \rightarrow 6CO_2(g) + 6H_2O(l)$$

Calculate the volume of oxygen required at STP to metabolize 0.025 kg of glucose ($C_6H_{12}O_6$). STP refers to standard temperature and pressure, that is, $T = 273$ K and $P = 1.00$ atm. Assume oxygen behaves ideally at STP.

From the stoichiometric equation, we see that 6 mols of O_2 are required for each mol of glucose. Therefore

$$V_{O_2} = \frac{n_{O_2} RT}{P} = \frac{\dfrac{6 \times 25\,\text{g}}{180.18\,\text{g mol}^{-1}} \times 8.206 \times 10^{-2}\,\text{L atm mol}^{-1}\,\text{K}^{-1} \times 273\,\text{K}}{1\,\text{atm}} = 18.6\,\text{L}$$

P1.18   A mixture of 2.10×10^{-3} g of O_2, 3.88×10^{-3} mol of N_2, and 5.25×10^{20} molecules of CO are placed into a vessel of volume 5.25 L at 12.5°C.

   a.   Calculate the total pressure in the vessel.

   b.   Calculate the mole fractions and partial pressures of each gas.

(a) $n_{O_2} = \dfrac{2.10 \times 10^{-3}\,\text{g}}{32.0\,\text{g mol}^{-1}} = 6.56 \times 10^{-5}\,\text{mol}$;

   $n_{CO} = \dfrac{5.25 \times 10^{20}\,\text{molecules}}{6.022 \times 10^{23}\,\text{molecules mol}^{-1}} = 8.71 \times 10^{-4}\,\text{mol}$

   $n_{total} = n_{O_2} + n_{N_2} + n_{CO} = 6.56 \times 10^{-5}\,\text{mol} + 3.88 \times 10^{-3}\,\text{mol} + 8.72 \times 10^{-4}\,\text{mol} = 4.82 \times 10^{-3}\,\text{mol}$

   $P_{total} = \dfrac{nRT}{V} = \dfrac{4.82 \times 10^{-3}\,\text{mol} \times 8.314 \times 10^{-2}\,\text{L bar mol}^{-1}\,\text{K}^{-1} \times (273.15 + 12.5)\,\text{K}}{5.25\,\text{L}} = 2.18 \times 10^{-2}\,\text{bar}$

(b) $x_{O_2} = \dfrac{6.56 \times 10^{-5}\,\text{mol}}{4.82 \times 10^{-3}\,\text{mol}} = 0.0136;$ $x_{N_2} = \dfrac{3.88 \times 10^{-3}\,\text{mol}}{4.82 \times 10^{-3}\,\text{mol}} = 0.805;$ $x_{CO} = \dfrac{8.71 \times 10^{-4}\,\text{mol}}{4.82 \times 10^{-3}\,\text{mol}} = 0.181$

   $P_{O_2} = x_{O_2} P_{total} = 0.0136 \times 2.18 \times 10^{-2}\,\text{bar} = 2.97 \times 10^{-4}\,\text{bar}$

   $P_{N_2} = x_{N_2} P_{total} = 0.805 \times 2.18 \times 10^{-2}\,\text{bar} = 1.76 \times 10^{-2}\,\text{bar}$

   $P_{CO} = x_{CO} P_{total} = 0.181 \times 2.18 \times 10^{-2}\,\text{bar} = 3.94 \times 10^{-3}\,\text{bar}$

P1.19   Calculate the pressure exerted by benzene for a molar volume of 2.00 L at 595 K using the Redlich–Kwong equation of state:

$$P = \frac{RT}{V_m - b} - \frac{a}{\sqrt{T}} \frac{1}{V_m(V_m + b)} = \frac{nRT}{V - nb} - \frac{n^2 a}{\sqrt{T}} \frac{1}{V(V + nb)}$$

The Redlich–Kwong parameters a and b for benzene are 452.0 bar dm^6 mol^{-2} K$^{1/2}$ and 0.08271 dm^3 mol^{-1}, respectively. Is the attractive or repulsive portion of the potential dominant under these conditions?

$$P = \frac{RT}{V_m - b} - \frac{a}{\sqrt{T}} \frac{1}{V_m(V_m + b)}$$

$$= \frac{8.314 \times 10^{-2}\,\text{bar dm}^3\,\text{mol}^{-1}\,\text{K}^{-1} \times 595\,\text{K}}{2.00\,\text{dm}^3\,\text{mol}^{-1} - 0.08271\,\text{dm}^3\,\text{mol}^{-1}} - \frac{452.0\,\text{bar dm}^6\,\text{mol}^{-2}\,\text{K}^{1/2}}{\sqrt{595\,\text{K}}}$$

$$\times \frac{1}{2.00\,\text{dm}^3\,\text{mol}^{-1} \times (2.00\,\text{dm}^3\,\text{mol}^{-1} + 0.08271\,\text{dm}^3\,\text{mol}^{-1})}$$

$$P = 21.4\,\text{bar}$$

$$P_{ideal} = \frac{RT}{V} = \frac{8.3145 \times 10^{-2} \times \text{L bar mol}^{-1}\,\text{K}^{-1} \times 595\,\text{K}}{2.00\,\text{L}} = 24.7\,\text{bar}$$

Because $P < P_{ideal}$, the attractive part of the potential dominates.

P1.21   An initial step in the biosynthesis of glucose $C_6H_{12}O_6$ is the carboxylation of pyruvic acid $CH_3COCOOH$ to form oxaloacetic acid $HOOCCOCH_2COOH$

$$CH_3COCOOH(s) + CO_2(g) \rightarrow HOOCCOCH_2COOH(s)$$

If you knew nothing else about the intervening reactions involved in glucose biosynthesis other than no further carboxylations occur, what volume of CO_2 is required to produce 1.10 g of glucose? Assume $P = 1$ atm and $T = 298.$ K.

From the stoichiometric equation,

$$n_{CO_2} = n_{glucose} = \frac{m_{glucose}}{M_{glucose}} = \frac{1.10\,g}{180.18\,g\,mol^{-1}} = 6.10 \times 10^{-3}\,mol$$

$$V_{CO_2} = \frac{n_{CO_2}RT}{P} = \frac{6.10 \times 10^{-3}\,mol \times 8.206 \times 10^{-2}\,L\,atm\,mol^{-1}\,K^{-1} \times 298\,K}{1\,atm} = 0.149\,L$$

P1.23   Assume that air has a mean molar mass of $28.9\,g\,mol^{-1}$ and that the atmosphere has a uniform temperature of $25.0°C$. Calculate the barometric pressure in Pa in Santa Fe, for which $z = 7000.$ ft. Use the information contained in Problem P1.20.

$$P = P^0 e^{-\frac{M_i g z}{RT}} = 10^5\,Pa\,exp\left(-\frac{28.9 \times 10^{-3}\,kg \times 9.81\,m\,s^{-2} \times 7000.\,ft \times 0.3048\,m\,ft^{-1}}{8.314\,J\,mol^{-1}\,K^{-1} \times 298.15\,K}\right) = 7.95 \times 10^4\,Pa$$

P1.29   A balloon filled with 11.50 L of Ar at $18.7°C$ and one atm rises to a height in the atmosphere where the pressure is 207 Torr and the temperature is $-32.4°C$. What is the final volume of the balloon? Assume that the pressure inside and outside the balloon have the same value.

$$V_f = \frac{P_i}{P_f}\frac{T_f}{T_i}V_i = \frac{760\,Torr \times (273.15 - 32.4)\,K}{207\,Torr \times (273.15 + 18.7)\,K} \times 11.5\,L = 34.8\,L$$

P1.30   Carbon monoxide competes with oxygen for binding sites on the transport protein hemoglobin. CO can be poisonous if inhaled in large quantities. A safe level of CO in air is 50. parts per million (ppm). When the CO level increases to 800. ppm, dizziness, nausea, and unconsciousness occur, followed by death. Assuming the partial pressure of oxygen in air at sea level is 0.20 atm, what proportion of CO to O_2 is fatal?

$$x_{O_2} = \frac{0.20\,atm}{1\,atm} = 2.0 \times 10^5\,ppm$$

$$\frac{x_{CO}}{x_{O_2}} = \frac{800.\,ppm}{2.0 \times 10^5\,ppm} = 4.0 \times 10^{-3}$$

P1.36   A glass bulb of volume 0.198 L contains 0.457 g of gas at 759.0 Torr and $134.0°C$. What is the molar mass of the gas?

$$n = \frac{m}{M} = \frac{PV}{RT}; \quad M = m\frac{RT}{PV}$$

$$M = 0.457\,g \times \frac{8.2057 \times 10^{-2}\,L\,atm\,mol^{-1}\,K^{-1} \times (273.15 + 134.0)\,K}{\frac{759}{760}\,atm \times 0.198\,L} = 77.2\,amu$$

2 Heat, Work, Internal Energy, Enthalpy, and the First Law of Thermodynamics

Numerical Problems

P2.1 A 3.75 mole sample of an ideal gas with $C_{V,m} = 3/2\,R$ initially at a temperature $T_i = 298$ K and $P_i = 1.00$ bar is enclosed in an adiabatic piston and cylinder assembly. The gas is compressed by placing a 725 kg mass on the piston of diameter 25.4 cm. Calculate the work done in this process and the distance that the piston travels. Assume that the mass of the piston is negligible.

We first calculate the external pressure and the initial volume.

$$P_{external} = P_{atm} + \frac{F}{A} = 10^5\,\text{Pa} + \frac{mg}{\pi r^2} = 10^5\,\text{Pa} + \frac{725\,\text{kg} \times 9.81\,\text{m s}^{-2}}{\pi \times (\frac{1}{2} \times 0.254\,\text{m})^2} = 2.40 \times 10^5\,\text{Pa}$$

$$V_i = \frac{nRT}{P_i} = \frac{3.75\,\text{mol} \times 8.314\,\text{J mol}^{-1}\text{K}^{-1} \times 298\,\text{K}}{10^5\,\text{Pa}} = 0.0929\,\text{m}^3 = 92.9\,\text{L}$$

Following Example Problem 2.6,

$$T_f = T_i \left(\frac{C_{V,m} + \dfrac{RP_{external}}{P_i}}{C_{V,m} + \dfrac{RP_{external}}{P_f}} \right) = 298\,\text{K} \times \left(\frac{12.47\,\text{J mol}^{-1}\text{K}^{-1} + \dfrac{8.314\,\text{J mol}^{-1}\text{K}^{-1} \times 2.40 \times 10^5\,\text{Pa}}{1.00 \times 10^5\,\text{Pa}}}{12.47\,\text{J mol}^{-1}\text{K}^{-1} + \dfrac{8.314\,\text{J mol}^{-1}\text{K}^{-1} \times 2.40 \times 10^5\,\text{Pa}}{2.40 \times 10^5\,\text{Pa}}} \right) = 465\,\text{K}$$

$$V_f = \frac{nRT}{P_f} = \frac{3.75\,\text{mol} \times 8.314\,\text{J mol}^{-1}\text{K}^{-1} \times 465\,\text{K}}{2.40 \times 10^5\,\text{Pa}} = 6.04 \times 10^{-2}\,\text{m}^3$$

$$w = -P_{external}(V_f - V_i) = -2.40 \times 10^5\,\text{Pa} \times (6.04 \times 10^{-2}\,\text{m}^3 - 9.29 \times 10^{-2}\,\text{m}^3) = 7.82 \times 10^3\,\text{J}$$

$$h = -\frac{V_f - V_i}{\pi r^2} = \frac{3.26 \times 10^{-2}\,\text{m}^3}{5.07 \times 10^{-2}\,\text{m}^2} = 0.642\,\text{m}$$

P2.3 A 2.50 mole sample of an ideal gas, for which $C_{V,m} = 3/2\,R$, is subjected to two successive changes in state: (1) From 25.0°C and $125. \times 10^3$ Pa, the gas is expanded isothermally against a constant pressure of 15.2×10^3 Pa to twice the initial volume. (2) At the end of the previous process, the gas is cooled at constant volume from 25.0°C to –29.0°C. Calculate q, w, ΔU, and ΔH for each of the stages. Also calculate q, w, ΔU, and ΔH for the complete process.

a.
$$V_i = \frac{nRT}{P_i} = \frac{2.50\,\text{mol} \times 8.314\,\text{J mol}^{-1}\text{K}^{-1} \times 298\,\text{K}}{125 \times 10^3\,\text{Pa}} = 4.96 \times 10^{-2}\,\text{m}^3$$

$$V_f = 2V_i = 9.92 \times 10^{-2}\,\text{m}^3$$

$$w = -P_{ext}(V_f - V_i) = -15.2 \times 10^3\,\text{Pa} \times (9.92 \times 10^{-2}\,\text{m}^3 - 4.96 \times 10^{-2}\,\text{m}^3) = -754\,\text{J}$$

ΔU and $\Delta H = 0$ because $\Delta T = 0$

$$q = -w = 754\,\text{J}$$

b. $\Delta U = nC_{V,m}(T_f - T_i) = 2.50 \text{ mol} \times 1.5 \times 8.314 \text{ J mol}^{-1}\text{K}^{-1} \times (244 \text{ K} - 298 \text{ K}) = -1.68 \times 10^3 \text{ J}$

$w = 0 \text{ because } \Delta V = 0$

$q = \Delta U = -1.68 \times 10^3 \text{ J}$

$\Delta H = nC_{P,m}(T_f - T_i) = n(C_{V,m} + R)(T_f - T_i)$

$\qquad = 2.50 \text{ mol} \times 2.5 \times 8.314 \text{ J mol}^{-1}\text{K}^{-1} \times (244 \text{ K} - 298 \text{ K})$

$\qquad = -2.81 \times 10^3 \text{ J}$

$\Delta U_{total} = 0 - 1.684 \times 10^3 \text{ J} = -1.68 \times 10^3 \text{ J}$

$w_{total} = 0 - 754 \text{ J} = -754 \text{ J}$

$q_{total} = 754 \text{ J} - 1.684 \times 10^3 \text{ J} = -930. \text{ J}$

$\Delta H_{total} = 0 - 2.81 \times 10^3 \text{ J} = -2.81 \times 10^3 \text{ J}$

P2.5 Count Rumford observed that using cannon boring machinery a single horse could heat 11.6 kg of ice water ($T = 273 \text{ K}$) to $T = 355 \text{ K}$ in 2.5 hours. Assuming the same rate of work, how high could a horse raise a 225 kg weight in 2.5 minutes? Assume the heat capacity of water is $4.18 \text{ J K}^{-1} \text{g}^{-1}$.

$$Rate = \frac{C_p m_{water} \Delta T}{time_1} = \frac{4.18 \text{ J K}^{-1} \text{g}^{-1} \times 11.6 \times 10^3 \text{ g} \times (355 - 273) \text{ K}}{2.5 \text{ hr} \times 3600 \text{ s hr}^{-1}} = 442 \text{ J s}^{-1}$$

$$h = \frac{Rate \times time_2}{m_{weight} g} = \frac{442 \text{ J s}^{-1} \times 150 \text{ s}}{225 \text{ kg} \times 9.81 \text{ m s}^{-2}} = 30. \text{ m}$$

P2.6 A 1.50 mole sample of an ideal gas at 28.5°C expands isothermally from an initial volume of 22.5 dm^3 to a final volume of 75.5 dm^3. Calculate w for this process (a) for expansion against a constant external pressure of 0.498×10^5 Pa and (b) for a reversible expansion.

a. $w = -P_{external}\Delta V = -0.498 \times 10^5 \text{ Pa} \times (75.5 - 22.5) \times 10^{-3} \text{ m}^3 = -2.64 \times 10^3 \text{ J}$

b. $w_{reversible} = -nRT \ln\dfrac{V_f}{V_i} = -1.50 \text{ mol} \times 8.314 \text{ J mol}^{-1}\text{K}^{-1} \times (273.15 + 28.5) \text{ K} \times \ln\dfrac{75.5 \text{ dm}^3}{22.5 \text{ dm}^3}$

$\qquad = -4.55 \times 10^3 \text{ J}$

P2.7 Calculate q, w, ΔU, and ΔH if 2.25 mol of an ideal gas with $C_{v,m} = 3/2\, R$ undergoes a reversible adiabatic expansion from an initial volume $V_i = 5.50 \text{ m}^3$ to a final volume $V_f = 25.0 \text{ m}^3$. The initial temperature is 275 K.

$q = 0$ because the process is adiabatic.

$$\frac{T_f}{T_i} = \left(\frac{V_f}{V_i}\right)^{1-\gamma}$$

$$T_f = \left(\frac{25.0 \text{ m}^3}{5.50 \text{ m}^3}\right)^{1-\frac{5}{3}} \times T_i = 100. \text{ K}$$

$$\Delta U = w = nC_{V,n}\Delta T = 2.25 \text{ mol} \times \frac{3 \times 8.314 \text{ J mol}^{-1}\text{K}^{-1}}{2} \times (100. \text{ K} - 300 \text{ K}) = -4.90 \times 10^3 \text{ J}$$

$$\Delta H = \Delta U + nR\Delta T = -4.904 \times 10^3 \text{ J} + 2.25 \text{ mol} \times 8.314 \text{ J mol}^{-1}\text{K}^{-1} \times (100. \text{ K} - 300 \text{ K})$$

$$\Delta H = -8.17 \times 10^3 \text{ J}$$

P2.10   A muscle fiber contracts by 3.5 cm and in doing so lifts a weight. Calculate the work performed by the fiber. Assume the muscle fiber obeys Hooke's law $F = -kx$ with a force constant, k, of $750.\,\text{N m}^{-1}$.

$$w = \frac{1}{2}kx^2 = \frac{1}{2} \times 750.\,\text{N m}^{-1} \times (3.5 \times 10^{-2}\,\text{m})^2 = 0.46\,\text{J}$$

This is the work done on the weight by the fiber. It is positive because the weight is higher than its initial position and has therefore gained energy. The work done on the fiber by the weight has the opposite sign.

P2.12   In the reversible adiabatic expansion of 1.75 mol of an ideal gas from an initial temperature of 27.0°C, the work done on the surroundings is 1300. J. If $C_{V,m} = 3/2R$, calculate q, w, ΔU, and ΔH.

$$q = 0 \text{ because the process is adiabatic}$$
$$\Delta U = w = -1300.\,\text{J}$$
$$\Delta U = nC_{V,m}(T_f - T_i)$$
$$T_f = \frac{\Delta U + nC_{V,m}T_i}{nC_{V,m}}$$
$$= \frac{-1300.\,\text{J} + 1.75\,\text{mol} \times 1.5 \times 8.314\,\text{J mol}^{-1}\,\text{K}^{-1} \times 300.\,\text{K}}{1.75\,\text{mol} \times 8.314\,\text{J mol}^{-1}\,\text{K}^{-1}}$$
$$= 241\,\text{K}$$
$$\Delta H = nC_{P,m}(T_f - T_i) = n(C_{V,m} + R)(T_f - T_i)$$
$$= 1.75\,\text{mol} \times 2.5 \times 8.314\,\text{J mol}^{-1}\,\text{K}^{-1}(241\,\text{K} - 300\,\text{K})$$
$$= -2.17 \times 10^3\,\text{J}$$

P2.14   A 1.25 mole sample of an ideal gas is expanded from 320. K and an initial pressure of 3.10 bar to a final pressure of 1.00 bar, and $C_{P,m} = 5/2R$. Calculate w for the following two cases:

a.   The expansion is isothermal and reversible.

b.   The expansion is adiabatic and reversible.

Without resorting to equations, explain why the result to part (b) is greater than or less than the result to part (a).

(a) $$w = -nRT\ln\frac{V_f}{V_i} = -nRT\ln\frac{P_i}{P_f}$$

$$= -1.25\,\text{mol} \times 8.314\,\text{J mol}^{-1}\,\text{K}^{-1} \times 320.\,\text{K} \times \ln\frac{3.10\,\text{bar}}{1.00\,\text{bar}} = -3.76 \times 10^3\,\text{J}$$

(b)   Because $q = 0$, $w = \Delta U$. In order to calculate ΔU, we first calculate T_f.

$$\frac{T_f}{T_i} = \left(\frac{V_f}{V_i}\right)^{1-\gamma} = \left(\frac{T_f}{T_i}\right)^{1-\gamma}\left(\frac{P_i}{P_f}\right)^{1-\gamma}; \quad \left(\frac{T_f}{T_i}\right)^{\gamma} = \left(\frac{P_i}{P_f}\right)^{1-\gamma}; \quad \frac{T_f}{T_i} = \left(\frac{P_i}{P_f}\right)^{\frac{1-\gamma}{\gamma}}$$

$$T_f = T_i \times \left(\frac{3.10\,\text{bar}}{1.00\,\text{bar}}\right)^{\frac{1-\frac{5}{3}}{\frac{5}{3}}} = 204\,\text{K}$$

$$w = \Delta U = nC_{V,m}\Delta T = 1.25\,\text{mol} \times \frac{3 \times 8.314\,\text{J mol}^{-1}\,\text{K}^{-1}}{2} \times (204\,\text{K} - 320.\,\text{K}) = -1.82 \times 10^3\,\text{J}$$

Less work is done on the surroundings in part (b) because in the adiabatic expansion, the temperature falls and therefore the final volume is less than that in part (a).

P2.16 A 2.25 mole sample of an ideal gas with $C_{V,m} = 3/2R$ initially at 310. K and 1.25×10^5 Pa undergoes a reversible adiabatic compression. At the end of the process, the pressure is 3.10×10^6 Pa. Calculate the final temperature of the gas. Calculate q, w, ΔU, and ΔH for this process.

$q = 0$ because the process is adiabatic.

$$T_f = T_i \left(\frac{P_i}{P_f} \right)^{\frac{1-C_P/C_V}{C_P/C_V}} = 325 \text{ K} \times \left(\frac{1.25 \times 10^5 \text{ Pa}}{3.10 \times 10^5 \text{ Pa}} \right)^{\frac{1-5/3}{5/3}} = 1.12 \times 10^3 \text{ K}$$

$$w = \Delta U = nC_{V,m}\Delta T = 2.25 \text{ mol} \times \frac{3 \times 8.314 \text{ J mol}^{-1}\text{K}^{-1}}{2} \times (1120 \text{ K} - 310. \text{ K}) = 22.7 \times 10^3 \text{ J}$$

$$\Delta H = \Delta U + \Delta(PV) = \Delta U + nR\Delta T$$

$$= 22.78 \times 10^3 \text{ J} + 2.25 \text{ mol} \times 8.314 \text{ J mol}^{-1}\text{K}^{-1} \times (1120 \text{ K} - 310. \text{ K})$$

$$\Delta H = 37.9 \times 10^3 \text{ J}$$

P2.17 A vessel containing 1.50 mol of an ideal gas with $P_i = 1.00$ bar and $C_{P,m} = 5/2R$ is in thermal contact with a water bath. Treat the vessel, gas, and water bath as being in thermal equilibrium, initially at 298 K, and as separated by adiabatic walls from the rest of the universe. The vessel, gas, and water bath have an average heat capacity of $C_P = 2450.$ J K^{-1}. The gas is compressed reversibly to $P_f = 20.5$ bar. What is the temperature of the system after thermal equilibrium has been established?

Assume initially that the temperature rise is so small that the reversible compression can be thought of as an isothermal reversible process. If the answer substantiates this assumption, it is valid.

$$w = -nRT_1 \ln \frac{V_f}{V_i} = -nRT_1 \ln \frac{P_i}{P_f}$$

$$= -1.50 \text{ mol} \times 8.314 \text{ J mol}^{-1}\text{K}^{-1} \times 312 \text{ K} \times \ln \frac{1.00 \text{ bar}}{20.5 \text{ bar}} = 11.2 \times 10^3 \text{ J}$$

$$\Delta U_{combined\ system} = C_P\Delta T$$

$$\Delta T = \frac{\Delta U_{combined\ system}}{C_P} = \frac{11.2 \times 10^3 \text{ J}}{2450 \text{ J K}^{-1}} = 3.05 \text{ K}$$

$$T_f = 301 \text{ K}$$

The result justifies the assumption.

P2.21 The heat capacity of solid lead oxide is given by

$$C_{P,m} = 44.35 + 1.47 \times 10^{-3}\frac{T}{K} \text{ in units of J K}^{-1}\text{ mol}^{-1}$$

Calculate the change in enthalpy of 1.75 mol of PbO(s) if it is cooled from 825 to 375 K at constant pressure.

$$\Delta H = n \int_{T_i}^{T_f} C_{p,m}dT$$

$$= 1.75 \text{ mol} \times \int_{825}^{375} \left(44.35 + 1.47 \times 10^{-3}\frac{T}{K} \right) d\left(\frac{T}{K} \right)$$

$$= 1.75 \text{ mol} \times \left(\begin{array}{c} 44.35 \times (375 \text{ K} - 825 \text{ K}) \\ + \left[\frac{1.47 \times 10^{-3}}{2}\left(\frac{T}{K} \right)^2 \right]_{825 \text{ K}}^{375 \text{ K}} \end{array} \right)$$

$$= -34.9 \times 10^3 \text{ J} - 695 \text{ J}$$

$$= -35.6 \times 10^3 \text{ J}$$

P2.25   A major league pitcher throws a baseball with a speed of 162 kilometers per hour. If the baseball weighs 235 grams and its heat capacity is $1.7 \text{ J g}^{-1} \text{ K}^{-1}$, calculate the temperature rise of the ball when it is stopped by the catcher's mitt. Assume no heat is transferred to the catcher's mitt and that the catcher's arm does not recoil when he or she catches the ball.

$$v = 162 \times 10^3 \text{ m hr}^{-1} \times \frac{\text{hr}}{3600 \text{ s}} = 45.0 \text{ m s}^{-1}$$

$$q_P = C_P \Delta T = \tfrac{1}{2} m v^2$$

$$\Delta T = \frac{\tfrac{1}{2} m v^2}{C_P m} = \frac{0.5 \times 0.235 \text{ kg} \times (45.0 \text{ m s}^{-1})^2}{1700 \text{ J kg}^{-1} \text{ K}^{-1}} = 0.60 \text{ K}$$

P2.26   A 2.50 mol sample of an ideal gas for which $C_{V,m} = 3/2R$ undergoes the following two-step process: (1) From an initial state of the gas described by $T = 13.1°C$ and $P = 1.75 \times 10^5$ Pa, the gas undergoes an isothermal expansion against a constant external pressure of 3.75×10^4 Pa until the volume has doubled. (2) Subsequently, the gas is cooled at constant volume. The temperature falls to $-23.6°C$. Calculate q, w, ΔU, and ΔH for each step and for the overall process.

a.   For the first step, $\Delta U = \Delta H = 0$ because the process is isothermal.

$$V_i = \frac{nRT_i}{P_i}$$

$$= \frac{2.50 \text{ mol} \times 8.314 \text{ J mol}^{-1}\text{K}^{-1} \times (273.15 + 13.1) \text{ K}}{1.75 \times 10^4 \text{ Pa}} = 0.0340 \text{ m}^3$$

$$w = -q = -P_{external}\Delta V = -3.75 \times 10^4 \text{ Pa} \times 2 \times 0.0340 \text{ m}^3$$

$$= -1.27 \times 10^3 \text{ J}$$

b.   For the second step, $w = 0$ because $\Delta V = 0$.

$$q = \Delta U = nC_V\Delta T = 2.50 \text{ mol} \times \frac{3 \times 8.314 \text{ J mol}^{-1}\text{K}^{-1}}{2} \times (-23.6°C - 13.1°C) = -1.14 \times 10^3 \text{ J}$$

$$\Delta H = \Delta U + \Delta(PV) = \Delta U + nR\Delta T = -1.14 \times 10^3 \text{ J}$$

$$+ 2.50 \text{ mol} \times 8.314 \text{ J mol}^{-1}\text{K}^{-1} \times (-23.6°C - 13.1°C)$$

$$\Delta H = -1.91 \times 10^3 \text{ J}$$

For the overall process,

$$w = -1.27 \times 10^3 \text{ J}, q = 130 \text{ J}$$

$$\Delta U = -1.14 \times 10^3 \text{ J, and } \Delta H = -1.91 \times 10^3 \text{ J}$$

P2.28   A 3.50 mole sample of an ideal gas with $C_{V,m} = 3/2R$ is expanded adiabatically against a constant external pressure of 1.45 bar. The initial temperature and pressure are $T_i = 310.$ K and $P_i = 15.2$ bar. The final pressure is $P_f = 1.45$ bar. Calculate q, w, ΔU, and ΔH for the process.

$$\Delta U = nC_{V,m}(T_f - T_i) = -P_{external}(V_f - V_i) = w$$

$q = 0$ because the process is adiabatic.

$$nC_{V,m}(T_f - T_i) = -nRP_{external}\left(\frac{T_f}{P_f} - \frac{T_i}{P_i}\right)$$

$$T_f\left(nC_{V,m} + \frac{nRP_{external}}{P_f}\right) = T_i\left(nC_{V,m} + \frac{nRP_{external}}{P_i}\right)$$

$$T_f = T_i \left(\frac{C_{V,m} + \dfrac{RP_{external}}{P_i}}{C_{V,m} + \dfrac{RP_{external}}{P_f}} \right) = 310.\,K \times \left(\frac{1.5 \times 8.314 \text{ J mol}^{-1}\text{K}^{-1} + \dfrac{8.314 \text{ J mol}^{-1}\text{K}^{-1} \times 1.45 \text{ bar}}{15.2 \text{ bar}}}{1.5 \times 8.314 \text{ J mol}^{-1}\text{K}^{-1} + \dfrac{8.314 \text{ J mol}^{-1}\text{K}^{-1} \times 1.45 \text{ bar}}{1.45 \text{ bar}}} \right)$$

$$T_f = 198 \text{ K}$$

$$\Delta U = w = nC_{V,n}\Delta T = 3.50 \text{ mol} \times \frac{3 \times 8.314 \text{ J mol}^{-1}\text{K}^{-1}}{2} \times (197 \text{ K} - 310.\ \text{K}) = -4.90 \times 10^3 \text{ J}$$

$$\Delta H = \Delta U + nR\Delta T = -4.90 \times 10^3 \text{ J} + 3.50 \text{ mol} \times 8.314 \text{ J mol}^{-1}\text{K}^{-1} \times (197 \text{ K} - 310.\ \text{K})$$

$$\Delta H = -8.16 \times 10^3 \text{ J}$$

P2.30   For 1.25 mol of an ideal gas, $P_{external} = P = 350. \times 10^3$ Pa. The temperature is changed from 135°C to 21.2°C, and $C_{V,m} = 3/2R$. Calculate q, w, ΔU, and ΔH.

$$\Delta U = nC_{V,m}\Delta T = 1.25 \text{ mol} \times \frac{3}{2} \times 8.314 \text{ J mol}^{-1}\text{K}^{-1} \times (408 \text{ K} - 294 \text{ K}) = -1.77 \times 10^3 \text{ J}$$

$$\Delta H = nC_{P,m}\Delta T = n(C_{V,m} + R)\Delta T$$

$$= 1.25 \text{ mol} \times \frac{5}{2} \times 8.314 \text{ J mol}^{-1}\text{K}^{-1} \times (408 \text{ K} - 294 \text{ K})$$

$$= -2.96 \times 10^3 \text{ J}$$

$$= q_P$$

$$w = \Delta U - q_P = -1.77 \times 10^3 + 2.96 \times 10^3 \text{ J} = 1.18 \times 10^3$$

P2.37   Calculate ΔH and ΔU for the transformation of 2.50 mol of an ideal gas from 19.0°C and 1.00 atm to 550.°C and 19.5 atm if

$$C_{P,m} = 20.9 + 0.042\frac{T}{K} \text{ in units of J K}^{-1}\text{ mol}^{-1}.$$

$$\Delta H = n\int_{T_i}^{T_f} C_{P,m}\, dT$$

$$= \int_{292\,K}^{823\,K} \left(20.9 + 0.042\frac{T'}{K} \right) dT'$$

$$= 20.9 \times (695 \text{ K} - 308 \text{ K})\text{J} + \left[0.021 T^2 \right]_{292\,K}^{823\,K} \text{J}$$

$$= 2.774 \times 10^4 \text{ J} + 3.109 \times 10^4 \text{ J}$$

$$= 5.88 \times 10^4 \text{ J}$$

$$\Delta U = \Delta H - \Delta(PV) = \Delta H - nR\Delta T$$

$$= 5.88 \times 10^4 \text{ J} - 8.314 \text{ J K}^{-1}\text{mol}^{-1} \times (823 - 292) \text{ K}$$

$$= 4.78 \times 10^4 \text{ J}$$

P2.38   A 1.75 mole sample of an ideal gas for which $C_{V,m} = 20.8 \text{ J K}^{-1} \text{mol}^{-1}$ is heated from an initial temperature of 21.2°C to a final temperature of 380.°C at constant volume. Calculate q, w, ΔU, and ΔH for this process.

$w = 0$ because $\Delta V = 0$.

$$\Delta U = q = nC_V \Delta T = 1.75 \text{ mol} \times 20.8 \text{ J mol}^{-1} \text{K}^{-1} \times 358.8 \text{ K} = 13.1 \times 10^3 \text{ J}$$

$$\Delta H = \Delta U + \Delta(PV) = \Delta U + nR\Delta T = 13.1 \times 10^3 \text{ J} + 1.75 \text{ mol} \times 8.314 \text{ J mol}^{-1} \text{K}^{-1} \times 358.8 \text{ K}$$

$$= 18.3 \times 10^3 \text{ J}$$

P2.42   DNA can be modeled as an elastic rod which can be twisted or bent. Suppose a DNA molecule of length L is bent such that it lies on the arc of a circle of radius R_c. The reversible work involved in bending DNA without twisting is $w_{bend} = \dfrac{BL}{2R_c^2}$ where B is the bending force constant. The DNA in a nucleosome particle is about 680. Å in length. Nucleosomal DNA is bent around a protein complex called the histone octamer into a circle of radius 55 Å. Calculate the reversible work involved in bending the DNA around the histone octamer if the force constant $B = 2.00 \times 10^{-28} \text{ J m}^{-1}$.

$$w_{bend} = \frac{BL}{2R_c^2} = \frac{2.00 \times 10^{-28} \text{ J m} \times 680. \times 10^{-10} \text{ m}}{2 \times (55 \times 10^{-10} \text{ m})^2} = 2.2 \times 10^{-19} \text{ J}$$

3 The Importance of State Functions: Internal Energy and Enthalpy

Numerical Problems

P3.2 Use the result of Problem P3.10 to show that $(\partial C_V / \partial V)_T$ for the van der Waals gas is zero.

We use the relationship

$$\left(\frac{\partial C_V}{\partial V}\right)_T = T\left(\frac{\partial^2 P}{\partial T^2}\right)_V$$

$$P = \frac{RT}{V_m - b} - \frac{a}{V_m^2}$$

$$\left(\frac{\partial P}{\partial T}\right)_V = \frac{R}{V_m - b}$$

$$\left(\frac{\partial^2 P}{\partial T^2}\right)_V = \left(\frac{\partial \frac{R}{V_m - b}}{\partial T}\right)_V = 0$$

therefore $\left(\dfrac{\partial C_V}{\partial V}\right)_T = T\left(\dfrac{\partial^2 P}{\partial T^2}\right)_V = T \times 0 = 0.$

P3.7 Integrate the expression $\beta = 1/V(\partial V/\partial T)_P$ assuming that β is independent of pressure. By doing so, obtain an expression for V as a function of T and β at constant P.

$$\beta = \frac{1}{V}\left(\frac{\partial V}{\partial T}\right)_P$$

$$\frac{dV}{V} = \beta dT$$

$$\int \frac{dV}{V} = \int \beta dT \text{ or } \ln \frac{V_f}{V_i} = \beta(T_f - T_i)$$

if β can be assumed constant in the temperature interval of interest.

P3.12 Calculate w, q, ΔH, and ΔU for the process in which 1.75 mol of water undergoes the transition $H_2O(l, 373\,\text{K}) \rightarrow H_2O(g, 610.\,\text{K})$ at 1 bar of pressure. The volume of liquid water at 373 K is $1.89 \times 10^{-5}\,\text{m}^3\,\text{mol}^{-1}$ and the molar volume of steam at 373 and 610. K is 3.03 and $5.06 \times 10^{-2}\,\text{m}^3\,\text{mol}^{-1}$, respectively. For steam, $C_{P,m}$ can be considered constant over the temperature interval of interest at 33.58 J mol^{-1} K^{-1}.

$q = \Delta H = n\Delta H_{vaporization} + nC_{P,m}^{steam}\Delta T$

$= 1.75\,\text{mol} \times 40656\,\text{J} + 1.75\,\text{mol} \times 33.58\,\text{J mol}^{-1}\,\text{K}^{-1} \times (610.\,\text{K} - 373\,\text{K}) = 8.51 \times 10^4\,\text{J}$

$$w = -P_{external}\Delta V = -10^5 \text{ Pa} \times (1.75 \times 5.06 \times 10^{-2} \text{ m}^3 - 1.75 \times 1.89 \times 10^{-5} \text{ m}^3)$$

$$= -8.85 \times 10^3 \text{ J}$$

$$\Delta U = w + q = -8.85 \times 10^3 \text{ J} + 8.51 \times 10^4 \text{ J} = 7.62 \times 10^4 \text{ J}$$

P3.16 The Joule coefficient is defined by $(\partial T/\partial V)_U = (1/C_V)[P - T(\partial P/\partial T)_V]$. Calculate the Joule coefficient for an ideal gas and for a van der Waals gas.

For an ideal gas

$$\left(\frac{\partial T}{\partial V}\right)_U = \frac{1}{C_{V,m}}\left[P - T\left(\frac{\partial}{\partial T}\frac{nRT}{V}\right)_V\right] = \frac{1}{C_{V,m}}\left[P - \frac{nRT}{V}\right] = 0$$

For a van der Waals gas,

$$\left(\frac{\partial T}{\partial V}\right)_U = \frac{1}{C_{V,m}}\left[P - T\left(\frac{\partial}{\partial T}\left[\frac{RT}{V_m - b} - \frac{a}{V_m^2}\right]\right)_V\right] = \frac{1}{C_{V,m}}\left[P - \frac{RT}{(V_m - b)}\right] = -\frac{1}{C_V}\frac{a}{V_m^2}$$

P3.18 Show that the expression $(\partial U/\partial V)_T = T(\partial P/\partial T)_V - P$ can be written in the form

$$\left(\frac{\partial U}{\partial V}\right)_T = T^2\left(\partial\left[\frac{P}{T}\right]\bigg/\partial T\right)_V = -\left(\partial\left[\frac{P}{T}\right]\bigg/\partial\left[\frac{1}{T}\right]\right)_V$$

$$\left(\frac{\partial U}{\partial V}\right)_T = T\left(\frac{\partial P}{\partial T}\right)_V - P$$

$$\left(\frac{\partial[P/T]}{\partial T}\right)_V = P\left(\frac{\partial[1/T]}{\partial T}\right)_V + \frac{1}{T}\left(\frac{\partial P}{\partial T}\right)_V$$

$$= -\frac{P}{T^2} + \frac{1}{T}\left(\frac{\partial P}{\partial T}\right)_V$$

$$\left(\frac{\partial P}{\partial T}\right)_V = T\left(\left(\frac{\partial[P/T]}{\partial T}\right)_V + \frac{P}{T^2}\right)$$

$$\left(\frac{\partial U}{\partial V}\right)_T = T^2\left(\left(\frac{\partial[P/T]}{\partial T}\right)_V + \frac{P}{T^2}\right) - P$$

$$= T^2\left(\frac{\partial[P/T]}{\partial T}\right)_V + P - P = T^2\left(\frac{\partial[P/T]}{\partial T}\right)_V$$

We now change the differentiation to the variable $1/T$.

$$\left(\frac{\partial[P/T]}{\partial T}\right)_V = \left(\frac{\partial[P/T]}{\partial[1/T]}\right)_V\left(\frac{\partial[1/T]}{\partial T}\right)_V = -\frac{1}{T^2}\left(\frac{\partial[P/T]}{\partial[1/T]}\right)_V$$

$$\left(\frac{\partial U}{\partial V}\right)_T = T^2\left(\frac{\partial[P/T]}{\partial T}\right)_V = T^2\left(-\frac{1}{T^2}\frac{\partial[P/T]}{\partial[1/T]}\right)_V = -\left(\frac{\partial[P/T]}{\partial[1/T]}\right)_V$$

P3.19 Derive an expression for the internal pressure of a gas that obeys the Bethelot equation of state, $P = \dfrac{RT}{V_m - b} - \dfrac{a}{TV_m^2}$.

The internal pressure of a gas is given by Equation (3.19)

$$\left(\frac{\partial U}{\partial V}\right)_T = T\left(\frac{\partial P}{\partial T}\right)_V - P$$

Using the Bethelot equation of state

$$\left(\frac{\partial P}{\partial T}\right)_V = \frac{R}{V_m - b} + \frac{a}{T^2 V_m^2}$$

$$\left(\frac{\partial U}{\partial V}\right)_T = \frac{RT}{V_m - b} + \frac{a}{TV_m^2} - \left(\frac{RT}{V_m - b} - \frac{a}{TV_m^2}\right) = \frac{2a}{TV_m^2}$$

P3.20   Because U is a state function, $(\partial/\partial V\,(\partial U/\partial T)_V)_T = (\partial/\partial T\,(\partial U/\partial V)_T)_V$. Using this relationship, show that $(\partial C_V/\partial V)_T = 0$ for an ideal gas.

For an ideal gas, by definition, $\left(\dfrac{\partial U}{\partial V}\right)_T = 0$. Because the order of differentiation can be changed for a state function,

$$\left(\frac{\partial}{\partial V}\left(\frac{\partial U}{\partial T}\right)_V\right)_T = \left(\frac{\partial C_V}{\partial V}\right)_T = \left(\frac{\partial}{\partial T}\left(\frac{\partial U}{\partial V}\right)_T\right)_V = 0$$

P3.22   Use $(\partial U/\partial V)_T = (\beta T - \kappa P)/\kappa$ to calculate $(\partial U/\partial V)_T$ for an ideal gas.

$$\beta = \frac{1}{V}\left(\frac{\partial V}{\partial T}\right)_P = \frac{1}{V}\frac{nR}{P}; \quad \kappa = -\frac{1}{V}\left(\frac{\partial V}{\partial P}\right)_T = \frac{nRT}{VP^2} = \frac{1}{P}$$

$$\left(\frac{\partial U}{\partial V}\right)_T = \frac{\beta T - \kappa P}{\kappa} = \frac{\dfrac{1}{V}\dfrac{nRT}{P} - 1}{\dfrac{1}{P}} = P(1-1) = 0$$

P3.24   A differential $dz = f(x, y)\,dx + g(x, y)\,dy$ is exact if the integral $\int f(x, y)\,dx + \int g(x, y)\,dy$ is independent of the path. Demonstrate that the differential $dz = 2xy\,dx + x^2\,dy$ is exact by integrating dz along the paths $(1,\,1) \to (1,\,8) \to (6,\,8)$ and $(1,\,1) \to (1,\,3) \to (4,\,3) \to (4,\,8) \to (6,\,8)$. The first number in each set of parentheses is the x coordinate, and the second number is the y coordinate.

$$\int dz = \int 2xy\,dx + \int x^2\,dy$$

Path 1

$$\int dz = 2\int_1^6 8x\,dx + 1\int_1^8 dy = 280 + 7 = 287$$

Path 2

$$\int dz = \int_1^3 dy + \int_1^4 6x\,dx + \int_3^8 16\,dy + \int_4^6 16x\,dx$$

$$= 2 + 45 + 80 + 160 = 287$$

P3.28   Use the relation

$$C_{P,m} - C_{V,m} = T\left(\frac{\partial V_m}{\partial T}\right)_P\left(\frac{\partial P}{\partial T}\right)_V$$

the cyclic rule and the van der Waals equation of state, to derive an equation for $C_{P,m} - C_{V,m}$ in terms of V_m, T, and the gas constants R, a, and b.

We use the cyclic rule to evaluate $\left(\dfrac{\partial V_m}{\partial T}\right)_P$.

$$\left(\frac{\partial V_m}{\partial T}\right)_P\left(\frac{\partial T}{\partial P}\right)_{V_m}\left(\frac{\partial P}{\partial V_m}\right)_T = -1$$

$$\left(\frac{\partial V_m}{\partial T}\right)_P = -\left(\frac{\partial P}{\partial T}\right)_{V_m}\left(\frac{\partial V_m}{\partial P}\right)_T$$

$$C_{P,m} - C_{V,m} = T\left(\frac{\partial V_m}{\partial T}\right)_P\left(\frac{\partial P}{\partial T}\right)_{V_m} = -T\left[\left(\frac{\partial P}{\partial T}\right)_{V_m}\right]^2\left(\frac{\partial V_m}{\partial P}\right)_T = -T\frac{\left[\left(\dfrac{\partial P}{\partial T}\right)_{V_m}\right]^2}{\left(\dfrac{\partial P}{\partial V_m}\right)_T}$$

$$P = \frac{RT}{V_m - b} - \frac{a}{V_m^2}$$

$$\left(\frac{\partial P}{\partial T}\right)_{V_m} = \frac{R}{V_m - b}$$

$$\left(\frac{\partial P}{\partial V_m}\right)_T = \frac{-RT}{(V_m - b)^2} + \frac{2a}{V_m^3} = \frac{-RTV_m^3 + 2a(V_m - b)}{V_m^3(V_m - b)^2}$$

$$C_{P,m} - C_{V,m} = -T\frac{\left(\frac{R}{V_m - b}\right)^2}{\frac{-RT}{(V_m - b)^2} + \frac{2a}{V_m^3}} = -T\frac{R}{-T + \frac{2a(V_m - b)^2}{RV_m^3}} = \frac{R}{1 - \frac{2a(V_m - b)^2}{RTV_m^3}}$$

In the ideal gas limit, $a = 0$, and $C_{P,m} - C_{V,m} = R$.

P3.31 This problem will give you practice in using the cyclic rule. Use the ideal gas law to obtain the three functions $P = f(V, T)$, $V = g(P, T)$, and $T = h(P, V)$. Show that the cyclic rule $(\partial P/\partial V)_T (\partial V/\partial T)_P (\partial T/\partial P)_V = -1$ is obeyed.

$$P = \frac{nRT}{V}; \quad V = \frac{nRT}{P}; \quad T = \frac{PV}{nR}$$

$$\left(\frac{\partial P}{\partial V}\right)_T = -\frac{nRT}{V^2}; \quad \left(\frac{\partial V}{\partial T}\right)_P = \frac{nR}{P}; \quad \left(\frac{\partial T}{\partial P}\right)_V = \frac{V}{nR}$$

$$\left(\frac{\partial P}{\partial V}\right)_T\left(\frac{\partial V}{\partial T}\right)_P\left(\frac{\partial T}{\partial P}\right)_V = \left(-\frac{nRT}{V^2}\right)\left(\frac{nR}{P}\right)\left(\frac{V}{nR}\right) = \frac{-nRT}{PV} = -1$$

P3.32 Regard the enthalpy as a function of T and P. Use the cyclic rule to obtain the expression

$$C_P = -\left(\frac{\partial H}{\partial P}\right)_T \bigg/ \left(\frac{\partial T}{\partial P}\right)_H$$

$$\left(\frac{\partial H}{\partial P}\right)_T\left(\frac{\partial P}{\partial T}\right)_H\left(\frac{\partial T}{\partial H}\right)_P = -1$$

$$C_P = \left(\frac{\partial H}{\partial T}\right)_P = -\left(\frac{\partial H}{\partial P}\right)_T\left(\frac{\partial P}{\partial T}\right)_H = -\frac{\left(\frac{\partial H}{\partial P}\right)_T}{\left(\frac{\partial T}{\partial P}\right)_H}$$

P3.36 For an ideal gas, $\left(\frac{\partial U}{\partial V}\right)_T$ and $\left(\frac{\partial H}{\partial P}\right)_T = 0$. Prove that C_V is independent of volume and C_P is independent of pressure.

$$C_V = \left(\frac{\partial U}{\partial T}\right)_V$$

$$\left(\frac{\partial C_V}{\partial V}\right)_T = \left(\frac{\partial\left(\frac{\partial U}{\partial T}\right)_V}{\partial V}\right)_T = \left(\frac{\partial\left(\frac{\partial U}{\partial V}\right)_T}{\partial T}\right)_V \quad \text{because } U \text{ is a state function.}$$

Because $\left(\frac{\partial U}{\partial V}\right)_T = 0$ for an ideal gas, $\left(\frac{\partial C_V}{\partial V}\right)_T = 0$

$$C_P = \left(\frac{\partial H}{\partial T}\right)_P$$

$$\left(\frac{\partial C_P}{\partial P}\right)_T = \left(\frac{\partial\left(\frac{\partial H}{\partial T}\right)_P}{\partial P}\right)_T = \left(\frac{\partial\left(\frac{\partial H}{\partial P}\right)_T}{\partial T}\right)_P \quad \text{because } H \text{ is a state function.}$$

Because $\left(\frac{\partial H}{\partial P}\right)_T = 0$ for an ideal gas, $\left(\frac{\partial C_P}{\partial P}\right)_T = 0$.

4   Thermochemistry

Numerical Problems

P4.2   At 1000. K, $\Delta H_R^\circ = -123.77 \text{ kJ mol}^{-1}$ for the reaction $N_2(g) + 3H_2(g) \rightarrow 2NH_3(g)$, with $C_{P,m} = 3.502\,R$, $3.466\,R$, and $4.217\,R$ for $N_2(g)$, $H_2(g)$, and $NH_3(g)$, respectively. Calculate ΔH_f° of $NH_3(g)$ at 450. K from this information. Assume that the heat capacities are independent of temperature.

$$\Delta H_R^\circ(450.\,K) = \Delta H_R^\circ(1000.\,K) + \int_{1000.\,K}^{450.\,K} \Delta C_P(T)\,dT$$

For this problem, the heat capacities are assumed to be independent of T.

$$\Delta H_R^\circ(450.\,K) = \Delta H_R^\circ(1000.\,K) + \Delta C_P \Delta T$$
$$= -123.77 \text{ kJ mol}^{-1} + [2C_{P,m}(NH_3, g) - C_{P,m}(N_2, g) - 3C_{P,m}(H_2, g)][-550.\,K]$$
$$= -123.77 \text{ kJ mol}^{-1} + 8.314 \text{ J mol}^{-1} K^{-1} \times [2 \times 4.217 - 3.502 - 3 \times 3.466][-550.\,K]$$
$$= -98.775 \text{ kJ mol}^{-1}$$

$$\Delta H_f^\circ(NH_3, g, 450.\,K) = \frac{1}{2}\Delta H_R^\circ(450.\,K) = -49.39 \text{ kJ mol}^{-1}$$

P4.3   A sample of K(s) of mass 2.740 g undergoes combustion in a constant volume calorimeter. The calorimeter constant is 1849 J K^{-1}, and the measured temperature rise in the inner water bath containing 1450. g of water is 1.60 K. Calculate ΔU_f° and ΔH_f° for K_2O.

$$2K(s) + \frac{1}{2}O_2(g) \rightarrow K_2O(s)$$

$$\Delta U_f^\circ = -\frac{M_s}{m_s}\left(\frac{m_{H_2O}}{M_{H_2O}}C_{H_2O,m}\Delta T + C_{calorimeter}\Delta T\right)$$

$$= -\frac{39.098 \text{ g mol}^{-1}}{2.740 \text{ g}} \times \frac{2 \text{ mol K}}{1 \text{ mol reaction}} \times \left(\begin{array}{c} \frac{1.450 \times 10^3 \text{ g}}{18.02 \text{ g mol}^{-1}} \times 75.3 \text{ J mol}^{-1} K^{-1} \times 1.60\,^\circ C \\ +1.849 \times 10^3 \text{ J}^\circ C^{-1} \times 1.60\,^\circ C \end{array}\right)$$

$$= -361 \text{ kJ mol}^{-1}$$

$$\Delta H_f^\circ = \Delta U_f^\circ + \Delta nRT$$

$$= -361 \text{ kJ mol}^{-1} - \frac{1}{2} \times 8.314 \text{ J K}^{-1}\text{mol}^{-1} \times 298.15 \text{ K} = -362 \text{ kJ mol}^{-1}$$

P4.10   The following data are a DSC scan of a solution of a T4 lysozyme mutant. From the data determine T_m. Determine also the excess heat capacity ΔC_P at $T = 308$ K. Determine also the intrinsic δC_P^{int} and transition δC_P^{trs} excess heat capacities $T = 308$ K. In your calculations use the extrapolated curves, shown as dashed lines in the DSC scan.

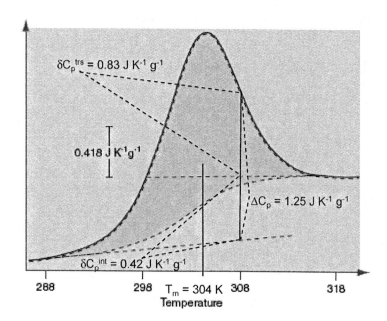

P4.11 At 298 K, $\Delta H_R^\circ = 131.28\,\text{kJ mol}^{-1}$ for the reaction $C(graphite) + H_2O(g) \rightarrow CO(g) + H_2(g)$, with $C_{P,m} = 8.53, 33.58, 29.12,$ and $28.82\,\text{J K}^{-1}\,\text{mol}^{-1}$ for graphite, $H_2O(g), CO(g),$ and $H_2(g)$, respectively. Calculate ΔH_R° at 240. °C from this information. Assume that the heat capacities are independent of temperature.

$$\Delta H_R^\circ(513\,\text{K}) = \Delta H_R^\circ(298\,\text{K}) + \int_{298\,\text{K}}^{513\,\text{K}} \Delta C_P(T)\,dT$$

For this problem, it is assumed that the heat capacities are independent of T.

$$\Delta H_R^\circ(513\,\text{K}) = \Delta H_R^\circ(298\,\text{K}) + [C_{P,m}(H_2, g) + C_{P,m}(CO, g) - C_{P,m}(C, graphite) - C_{P,m}(H_2O, g)]\Delta T$$

$$= 131.28\,\text{kJ mol}^{-1} + [28.82 + 29.12 - 8.53 - 33.58]\,\text{J mol}^{-1}\,\text{K}^{-1} \times 215\,\text{K}$$

$$= 134.68\,\text{kJ mol}^{-1}$$

P4.12 Consider the reaction $TiO_2(s) + 2\,C(graphite) + 2\,Cl_2(g) \rightarrow 2\,CO(g) + TiCl_4(l)$ for which $\Delta H_{R,298\,K}^\circ = -80.\,\text{kJ mol}^{-1}$. Given the following data at 25°C, (a) calculate ΔH_R° at 135.8°C, the boiling point of $TiCl_4$, and (b) calculate ΔH_f° for $TiCl_4\,(l)$ at 25°C:

Substance	$TiO_2(s)$	$Cl_2(g)$	$C(graphite)$	$CO(g)$	$TiCl_4(l)$
$\Delta H_f^\circ(\text{kJ mol}^{-1})$	−945			−110.5	
$C_{P,m}(\text{J K}^{-1}\,\text{mol}^{-1})$	55.06	33.91	8.53	29.12	145.2

Assume that the heat capacities are independent of temperature.

a.   Calculate ΔH_R° at 135.8°C, the boiling point of $TiCl_4$.

b.   Calculate ΔH_f° for $TiCl_4(l)$ at 25°C.

Assume that the heat capacities are independent of temperature.

(a) $$\Delta H_R^\circ(409.0\,\text{K}) = \Delta H_R^\circ(298\,\text{K}) + \int_{298\,\text{K}}^{409.0\,\text{K}} \Delta C_{P,m}\,dT$$

In this case, the heat capacities are assumed to be independent of T.

$$\Delta H_R^\circ(409.0\ K) = \Delta H_R^\circ(298\ K) + \Delta C_{P,m}[409.0\ K - 298\ K]$$

$$= -80.\ kJ\ mol^{-1} + [C_{P,m}(TiCl_4, l) + 2C_{P,m}(CO, g) - C_{P,m}(TiO_2, s)$$

$$- 2C_{P,m}(graphite, s) - 2C_{P,m}(Cl_2, g)][409.0\ K - 298\ K]$$

$$= -80.\ kJ\ mol^{-1} + [145.2 + 2\times 29.12 - 55.06 - 2\times 8.53 - 2\times 33.91][409.0\ K - 298\ K]$$

$$= -73.0\ kJ\ mol^{-1}$$

(b) $TiO_2(s) + 2C(graphite) + 2Cl_2(g) \rightarrow 2CO(g) + TiCl_4(l)$

$$\Delta H_R^\circ = -80.\ kJ\ mol^{-1} = 2\Delta H_f^\circ(CO, g) + \Delta H_f^\circ(TiCl_4, l) - \Delta H_f^\circ(TiO_2, s)$$

$$\Delta H_f^\circ(TiCl_4, l) = \Delta H_f^\circ(TiO_2, s) - 2\Delta H_f^\circ(CO, g) - 80.\ kJ\ mol^{-1}$$

$$= -945\ kJ\ mol^{-1} + 2\times 110.5\ kJ\ mol^{-1} - 80.\ kJ\ mol^{-1}$$

$$= -804\ kJ\ mol^{-1}$$

P4.13   Calculate ΔH_R° and ΔU_R° for the oxidation of benzene (g). Also calculate

$$\frac{\Delta H_R^\circ - \Delta U_R^\circ}{\Delta H_R^\circ}$$

$$\frac{15}{2}O_2(g) + C_6H_6(l) \rightarrow 3H_2O(l) + 6CO_2(g)$$

From the data tables,

$$\Delta H_{combustion}^\circ = 3\Delta H_f^\circ(H_2O, l) + 6\Delta H_f^\circ(CO_2, g) - \Delta H_f^\circ(C_6H_6, l)$$

$$= -3\times 285.8\ kJ\ mol^{-1} - 6\times 393.5\ kJ\ mol^{-1} - 49.1\ kJ\ mol^{-1}$$

$$= -3268\ kJ\ mol^{-1}$$

$$\Delta U_R^\circ = \Delta H_R^\circ - \Delta nRT = -3268\ kJ\ mol^{-1} + 1.5\times 8.314\ J\ K^{-1}\ mol^{-1}\times 298.15\ K$$

$$= -3264\ kJ\ mol^{-1}$$

$$\frac{\Delta H_R^\circ - \Delta U_R^\circ}{\Delta H_R^\circ} = \frac{-3268\ kJ\ mol^{-1} + 3264\ kJ\ mol^{-1}}{-3268\ kJ\ mol^{-1}} = 0.0122$$

P4.23   Calculate ΔH_R° at 675 K for the reaction $4NH_3(g) + 6NO(g) \rightarrow 5N_2(g) + 6H_2O(g)$ using the temperature dependence of the heat capacities from the data tables. Compare your result with ΔH_R° at 298.15 K. Is the difference large or small? Why?

$$\Delta H_R^\circ(675\ K) = \Delta H_R^\circ(298.15\ K) + \int_{298.15}^{675} \Delta C_P\left(\frac{T}{K}\right)d\frac{T}{K}$$

$$\Delta C_P = 5C_{P,m}(N_2, g) + 6C_{P,m}(H_2O, g) - 4C_{P,m}(NH_3, g) - 6C_{P,m}(NO, g)$$

$$= \begin{bmatrix} (5\times 30.81 + 6\times 33.80 - 4\times 29.29 - 6\times 33.58) \\ -(5\times 0.01187 + 6\times 0.00795 + 4\times 0.01103 - 6\times 0.02593)\dfrac{T}{K} \\ +(5\times 2.3968 + 6\times 2.8228 - 4\times 4.2446 - 6\times 5.3326)\times 10^{-5}\dfrac{T^2}{K^2} \\ -(5\times 1.0176 + 6\times 1.3115 - 4\times 2.7706 - 6\times 2.7744)\times 10^{-8}\dfrac{T^3}{K^3} \end{bmatrix} J\ K^{-1}\ mol^{-1}$$

$$= \begin{bmatrix} 38.21 + 0.00441\dfrac{T}{K} - 2.0053\times 10^{-4}\dfrac{T^2}{K^2} + 1.4772\times 10^{-7}\dfrac{T^3}{K^3} \end{bmatrix} J\ K^{-1}\ mol^{-1}$$

$$\int_{298.15}^{675} \Delta C_P\left(\frac{T}{K}\right)d\frac{T}{K} = \left[\int_{298.15}^{675}\left(38.21 + 0.00441\frac{T}{K} - 2.0053\times10^{-4}\frac{T^2}{K^2} + 1.4772\times10^{-8}\frac{T^3}{K^3}\right)d\frac{T}{K}\right]J\,mol^{-1}$$

$$= (14.400 + 0.8086 - 18.78 + 7.374)kJ\,mol^{-1} = 3.80\,kJ\,mol^{-1}$$

$$\Delta H_R^\circ(298.15\,K) = 5\Delta H_f^\circ(N_2, g) + 6\Delta H_f^\circ(H_2O, g) - 4\Delta H_f^\circ(NH_3, g) - 6\Delta H_f^\circ(NO, g)$$

$$\Delta H_R^\circ(298.15\,K) = -6\times241.8\,kJ\,mol^{-1} + 4\times45.9\,kJ\,mol^{-1} - 6\times91.3\,kJ\,mol^{-1} = -1815\,kJ\,mol^{-1}$$

$$\Delta H_R^\circ(675\,K) = -1815\,kJ\,mol^{-1} + 3.80\,kJ\,mol^{-1} = -1811\,kJ\,mol^{-1}$$

The difference is small, not because the heat capacities of reactants and products are small, but because the difference in heat capacities of reactants and products is small.

P4.24   From the following data at 298.15 K as well as data in Table 4.1 (Appendix B, Data Tables), calculate the standard enthalpy of formation of $H_2S(g)$ and of $FeS_2(s)$:

	$\Delta H_R^\circ(kJ\,mol^{-1})$
$Fe(s) + 2H_2S(g) \rightarrow FeS_2(s) + 2H_2(g)$	-137.0
$H_2S(g) + \frac{3}{2}O_2(g) \rightarrow H_2O(l) + SO_2(g)$	-562.0

	$\Delta H_R^\circ(kJ\,mol^{-1})$
$H_2O(l) + SO_2(g) \rightarrow H_2S(g) + \frac{3}{2}O_2(g)$	562.0
$S(s) + O_2(g) \rightarrow SO_2(g)$	-296.8
$H_2(g) + \frac{1}{2}O_2(g) \rightarrow H_2O(l)$	-285.8
$H_2(g) + S(s) \rightarrow H_2S(g)$	$\Delta H_f^\circ = -20.6\,kJ\,mol^{-1}$

	$\Delta H_R^\circ(kJ\,mol^{-1})$
$Fe(s) + 2H_2S(g) \rightarrow FeS_2(s) + 2H_2(g)$	-137.0
$2H_2(g) + 2S(s) \rightarrow 2H_2S(g)$	-2×20.6
$Fe(s) + 2S(s) \rightarrow FeS_2(s)$	$\Delta H_f^\circ = -178.2\,kJ\,mol^{-1}$

P4.27   Calculate ΔH for the process in which $Cl_2(g)$ initially at 298.15 K at 1 bar is heated to 690. K at 1 bar. Use the temperature-dependent heat capacities in the data tables. How large is the relative error if the molar heat capacity is assumed to be constant at its value of 298.15 K over the temperature interval?

$$\Delta H = \int_{298.15}^{690.} C_{P,m}\left(\frac{T}{K}\right)d\frac{T}{K}$$

$$= \left[\int_{298.15}^{690.}\left(22.85 + 0.06543\frac{T}{K} - 1.2517\times10^{-4}\frac{T^2}{K^2} + 1.1484\times10^{-7}\frac{T^3}{K^3}\right)d\frac{T}{K}\right]J\,K^{-1}\,mol^{-1}$$

$$= (8954 + 12668 - 12601 + 6281)\,J\,mol^{-1} = 15.3\,kJ\,mol^{-1}$$

If it is assumed that the heat capacity is constant at its value at 298 K,

$$\Delta H^\circ \approx \left[\int_{298.15}^{690}(33.95)d\frac{T}{K}\right]J\,K^{-1}\,mol^{-1} = 13.3\,kJ\,mol^{-1}$$

$$Error = 100\times\frac{15.3\,kJ\,mol^{-1} - 13.3\,kJ\,mol^{-1}}{15.3\,kJ\,mol^{-1}} = 13.1\%$$

P4.28   From the following data at 298.15 K calculate the standard enthalpy of formation of $FeO(s)$ and of $Fe_2O_3(s)$:

$$\Delta H_R^\circ (\text{kJ mol}^{-1})$$

$Fe_2O_3(s) + 3C(graphite) \rightarrow 2Fe(s) + 3CO(g)$	492.6
$FeO(s) + C(graphite) \rightarrow Fe(s) + CO(g)$	155.8
$C(graphite) + O_2(g) \rightarrow CO_2(g)$	−393.51
$CO(g) + \dfrac{1}{2}O_2(g) \rightarrow CO_2(g)$	−282.98

$$\Delta H_R^\circ (\text{kJ mol}^{-1})$$

$Fe(s) + CO(g) \rightarrow FeO(s) + C(graphite)$	−155.8
$CO_2(s) \rightarrow CO(g) + \dfrac{1}{2}O_2(g)$	282.98
$C(graphite) + O_2(g) \rightarrow CO_2(g)$	−393.51

$Fe(s) + \dfrac{1}{2}O_2(g) \rightarrow FeO(s)$	$\Delta H_f^\circ = -266.3 \text{ KJ mol}^{-1}$

$$\Delta H_R^\circ (\text{kJ mol}^{-1})$$

$2Fe(s) + 3CO(g) \rightarrow Fe_2O_3(s) + 3C(graphite)$	−492.6
$3C(graphite) + 3O_2(g) \rightarrow 3CO_2(g)$	-3×393.51
$3CO_2(g) \rightarrow 3CO(g) + \dfrac{3}{2}O_2(g)$	3×282.98

$2Fe(s) + \dfrac{3}{2}O_2(g) \rightarrow Fe_2O_3(s)$	$\Delta H_f^\circ = -824.2 \text{ kJ mol}^{-1}$

P4.30   Use the average bond energies in Table 4.3 to estimate ΔU for the reaction $C_2H_4(g) + H_2(g) \rightarrow C_2H_6(g)$. Also calculate ΔU_R° from the tabulated values of ΔH_f° for reactant and products (Appendix B, Data Tables). Calculate the percent error in estimating ΔU_R° from the average bond energies for this reaction.

$$\Delta U_R = -[(\text{C—C bond energy} + 6\text{C—H bond energy}) - \text{H—H bond energy}$$
$$- (\text{C}{=}\text{C bond energy} - 4\text{C—H bond energy})]$$
$$\Delta U_R = -[(346 \text{ kJ mol}^{-1} + 6 \times 411 \text{ kJ mol}^{-1}) - 432 \text{ kJ mol}^{-1} - (602 \text{ kJ mol}^{-1}$$
$$- 4 \times 411 \text{ kJ mol}^{-1})] = -134 \text{ kJ mol}^{-1}$$

Using the data tables,

$$\Delta H_R^\circ (298.15 \text{ K}) = \Delta H_f^\circ(C_2H_6, g) - \Delta H_f^\circ(C_2H_4, g) - \Delta H_f^\circ(H_2, g)$$
$$\Delta H_R^\circ (298.15 \text{ K}) = -84.0 \text{ kJ mol}^{-1} - 52.4 \text{ kJ mol}^{-1} = -136.4 \text{ kJ mol}^{-1}$$
$$\Delta U_R^\circ (298.15 \text{ K}) = \Delta H_R^\circ (298.15 \text{ K}) - \Delta n R T$$
$$= -136.4 \text{ kJ mol}^{-1} + 8.314 \text{ J mol}^{-1} \text{ K}^{-1} \times 298.15 \text{ K} = -133.9 \text{ kJ mol}^{-1}$$
$$\text{Relative Error} = 100 \times \frac{+134 \text{ kJ mol}^{-1} - 133.9 \text{ kJ mol}^{-1}}{-133.9 \text{ kJ mol}^{-1}} \approx 0\%$$

P4.33 A camper stranded in snowy weather loses heat by wind convection. The camper is packing emergency rations consisting of 58% sucrose, 31% fat, and 11% protein by weight. Using the data provided in Problem P4.32 and assuming the fat content of the rations can be treated with palmitic acid data and the protein content similarly by the protein data in Problem P4.32, how much emergency rations must the camper consume in order to compensate for a reduction in body temperature of 3.5 K? Assume the heat capacity of the body equals that of water. Assume the camper weighs 67 kg. State any additional assumptions.

At constant pressure $q = \Delta H$. The composition of the emergency rations means that 1 kg of the rations contains the following number of moles of sucrose, fat, and protein:

$$n_{succrose} = \frac{m}{M} = \frac{(0.58\text{ kg})}{(342.3\text{ g mol}^{-1})} = 1.694\text{ mol}$$

$$n_{fat} = \frac{m}{M} = \frac{(0.31\text{ kg})}{(256.43\text{ g mol}^{-1})} = 1.209\text{ mol}$$

$$n_{protein} = \frac{m}{M} = \frac{(0.11\text{ kg})}{(88.30\text{ g mol}^{-1})} = 1.246\text{ mol}$$

Therefore, the enthalpy of combustion for 1 kg of rations is:

$$\Delta H^{\circ}_{combustion,1\,kg} = (1.694\text{ mol}) \times (-5647\text{ kJ mol}^{-1}) + (1.209\text{ mol}) \times (-10035\text{ kJ mol}^{-1})$$

$$+ (1.246\text{ mol}) \times (-22\text{ kJ mol}^{-1}) = -24120\text{ kJ}$$

The heat the stranded camper loses is given by:

$$q_{lost} = n_{H_2O}C_{p,m}\Delta T = \frac{m_{H_2O}}{M_{H_2O}} \times C_{p,m} \times \Delta T = \frac{(67\text{ kg})}{(18.01\text{ g mol}^{-1})} \times (75.3\text{ J K}^{-1}\text{ mol}^{-1}) \times (3.5\text{ K})$$

$$= 980\text{ kJ}$$

Finally, the mass of rations that needs to be consumed to produce the lost amount of heat assuming the body consists of 90% water is then:

$$m_{rations} = 0.9 \times \frac{(1\text{ kg}) \times (980\text{ kJ})}{(24120\text{ kJ})} = 49\text{ g}$$

5 Entropy and the Second and Third Laws of Thermodynamics

Numerical Problems

P5.5 One mole of $H_2O(l)$ is compressed from a state described by $P = 1.00$ bar and $T = 350.$ K to a state described by $P = 590.$ bar and $T = 750.$ K. In addition, $\beta = 2.07 \times 10^{-4}$ K^{-1}, and the density can be assumed to be constant at the value 997 kg m^{-3}. Calculate ΔS for this transformation, assuming that $\kappa = 0$.

From Equation (5.24),

$$\Delta S = \int_{T_i}^{T_f} \frac{C_P}{T}\,dT - \int_{P_i}^{P_f} V\beta\,dP \approx nC_{P,m}\,\ln\frac{T_f}{T_i} - nV_{m,i}\beta(P_f - P_i)$$

$$= 1\,\text{mol} \times 75.3\,\text{J mol}^{-1}\text{K}^{-1} \times \ln\frac{750.\,\text{K}}{350.\,\text{K}} - 1\,\text{mol} \times \frac{18.02 \times 10^{-3}\,\text{kg mol}^{-1}}{997\,\text{kg m}^{-3}}$$

$$\times 2.07 \times 10^{-4}\,\text{K}^{-1} \times 589\,\text{bar} \times 10^5\,\text{Pa bar}^{-1}$$

$$= 57.4\,\text{J K}^{-1} - 0.220\,\text{J K}^{-1} = 57.2\,\text{J K}^{-1}$$

P5.7 Consider the reversible Carnot cycle shown in Figure 5.2 with 1.25 mol of an ideal gas with $C_V = 5/2R$ as the working substance. The initial isothermal expansion occurs at the hot reservoir temperature of $T_{hot} = 740.$ K from an initial volume of 3.75 L (V_a) to a volume of 12.8 L (V_b). The system then undergoes an adiabatic expansion until the temperature falls to $T_{cold} = 310.$ K. The system then undergoes an isothermal compression and a subsequent adiabatic compression until the initial state described by $T_a = 740.$ K and $V_a = 3.75$ L is reached.

a. Calculate V_c and V_d.

$$V_c = V_b\left(\frac{T_c}{T_b}\right)^{\frac{1}{1-\gamma}} = 12.8\,\text{L} \times \left(\frac{310.\,\text{K}}{740.\,\text{K}}\right)^{\frac{1}{1-1.4}} = 113\,\text{L}$$

$$V_d = V_a\left(\frac{T_d}{T_a}\right)^{\frac{1}{1-\gamma}} = 3.75\,\text{L} \times \left(\frac{310.\,\text{K}}{740.\,\text{K}}\right)^{\frac{1}{1-1.4}} = 33.0\,\text{L}$$

b. Calculate w for each step in the cycle and for the total cycle.

$$w_{ab} = -nRT_a\ln\frac{V_b}{V_a} = 1.25\,\text{mol} \times 8.314\,\text{J mol}^{-1}\,\text{K}^{-1} \times \ln\frac{12.8\,\text{L}}{3.75\,\text{L}} = -9.44 \times 10^3\,\text{J}$$

$$w_{bc} = nC_{V,m}(T_c - T_b) = 1.25\,\text{mol} \times 2.5 \times 8.314\,\text{J mol}^{-1}\,\text{K}^{-1} \times (310.\,\text{K} - 740.\,\text{K}) = -11.2 \times 10^3\,\text{J}$$

$$w_{cd} = -nRT_c \ln\frac{V_d}{V_c} = 1.25 \text{ mol} \times 8.314 \text{ J mol}^{-1} \text{ K}^{-1} \times \ln\frac{33.0 \text{ L}}{113 \text{ L}} = 3.96 \times 10^3 \text{ J}$$

$$w_{da} = nC_{V,m}(T_a - T_d) = 1.75 \text{ mol} \times \frac{5}{2} \times 8.314 \text{ J mol}^{-1} \text{ K}^{-1} \times (740. \text{ K} - 310. \text{ K}) = +11.2 \times 10^3 \text{ J}$$

$$w_{total} = -9.44 \times 10^3 \text{ J} - 11.2 \times 10^3 \text{ J} + 3.96 \times 10^3 \text{ J} + 11.2 \times 10^3 \text{ J} = -5.49 \times 10^3 \text{ J}$$

c. Calculate ε and the amount of heat that is extracted from the hot reservoir to do 1.00 kJ of work in the surroundings.

$$\varepsilon = 1 - \frac{T_{cold}}{T_{hot}} = 1 - \frac{310. \text{ K}}{740. \text{ K}} = 0.581 \quad q = \frac{w}{\varepsilon} = \frac{1.00 \times 10^3 \text{ J}}{0.581} = 1.72 \times 10^3 \text{ J}$$

P5.8   The average heat evolved by the oxidation of foodstuffs in an average adult per hour per kilogram of body weight is $7.20 \text{ kJ kg}^{-1} \text{ hr}^{-1}$. Assume the weight of an average adult is 62.0 kg. Suppose the total heat evolved by this oxidation is transferred into the surroundings over a period lasting one week. Calculate the entropy change of the surroundings associated with this heat transfer. Assume the surroundings are at $T = 293 \text{ K}$.

$$q(\text{per day, 62.0 kg}) = 7.20 \text{ kJ kg}^{-1} \text{ hr}^{-1} \times 24 \text{ h day}^{-1} \times 7 \text{ day week}^{-1} \times 62.0 \text{ kg}$$

$$= 44.6 \times 10^4 \text{ kJ week}^{-1}$$

$$\Delta S = \frac{q}{T} = \frac{44.6 \times 10^4 \text{ kJ day}^{-1}}{293 \text{ K}} = 256 \text{ kJ K}^{-1} \text{ week}^{-1}$$

P5.9   Calculate ΔS, ΔS_{total}, and $\Delta S_{surroundings}$ when the volume of 150. g of CO initially at 273 K and 1.00 bar increases by a factor of two in (a) an adiabatic reversible expansion, (b) an expansion against $P_{external} = 0$, and (c) an isothermal reversible expansion. Take $C_{P,m}$ to be constant at the value 29.14 J mol^{-1} K^{-1} and assume ideal gas behavior. State whether each process is spontaneous. The temperature of the surroundings is 273 K.

a. An adiabatic reversible expansion $\Delta S_{surroundings} = 0$ because $q = 0$. $\Delta S = 0$ because the process is reversible. $\Delta S_{total} = \Delta S + \Delta S_{surroundings} = 0$. The process is not spontaneous.

b. An expansion against $P_{external} = 0$. ΔT and $w = 0$. Therefore $\Delta U = q = 0$.

$$\Delta S = nR \ln\frac{V_f}{V_i} = \frac{150. \text{ g}}{28.01 \text{ g mol}^{-1}} \times 8.314 \text{ J mol}^{-1} \text{ K}^{-1} \times \ln 2 = 30.9 \text{ J K}^{-1}$$

$\Delta S_{total} = \Delta S + \Delta S_{surroundings} = 30.9 \text{ J K}^{-1} + 0 = 30.9 \text{ J K}^{-1}$. The process is spontaneous.

c. An isothermal reversible expansion $\Delta T = 0$. Therefore $\Delta U = 0$.

$$w = -q = -nRT \ln\frac{V_f}{V_i} = -\frac{150. \text{ g}}{28.01 \text{ g mol}^{-1}} \text{ mol} \times 8.314 \text{ J mol}^{-1} \text{ K}^{-1} \times 273 \text{ K} \times \ln 2 = -8.43 \times 10^3 \text{ J}$$

$$\Delta S = \frac{q_{reversible}}{T} = \frac{8.43 \times 10^3 \text{ J}}{273 \text{ K}} = 30.9 \text{ J K}^{-1}$$

$$\Delta S_{surroundings} = \frac{-q}{T} = \frac{-8.43 \times 10^3 \text{ J}}{273 \text{ K}} = -30.9 \text{ J K}^{-1}$$

$\Delta S_{total} = \Delta S + \Delta S_{surroundings} = 30.9 \text{ J K}^{-1} - 30.9 \text{ J K}^{-1} = 0$. The system and surroundings are at equilibrium.

P5.10   The maximum theoretical efficiency of an internal combustion engine is achieved in a reversible Carnot cycle. Assume that the engine is operating in the Otto cycle and that $C_{V,m} = 5/2 R$ for the fuel–air mixture initially at 273 K (the temperature of the cold reservoir). The mixture is compressed by a factor of 6.9 in the adiabatic compression step. What is the maximum theoretical efficiency of this engine? How much would the efficiency increase if the compression ratio could be increased to 15? Do you see a problem in doing so?

$$T_f = T_i \left(\frac{V_f}{V_i}\right)^{1-\gamma} = T_i \left(\frac{1}{7.5}\right)^{1-\frac{7}{5}} = 273\,\text{K} \times \left(\frac{1}{6.9}\right)^{-0.4} = 591\,\text{K}$$

$$\varepsilon = 1 - \frac{T_{cold}}{T_{hot}} = 1 - \frac{273\,\text{K}}{591\,\text{K}} = 0.538$$

$$T_f = T_i \left(\frac{V_f}{V_i}\right)^{1-\gamma} = T_i \left(\frac{1}{15}\right)^{1-\frac{7}{5}} = 298\,\text{K} \times \left(\frac{1}{15}\right)^{-0.4} = 806\,\text{K}$$

$$\varepsilon = 1 - \frac{T_{cold}}{T_{hot}} = 1 - \frac{273\,\text{K}}{806\,\text{K}} = 0.661$$

It would be difficult to avoid ignition of the fuel–air mixture before the compression was complete.

P5.13   Calculate ΔS for the isothermal compression of 1.75 mol of Cu(s) from 2.15 bar to 1250. bar at 298 K. $\beta = 0.492 \times 10^{-4}\,\text{K}^{-1}$, $\kappa = 0.78 \times 10^{-6}\,\text{bar}^{-1}$, and the density is 8.92 g cm^{-3}. Repeat the calculation assuming that $\kappa = 0$.

$$\Delta S = -\int_{P_i}^{P_f} V_i (1 - \kappa P)\beta\, dP$$

$$= -\int_{2.15 \times 10^5}^{1.25 \times 10^8} \frac{63.55 \times 10^{-3}\,\text{kg mol}^{-1}}{8.92 \times 10^3\,\text{kg m}^3}(1 - 0.780 \times 10^{-11}\,\text{Pa}^{-1} \times P)0.492 \times 10^{-4}\,\text{K}^{-1}dP$$

$$= -0.0765\,\text{J K}^{-1}$$

Repeating the calculation for $\kappa = 0$, we see no change because $1 \gg \kappa P$.

P5.14   Calculate $\Delta S°$ for the reaction $3H_2(g) + N_2(g) \rightarrow 2NH_3(g)$ at 725 K. Omit terms in the temperature-dependent heat capacities higher than T^2/K^2.

From Table 2.4,

$$C_P°(H_2, g) = 22.66 + 4.38 \times 10^{-2}\frac{T}{\text{K}} - 1.0835 \times 10^{-4}\frac{T^2}{\text{K}^2}\,\text{J K}^{-1}\,\text{mol}^{-1}$$

$$C_P°(N_2, g) = 30.81 - 1.187 \times 10^{-2}\frac{T}{\text{K}} + 2.3968 \times 10^{-5}\frac{T^2}{\text{K}^2}\,\text{J K}^{-1}\,\text{mol}^{-1}$$

$$C_P°(NH_3, g) = 29.29 + 1.103 \times 10^{-2}\frac{T}{\text{K}} + 4.2446 \times 10^{-5}\frac{T^2}{\text{K}^2}\,\text{J K}^{-1}\,\text{mol}^{-1}$$

$$\Delta C_P° = 2\left(29.29 + 1.103 \times 10^{-2}\frac{T}{\text{K}} + 4.2446 \times 10^{-5}\frac{T^2}{\text{K}^2}\,\text{J K}^{-1}\,\text{mol}^{-1}\right)$$

$$- \left(30.81 - 1.187 \times 10^{-2}\frac{T}{\text{K}} + 2.3968 \times 10^{-5}\frac{T^2}{\text{K}^2}\,\text{J K}^{-1}\,\text{mol}^{-1}\right)$$

$$- 3\left(22.66 + 4.38 \times 10^{-2}\frac{T}{\text{K}} - 1.0835 \times 10^{-4}\frac{T^2}{\text{K}^2}\,\text{J K}^{-1}\,\text{mol}^{-1}\right)$$

$$\Delta C_P° = -40.21 - 0.0975\frac{T}{\text{K}} + 3.860 \times 10^{-4}\frac{T^2}{\text{K}^2}\,\text{J K}^{-1}\,\text{mol}^{-1}$$

$$\Delta S° = 2S_{298.15}°(NH_3, g) - S_{298.15}°(N_2, g) - 3S_{298.15}°(H_2, g)$$

$$= 2 \times 192.8\,\text{J K}^{-1}\,\text{mol}^{-1} - 191.6\,\text{J K}^{-1}\,\text{mol}^{-1} - 3 \times 130.7\,\text{J K}^{-1}\,\text{mol}^{-1}$$

$$= -198.1\,\text{J K}^{-1}\,\text{mol}^{-1}$$

$$\Delta S_T^\circ = \Delta S_{298.15}^\circ + \int_{298.15}^{T} \frac{\Delta C_p^\circ}{T'} dT'$$

$$= -198.1 \, \text{J K}^{-1} \, \text{mol}^{-1} + \int_{298.15}^{725} \frac{\left(-40.21 - 0.0975\frac{T}{\text{K}} + 3.860 \times 10^{-4} \frac{T^2}{\text{K}^2}\right)}{\frac{T}{\text{K}}} d\frac{T}{\text{K}} \, \text{J K}^{-1} \, \text{mol}^{-1}$$

$$= -198.1 \, \text{J K}^{-1} \, \text{mol}^{-1} - 31.34 \, \text{J K}^{-1} \, \text{mol}^{-1} - 41.62 \, \text{J K}^{-1} \, \text{mol}^{-1} + 84.28 \, \text{J K}^{-1} \, \text{mol}^{-1}$$

$$= -191.2 \, \text{J K}^{-1} \, \text{mol}^{-1}$$

P5.17 The interior of a refrigerator is typically held at 36°F and the interior of a freezer is typically held at 0.00°F. If the room temperature is 65°F, by what factor is it more expensive to extract the same amount of heat from the freezer than from the refrigerator? Assume that the theoretical limit for the performance of a reversible refrigerator is valid in this case.

From Equation (5.44)

$$\eta_r = \frac{T_{cold}}{T_{hot} - T_{cold}}$$

$$T_{room} = \frac{5}{9}(65 - 32) + 273.15 = 291.5 \, \text{K}$$

$$T_{freezer} = \frac{5}{9}(0.00 - 32) + 273.15 = 255.4 \, \text{K}$$

$$T_{refrigerator} = \frac{5}{9}(36 - 32) + 273.15 = 275.4 \, \text{K}$$

For the freezer $\eta_r = \dfrac{255.4 \, \text{K}}{291.5 \, \text{K} - 255.4 \, \text{K}} = 7.1.$

For the refrigerator $\eta_r = \dfrac{275.4 \, \text{K}}{291.5 \, \text{K} - 275.4 \, \text{K}} = 17.$

The freezer is more expensive to operate than the refrigerator by the ratio $17/7.1 = 2.4$.

P5.19 At the transition temperature of 95.4°C, the enthalpy of transition from rhombic to monoclinic sulfur is $0.38 \, \text{kJ mol}^{-1}$.

    a.   Calculate the entropy of transition under these conditions.

    b.   At its melting point, 119°C, the enthalpy of fusion of monoclinic sulfur is $1.23 \, \text{kJ mol}^{-1}$. Calculate the entropy of fusion.

    c.   The values given in parts (a) and (b) are for 1 mol of sulfur; however, in crystalline and liquid sulfur, the molecule is present as S_8. Convert the values of the enthalpy and entropy of fusion in parts (a) and (b) to those appropriate for S_8.

(a) $\Delta S_{transition} = \dfrac{\Delta H_{transition}}{T_{transition}} = \dfrac{0.38 \, \text{kJ mol}^{-1}}{(273.15 + 95.4) \, \text{K}} = 1.0 \, \text{J K}^{-1} \, \text{mol}^{-1}$

(b) $\Delta S_{fusion} = \dfrac{\Delta H_{fusion}}{T_{fusion}} = \dfrac{1.23 \, \text{kJ mol}^{-1}}{(273.15 + 119) \, \text{K}} = 3.14 \, \text{J K}^{-1} \, \text{mol}^{-1}$

(c)   Each of the ΔS in parts (a) and (b) should be multiplied by 8.

    $\Delta S_{transition} = 8.24 \, \text{J K}^{-1} \, \text{mol}^{-1}$

    $\Delta S_{fusion} = 25.1 \, \text{J K}^{-1} \text{mol}^{-1}$

P5.22 Calculate ΔH and ΔS if the temperature of 1.75 moles of Hg(l) is increased from 0.00°C to 75.00°C at 1 bar. Over this temperature range, $C_{P,m} = 30.093 - 4.944 \times 10^{-3} \, T$ J mol^{-1} K^{-1}.

$$\Delta H = 1.75 \text{ mol} \times \int_{273.15 \text{ K}}^{348.15 \text{ K}} (30.093 - 4.944 \times 10^{-3} T) \text{ J mol}^{-1} \text{ K}^{-1} \, dT$$

$$= 1.75 \times (30.093(348.15 \text{ K} - 273.15 \text{ K}) - 2.472 \times 10^{-2}(348.15 \text{ K} - 273.15 \text{ K})^2) \text{ J}$$

$$= 3.75 \times 10^3 \text{ J}$$

$$\Delta S = 1.75 \text{ mol} \times \int_{273.15 \text{ K}}^{348.15 \text{ K}} \left(\frac{30.093 - 4.944 \times 10^{-3} T \text{ J mol}^{-1} \text{ K}^{-1}}{T} \right) dT$$

$$= 1.75 \times \left(-4.944 \times 10^{-3}(T_f - T_i) + 30.093 \ln \frac{T_f}{T_i} \text{ J K}^{-1} \right) = 12.1 \text{ J K}^{-1}$$

P5.23 Calculate ΔS if the temperature of 2.50 mol of an ideal gas with $C_V = 5/2R$ is increased from 160. to 675 K under conditions of (a) constant pressure and (b) constant volume.

a. at constant pressure

$$\Delta S = nC_{P,m} \ln \frac{T_f}{T_i} = 2.50 \text{ mol} \times \left(\frac{5}{2} + 1 \right) \times 8.314 \text{ J mol}^{-1} \text{ K}^{-1} \times \ln \frac{675 \text{ K}}{160. \text{ K}} = 105 \text{ J K}^{-1}$$

b. at constant volume

$$\Delta S = nC_{V,m} \ln \frac{T_f}{T_i} = 2.50 \text{ mol} \times \frac{5}{2} \times 8.314 \text{ J mol}^{-1} \text{ K}^{-1} \times \ln \frac{675 \text{ K}}{160. \text{ K}} = 74.8 \text{ J K}^{-1}$$

P5.25 Calculate ΔS_R° for the reaction $H_2(g) + Cl_2(g) \rightarrow 2 \, HCl(g)$ at 870. K. Omit terms in the temperature-dependent heat capacities higher than T^2/K^2.

From Table 2.4,

$$C_P^{\circ}(H_2, g) = 22.66 + 4.38 \times 10^{-2} \frac{T}{K} - 1.0835 \times 10^{-4} \frac{T^2}{K^2} \text{ J K}^{-1} \text{ mol}^{-1}$$

$$C_P^{\circ}(Cl_2, g) = 22.85 + 6.543 \times 10^{-2} \frac{T}{K} - 1.2517 \times 10^{-4} \frac{T^2}{K^2} \text{ J K}^{-1} \text{ mol}^{-1}$$

$$C_P^{\circ}(HCl, g) = 29.81 - 4.12 \times 10^{-3} \frac{T}{K} + 6.2231 \times 10^{-6} \frac{T^2}{K^2} \text{ J K}^{-1} \text{ mol}^{-1}$$

$$\Delta C_P^{\circ} = 2 \left(29.81 - 4.12 \times 10^{-3} \frac{T}{K} + 6.2231 \times 10^{-6} \frac{T^2}{K^2} \text{ J K}^{-1} \text{ mol}^{-1} \right)$$

$$- \left(22.66 + 4.38 \times 10^{-2} \frac{T}{K} - 1.0835 \times 10^{-4} \frac{T^2}{K^2} \text{ J K}^{-1} \text{ mol}^{-1} \right)$$

$$- \left(22.85 + 6.543 \times 10^{-2} \frac{T}{K} - 1.2517 \times 10^{-4} \frac{T^2}{K^2} \text{ J K}^{-1} \text{ mol}^{-1} \right)$$

$$\Delta C_P^{\circ} = 14.11 - 0.117 \frac{T}{K} + 2.460 \times 10^{-4} \frac{T^2}{K^2} \text{ J K}^{-1} \text{ mol}^{-1}$$

$$\Delta S^{\circ} = 2 S_{298.15}^{\circ}(HCl, g) - S_{298.15}^{\circ}(Cl_2, g) - S_{298.15}^{\circ}(H_2, g)$$

$$= 2 \times 186.9 \text{ J K}^{-1} \text{ mol}^{-1} - 223.1 \text{ J K}^{-1} \text{ mol}^{-1} - 130.7 \text{ J K}^{-1} \text{ mol}^{-1} = 20.0 \text{ J K}^{-1} \text{ mol}^{-1}$$

$$\Delta S_T^{\circ} = \Delta S_{298.15}^{\circ} + \int_{298.15}^{T} \frac{\Delta C_p^{\circ}}{T'} \, dT'$$

$$= 20.0 \text{ J K}^{-1} \text{ mol}^{-1} + \int_{298.15}^{870.} \frac{\left(14.11 - 0.117 \frac{T}{K} + 2.460 \times 10^{-4} \frac{T^2}{K^2} \text{ J K}^{-1} \text{ mol}^{-1} \right)}{\frac{T}{K}} \, d\frac{T}{K} \text{ J K}^{-1} \text{ mol}^{-1}$$

$$= 20.0 \text{ J K}^{-1} \text{ mol}^{-1} + 15.11 \text{ J K}^{-1} \text{ mol}^{-1} - 66.91 \text{ J K}^{-1} \text{ mol}^{-1} + 82.16 \text{ J K}^{-1} \text{ mol}^{-1} = 50.1 \text{ J K}^{-1} \text{ mol}^{-1}$$

P5.28 The amino acid glycine dimerizes to form the dipeptide glycylglycine according to the reaction

$$2\text{Glycine}(s) \rightarrow \text{Glycylglycine}(s) + \text{H}_2\text{O}(l)$$

Calculate ΔS, $\Delta S_{surroundings}$, and $\Delta S_{universe}$ at $T = 298$ K. Useful thermodynamic data follow:

	Glycine	Glycylglycine	Water
ΔH_f° (kJ mol^{-1})	−537.2	−746.0	−285.8
S_m° (J K^{-1} mol^{-1})	103.5	190.0	70.0

$$\Delta S_R^{\circ} = -2 \times 103.5 \text{ J K}^{-1} \text{ mol}^{-1} + 190.0 \text{ J K}^{-1} \text{ mol}^{-1} + 70.0 \text{ J K}^{-1} \text{ mol}^{-1} = 53.0 \text{ J K}^{-1} \text{ mol}^{-1}$$

$$\Delta H_R^{\circ} = 2 \times 537.2 \text{ kJ mol}^{-1} - 746 \text{ kJ mol}^{-1} - 285.8 \text{ kJ mol}^{-1} = 42.6 \text{ kJ mol}^{-1}$$

$$\Delta S_{surroundings} = \frac{-\Delta H_R^{\circ}}{T} = \frac{-42.6 \text{ kJ mol}^{-1}}{298.0 \text{ K}} = -143 \text{ J K}^{-1} \text{ mol}^{-1}$$

$$\Delta S_{total} = -142.95 \text{ J K}^{-1} \text{ mol}^{-1} + 53.0 \text{ J K}^{-1} \text{ mol}^{-1} = -90.0 \text{ J K}^{-1} \text{ mol}^{-1}$$

P5.31 The following heat capacity data have been reported for L-alanine:

T(K)	10.	20.	40.	60.	80.	100.	140.	180.	220.	260.	300.
$C^{\circ}_{P,m}$ (J K^{-1} mol^{-1})	0.49	3.85	17.45	30.99	42.59	52.50	68.93	83.14	96.14	109.6	122.7

By a graphical treatment, obtain the molar entropy of L-alanine at $T = 300$. K. You can perform the integration numerically using either a spread sheet program or a curve-fitting routine and a graphing calculator (see Example Problem 5.9).

The data is graphed below with the heat capacity and temperature on the vertical and horizontal axes, respectively. The best fit to the data has the form

$$C_{P,m} = -9.160 + 0.7761 \, T - 0.001951 \, T^2 + 0.000002764 \, T^3$$

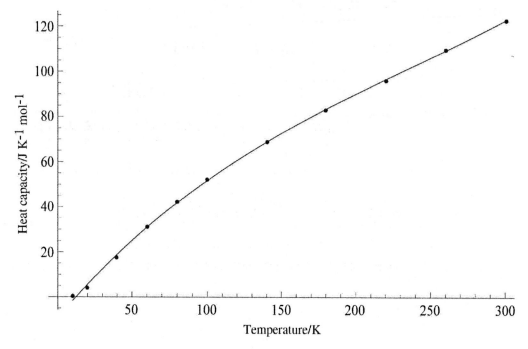

To obtain the entropy at 300. K, we evaluate the integral

$$S°(300. \text{ K}) = \int\limits_{10}^{300} \frac{-9.160 + 0.7761\,T - 0.001951\,T^2 + 0.000002764\,T^3}{T}\,dT$$

$$= [0.7761\,T - 0.0009756\,T^2 + 9.213 \times 10^{-7}\,T^3 - 9.160\,\text{Log}[T]]_{10.\text{ K}}^{300.\text{ K}}$$

$$= 131.1\,\text{J mol}^{-1}\,\text{K}^{-1}$$

We have neglected the contribution to S from temperatures below 10. K as the data is not available. This contribution is very small.

P5.35   Between 0°C and 100°C, the heat capacity of Hg(l) is given by

$$\frac{C_{P,m}(\text{Hg},\,l)}{\text{J K}^{-1}\,\text{mol}^{-1}} = 30.093 - 4.944 \times 10^{-3}\frac{T}{\text{K}}$$

Calculate ΔH and ΔS if 2.25 moles of Hg(l) is raised in temperature from 0.00° to 88.0°C at constant P.

$$\Delta H = n \int\limits_{273.15}^{361.15} C_{P,m}\,d[T/K]$$

$$= 2.25\,\text{mol} \times [30.093 \times (361.15 - 273.15) - 2.472 \times 10^{-3}(361.15^2 - 273.15^2)]\text{J mol}^{-1}$$

$$= 5.65 \times 10^3\,\text{J}$$

$$\Delta S = n \int\limits_{273.15}^{361.15} \frac{C_{P,m}}{[T/K]}\,d[T/K]$$

$$= 2.25\,\text{mol} \times \left[30.093\,\ln\frac{T_f}{T_i} - 4.944 \times 10^{-3}(348.15 - 273.15)\right]\text{J K}^{-1}\,\text{mol}^{-1}$$

$$= 17.9\,\text{J K}^{-1}$$

P5.38   The heat capacity of α-quartz is given by

$$\frac{C_{P,m}(\alpha\text{-quartz},\,s)}{\text{J K}^{-1}\,\text{mol}^{-1}} = 46.94 + 34.31 \times 10^{-3}\frac{T}{\text{K}} - 11.30 \times 10^5\frac{T^2}{\text{K}^2}$$

The coefficient of thermal expansion is given by $\beta = 0.3530 \times 10^{-4}\,\text{K}^{-1}$ and $V_m = 22.6\,\text{cm}^3\,\text{mol}^{-1}$. Calculate ΔS_m for the transformation α-quartz (15.0°C, 1 atm) $\rightarrow \alpha$-quartz (420.°C, 925 atm). From Equations (5.23) and (5.24)

$$\Delta S_m = \int\limits_{T_i}^{T_f} C_{P,m}\frac{dT}{T} - V\beta(P_f - P_i)$$

$$= \int\limits_{288.15}^{693.15} \frac{\left(46.94 + 34.31 \times 10^{-3}\dfrac{T}{\text{K}} - 11.3 \times 10^{-5}\left(\dfrac{T}{\text{K}}\right)^2\right)}{T/K}\,d\frac{T}{\text{K}}\,\text{J K}^{-1}\,\text{mol}^{-1} - V\beta(P_f - P_i)$$

$$= \left[\begin{array}{l} 46.94 \times \ln\dfrac{693.15\,\text{K}}{288.15} + 34.31 \times 10^{-3} \times (693.15 - 288.15) \\[2mm] -5.65 \times 10^{-5} \times (693.15^2 - 288.15^2) \end{array}\right]\text{J K}^{-1}\,\text{mol}^{-1}$$

$$- 22.6\,\text{cm}^3\,\text{mol}^{-1} \times \frac{1\,\text{m}^3}{10^6\,\text{cm}^3} \times 0.3530 \times 10^{-4}\,\text{K}^{-1} \times 924\,\text{atm} \times \frac{1.0125 \times 10^5\,\text{Pa}}{\text{atm}}$$

$$= 41.20\,\text{J K}^{-1}\,\text{mol}^{-1} + 13.896\,\text{J K}^{-1}\,\text{mol}^{-1} - 22.455\,\text{J K}^{-1}\,\text{mol}^{-1} - 0.0746\,\text{J K}^{-1}\,\text{mol}^{-1}$$

$$= 32.6\,\text{J K}^{-1}\,\text{mol}^{-1}$$

P5.39 a. Calculate ΔS if 1.00 mol of liquid water is heated from 0.00°C to 10.0°C under constant pressure if $C_{P,m} = 75.3 \, \text{J K}^{-1} \, \text{mol}^{-1}$.

   b. The melting point of water at the pressure of interest is 0.00°C and the enthalpy of fusion is 6.010 kJ mol^{-1}. The boiling point is 100.°C and the enthalpy of vaporization is 40.65 kJ mol^{-1}. Calculate ΔS for the transformation $H_2O(s, 0.00°C) \rightarrow H_2O(g, 100.°C)$.

   (a) The heat input is the same for a reversible and an irreversible process.

   $$dq = dq_{reversible} = nC_{P,m} \, dT$$

   $$\Delta S = n \int \frac{C_{P,m}}{T} \, dT = nC_{P,m} \ln \frac{T_f}{T_i}$$

   $$= 1 \, \text{mol} \times 75.3 \, \text{J mol}^{-1} \, \text{K}^{-1} \ln \frac{283.15 \, \text{K}}{273.15 \, \text{K}}$$

   $$= 2.71 \, \text{J K}^{-1}$$

   (b) $$\Delta S_{fusion} = \frac{\Delta H_{fusion}}{T_{fusion}} = \frac{1 \, \text{mol} \times 6010 \, \text{J mol}^{-1}}{273.15 \, \text{K}} = 22.00 \, \text{J K}^{-1}$$

   $$\Delta S_{vaporization} = \frac{\Delta H_{vaporization}}{T_{vaporization}} = \frac{1 \, \text{mol} \times 40650. \, \text{J mol}^{-1}}{373.15 \, \text{K}} = 108.94 \, \text{J K}^{-1}$$

   $$\Delta S_{total} = \Delta S_{fusion} + \Delta S_{vaporization} + \Delta S_{heating} = (22.00 + 108.94 + 2.71) \, \text{J K}^{-1} = 133.65 \, \text{J K}^{-1}$$

P5.42 The mean solar flux at the Earth's surface is ~2.00 J cm^{-2} min^{-1}. In a nonfocusing solar collector, the temperature reaches a value of 79.5°C. A heat engine is operated using the collector as the hot reservoir and a cold reservoir at 298 K. Calculate the area of the collector needed to produce 1000. watts. Assume that the engine operates at the maximum Carnot efficiency.

$$\varepsilon = 1 - \frac{T_{hot}}{T_{cold}} = 1 - \frac{298 \, \text{K}}{352.65 \, \text{K}} = 0.155$$

The area required for the solar panal is

$$\frac{\text{power}}{\text{efficiency} \times \text{flux}} = \frac{1000. \, \text{J s}^{-1}}{0.155 \times 2.00 \, \text{J cm}^{-2} \, \text{min}^{-1} \times 1 \, \text{min}/60 \, \text{s} \times 10^4 \, \text{cm}^2/\text{m}^2} = 19.4 \, \text{m}^2$$

6 Chemical Equilibrium

Numerical Problems

P6.1   Calculate ΔA_R° and ΔG_R° for the reaction $C_6H_6(l) + \frac{15}{2}O_2(g) \rightarrow 6CO_2(g) + 3H_2O(l)$ at 298 K from the combustion enthalpy of benzene and the entropies of the reactants and products.

All gaseous reactants and products are treated as ideal gases.

$$\Delta G_{combustion}^\circ = \Delta H_{combustion}^\circ - T\Delta S_{combustion}^\circ$$

$$\Delta S_{combustion}^\circ = 6S^\circ(CO_2, g) + 3S^\circ(H_2O, l) - S^\circ(C_6H_6, l) - 15/2 S^\circ(O_2, g)$$

$$= 6 \times 213.8 \text{ J mol}^{-1}\text{K}^{-1} + 3 \times 70.0 \text{ J mol}^{-1}\text{K}^{-1} - 173.4 \text{ J mol}^{-1}\text{K}^{-1} - 15/2 \times 205.2 \text{ J mol}^{-1}\text{K}^{-1}$$

$$= -219.6 \text{ J mol}^{-1}\text{K}^{-1}$$

$$\Delta G_{combustion}^\circ = -3268 \times 10^3 \text{ kJ mol}^{-1} - 298.15 \text{ K} \times (-219.6 \text{ J mol}^{-1}\text{K}^{-1}) = -3203 \times 10^3 \text{ kJ mol}^{-1}$$

$$\Delta A_{combustion}^\circ = \Delta U_{combustion}^\circ - T\Delta S_{combustion}^\circ$$

$$= \Delta H_{combustion}^\circ - \Delta(PV)_{combustion} - T\Delta S_{combustion}^\circ$$

$$= \Delta G_{combustion}^\circ + T\Delta S_{combustion}^\circ - \Delta(PV) - T\Delta S_{combustion}^\circ$$

$$= \Delta G_{combustion}^\circ - \Delta n RT$$

where Δn is the change in the number of moles of gas phase species in the reaction

$$\Delta A_{combustion}^\circ = -3203 \times 10^3 \text{ kJ mol}^{-1} + 1.5 \times 8.314 \text{ J mol}^{-1}\text{K}^{-1} \times 298.15 \text{ K}$$

$$= -3199 \times 10^3 \text{ kJ mol}^{-1}$$

The change in PV for the liquid can be neglected because liquids are essentially incompressible over the pressure range in this problem.

P6.3   A sample containing 2.75 moles of N_2 and 6.25 mol of H_2 is placed in a reaction vessel and brought to equilibrium at 52.0 bar and 690. K in the reaction $\frac{1}{2}N_2(g) + \frac{3}{2}H_2(g) \rightleftharpoons NH_3(g)$.

     a.   Calculate K_P at this temperature.

     b.   Set up an equation relating K_P and the extent of reaction as in Example Problem 6.10.

     c.   Using numerical equation solving software, calculate the number of moles of each species present at equilibrium.

     (a) $\Delta H_R^\circ = \Delta H_f^\circ(NH_3, g) = -45.9 \times 10^3 \text{ J mol}^{-1}$

$$\Delta G_R^\circ(298.15 \text{ K}) = \Delta G_f^\circ(NH_3, g) = -16.5 \times 10^3 \text{ J mol}^{-1}$$

$$\ln K_P(T_f) = -\frac{\Delta G_R^\circ(298.15 \text{ K})}{R \times 298.15 \text{ K}} - \frac{\Delta H_R^\circ}{R}\left(\frac{1}{T_f} - \frac{1}{298.15 \text{ K}}\right)$$

$$\ln K_P(690. \text{ K}) = \frac{16.5 \times 10^3 \text{ J mol}^{-1}}{8.314 \text{ J mol}^{-1}\text{K}^{-1} \times 298.15 \text{ K}} + \frac{45.9 \times 10^3 \text{ J mol}^{-1}}{8.314 \text{ J mol}^{-1}\text{K}^{-1}} \times \left(\frac{1}{690. \text{ K}} - \frac{1}{298.15 \text{ K}}\right)$$

$$= -3.85931$$

$$K_P(690. \text{ K}) = 2.11 \times 10^{-2}$$

(b)
$$\tfrac{1}{2}N_2(g) \quad + \quad \tfrac{3}{2}H_2(g) \quad \leftrightarrow \quad NH_3(g)$$

	$\tfrac{1}{2}N_2(g)$	$\tfrac{3}{2}H_2(g)$	$NH_3(g)$
Initial number of moles	2.75	6.25	0
Moles present at equilibrium	$2.75 - \xi$	$6.25 - 3\xi$	2ξ
Mole fraction present at equilibrium	$\dfrac{2.75 - \xi}{9.00 - 2\xi}$	$\dfrac{6.25 - 3\xi}{9.00 - 2\xi}$	$\dfrac{2\xi}{9.00 - 2\xi}$
Partial pressure at Equilibrium, $P_i = x_i P$	$\dfrac{2.75 - \xi}{9.00 - 2\xi}P$	$\dfrac{6.25 - 3\xi}{9.00 - 2\xi}P$	$\left(\dfrac{2\xi}{9.00 - 2\xi}\right)P$

We next express K_P in terms of n_0, ξ, and P.

$$K_P(T) = \frac{\left(\dfrac{P_{NH_3}^{eq}}{P^\circ}\right)}{\left(\dfrac{P_{N_2}^{eq}}{P^\circ}\right)^{\tfrac{1}{2}}\left(\dfrac{P_{H_2}^{eq}}{P^\circ}\right)^{\tfrac{3}{2}}} = \frac{\left(\dfrac{2\xi}{9.00 - 2\xi}\right)\dfrac{P}{P^\circ}}{\left(\left(\dfrac{2.75 - \xi}{9.00 - 2\xi}\right)\dfrac{P}{P^\circ}\right)^{\tfrac{1}{2}}\left(\left(\dfrac{6.25 - 3\xi}{9.00 - 2\xi}\right)\dfrac{P}{P^\circ}\right)^{\tfrac{3}{2}}}$$

The following equation can be solved numerically using a program such as Mathematica.

$$K_P(T) = \frac{\left(\dfrac{2\xi}{9.00 - 2\xi}\right)52.0}{\left(\left(\dfrac{2.75 - \xi}{9.00 - 2\xi}\right)52.0\right)^{\tfrac{1}{2}}\left(\left(\dfrac{6.25 - 3\xi}{9.00 - 2\xi}\right)52.0\right)^{\tfrac{3}{2}}} = 2.11 \times 10^{-2}$$

The physically meaningful root of the cubic equation is $\xi = 0.7902$. Therefore, there are 1.96 moles of $N_2(g)$, 3.88 moles of $H_2(g)$, and 1.58 moles of $NH_3(g)$ at equilibrium.

P6.4   Consider the equilibrium $NO_2(g) \rightleftharpoons NO(g) + \tfrac{1}{2}O_2(g)$. One mole of $NO_2(g)$ is placed in a vessel and allowed to come to equilibrium at a total pressure of 1 bar. An analysis of the contents of the vessel gives the following results:

T	700. K	800. K
P_{NO}/P_{NO_2}	0.872	2.50

a.   Calculate K_P at 700. and 800. K.

b.   Calculate ΔG_R° and ΔH_R° for this reaction at 298.15 K, using only the data in the problem. Assume that ΔH_R° is independent of temperature.

c.   Calculate ΔG_R° and ΔH_R° using the data tables and compare your answer with that obtained in part (b).

(a) $NO_2(g) \rightleftharpoons NO(g) + \dfrac{1}{2}O_2(g)$

$$K_P = \frac{(P_{NO}/P^\circ)(P_{O_2}/P^\circ)^{1/2}}{P_{NO_2}/P^\circ}$$

$$\text{At } 700.\,K, \frac{P_{NO}}{P_{NO_2}} = 0.872 \text{ and } P_{O_2} = \frac{1}{2}P_{NO}$$

$$P_{total} = P_{NO} + P_{NO_2} + P_{O_2} = 1 \text{ bar}$$

$$1 \text{ bar} = 0.872\,P_{NO_2} + P_{NO_2} + 0.436\,P_{NO_2}$$

$$P_{NO_2} = 0.4333 \text{ bar}$$

$$K_P = \frac{(0.872 \times 0.4333) \times \sqrt{0.436 \times 0.4333}}{0.4333} = 0.379$$

At 800. K, $\dfrac{P_{NO}}{P_{NO_2}} = 2.50$ and $P_{O_2} = \dfrac{1}{2} P_{NO}$

$$P_{Total} = P_{NO} + P_{NO_2} + P_{O_2}$$

$$1 \text{ bar} = 2.50 \, P_{NO_2} + P_{NO_2} + 1.25 \, P_{NO_2}$$

$$P_{NO_2} = 0.2105 \text{ bar}$$

$$K_P = \frac{(2.50 \times 0.2105) \times \sqrt{1.25 \times 0.0.2105}}{0.2105} = 1.28$$

(b)   Assuming that ΔH_R° is independent of temperature,

$$\ln \frac{K_P(800.\,\text{K})}{K_P(700.\,\text{K})} = \frac{-\Delta H_R^\circ}{R}\left(\frac{1}{800.\,\text{K}} - \frac{1}{700.\,\text{K}}\right)$$

$$\Delta H_R^\circ = -\frac{R \times \ln\left(\dfrac{K_P(800.\,\text{K})}{K_P(700.\,\text{K})}\right)}{\left(\dfrac{1}{800.\,\text{K}} - \dfrac{1}{700.\,\text{K}}\right)} = 56.7 \times 10^3 \text{ J mol}^{-1}$$

$$\ln K_P(298.15 \text{ K}) = \ln K_P(700.\,\text{K}) - \frac{\Delta H_R^\circ}{R} \times \left(\frac{1}{298.15 \text{ K}} - \frac{1}{700.\,\text{K}}\right) = -14.1141$$

$$\Delta G_R^\circ(298.15 \text{ K}) = -RT \ln K_P(298.15 \text{ K})$$

$$= -8.314 \text{ J mol}^{-1} \text{ K}^{-1} \times 298.15 \text{ K} \times (-14.1141)$$

$$= 35.0 \times 10^3 \text{ J mol}^{-1}$$

(c) $\Delta H_R^\circ = \Delta H_f^\circ(NO, g) - \Delta H_f^\circ(NO_2, g)$

$$= 91.3 \times 10^3 \text{ J mol}^{-1} - 33.2 \times 10^3 \text{ J mol}^{-1} = 58.1 \times 10^3 \text{ J mol}^{-1}$$

P6.7   The pressure dependence of G is quite different for gases and condensed phases. Calculate ΔG_m for the processes (C, *solid, graphite*, 1 bar, 298.15 K) $\to$ (C, *solid, graphite*, 325 bar, 298.15 K) and (He, *g*, 1 bar, 298.15 K) $\to$ (He, *g*, 325 bar, 298.15 K). By what factor is ΔG_m greater for He than for graphite?

For a solid or liquid, we can assume that the volume is indepent of pressure over a limited range in P.

$$\Delta G = \int_{P_i}^{P_f} V \, dP = V(P_f - P_i)$$

$$\Delta G_m(C, s, 325 \text{ bar}) = \Delta G_m(C, s, 1 \text{ bar}) + V_m(P_f - P_i) = G_m(C, s\, 1 \text{ bar}) + \frac{M}{\rho}(P_f - P_i)$$

$$= 0 + \frac{12.011 \times 10^{-3} \text{ kg/mol}}{2250 \text{ kg m}^{-3}} \times 324.0 \times 10^5 \text{ Pa} = 173 \text{ J mol}^{-1}$$

Treating He as an ideal gas,

$$G_m(\text{He}, g, 325 \text{ bar}) = G_m(\text{He}, g, 1 \text{ bar}) + \int_{P_i}^{P_f} V \, dP$$

$$= 0 + RT \ln \frac{P_f}{P_i} = 1 \text{ mole} \times 8.314 \text{ J mol}^{-1} \text{ K}^{-1} \times 298.15 \text{ K} \times \ln \frac{325 \text{ bar}}{1 \text{ bar}} = 14.3 \times 10^3 \text{ J mol}^{-1}$$

This result is a factor of 82.9 greater than that for graphite.

P6.10   Calculate K_P at 600. K for the reaction $N_2O_4(l) \rightleftharpoons 2NO_2(g)$ assuming that ΔH_R° is constant over the interval 298–725 K.

$$\Delta H_R^\circ = 2\Delta H_f^\circ(NO_2, g) - \Delta H_f^\circ(N_2O_4, l)$$

$$= 2 \times 33.2 \times 10^3 \text{ J mol}^{-1} + 19.5 \times 10^3 \text{ J mol}^{-1} = 85.9 \times 10^3 \text{ J mol}^{-1}$$

$$\Delta G_R^\circ = 2\Delta G_f^\circ(NO_2, g) - \Delta G_f^\circ(N_2O_4, l)$$

$$= 2 \times 51.3 \times 10^3 \text{ J mol}^{-1} - 97.5 \times 10^3 \text{ J mol}^{-1} = 5.1 \times 10^3 \text{ J mol}^{-1}$$

We use the Gibbs–Helmholtz equation to relate the Gibbs reaction energy at two temperatures.

$$\ln K_P(T_f) = -\frac{\Delta G_R^\circ(298.15 \text{ K})}{R \times 298.15 \text{ K}} - \frac{\Delta H_R^\circ}{R}\left(\frac{1}{T_f} - \frac{1}{298.15 \text{ K}}\right)$$

$$\ln K_P(600. \text{ K}) = -\frac{5.10 \times 10^3 \text{ J mol}^{-1}}{8.314 \text{ J K}^{-1} \text{ mol}^{-1} \times 298.15 \text{ K}} - \frac{85.9 \times 10^3 \text{ J mol}^{-1}}{8.314 \text{ J K}^{-1} \text{ mol}^{-1}} \times \left(\frac{1}{600. \text{ K}} - \frac{1}{298.15 \text{ K}}\right)$$

$$\ln K_P(600. \text{ K}) = 15.3762$$

$$K_P(600. \text{ K}) = 4.76 \times 10^6$$

P6.12   For the reaction $C(graphite) + H_2O(g) \rightleftharpoons CO(g) + H_2(g)$, $\Delta H_R^\circ = 131.28 \text{ kJ mol}^{-1}$ at 298.15 K. Use the values of $C_{P,m}$ at 298.15 K in the data tables to calculate ΔH_R° at 125.0°C.

$$\Delta H(T_2) = \Delta H(T_1) + \Delta C_{P,m}(T_2 - T_1)$$

$$\Delta H(398.15 \text{ K}) = 131.28 \text{ kJ mol}^{-1}$$

$$+ 100. \text{ K} \times \left(\begin{array}{l} 29.14 \text{ J K}^{-1} \text{ mol}^{-1} + 28.84 \text{ J K}^{-1} \text{ mol}^{-1} - 8.52 \text{ J K}^{-1} \text{ mol}^{-1} \\ -33.59 \text{ J K}^{-1} \text{ mol}^{-1} \end{array}\right)$$

$$= 132.9 \text{ kJ mol}^{-1}$$

P6.13 $Ca(HCO_3)_2(s)$ decomposes at elevated temperatures according to the stoichiometric equation $Ca(HCO_3)_2(s) \rightleftharpoons CaCO_3(s) + H_2O(g) + CO_2(g)$.

a.   If pure $Ca(HCO_3)_2(s)$ is put into a sealed vessel, the air is pumped out, and the vessel and its contents are heated, the total pressure is 0.290 bar. Determine K_P under these conditions.

b.   If the vessel also contains 0.120 bar $H_2O(g)$ at the final temperature, what is the partial pressure of $CO_2(g)$ at equilibrium?

(a) $Ca(HCO_3)_2(s) \rightleftharpoons CaCO_3(s) + H_2O(g) \quad + \quad CO_2(g)$

Partial pressure at equilibrium, $P_i = x_i P$ ξP ξP

The total pressure is made up of equal partial pressures of $H_2O(g)$ and $CO_2(g)$.

$$K_P = \frac{P_{H_2O}}{P^\circ}\frac{P_{CO_2}}{P^\circ} = \left(\frac{P_{H_2O}}{P^\circ}\right)^2 = \left(\frac{0.290}{2}\right)^2 = 0.0210$$

(b)   If one of the products is originally present

$$Ca(HCO_3)_2(s) \leftrightarrow CaCO_3(s) + H_2O(g) \quad + \quad CO_2(g)$$

Partial pressure at equilibrium, $P_i = x_i P$ $\xi P + P_i$ ξP

$$K_P = \frac{P_{H_2O}}{P^\circ}\frac{P_{CO_2}}{P^\circ} = \left(\frac{P + P_i}{P^\circ}\right)\left(\frac{P}{P^\circ}\right) = \left(\frac{P}{P^\circ} + 0.120\right)\left(\frac{P}{P^\circ}\right) = 0.02103$$

$$\frac{P}{P^\circ} = \frac{P_{CO_2}}{P^\circ} = 0.0969; \quad P_{CO_2} = 0.0969 \text{ bar}$$

P6.16   Collagen is the most abundant protein in the mammalian body. It is a fibrous protein that serves to strengthen and support tissues. Suppose a collagen fiber can be stretched reversibly with a force constant of $k = 10.0 \text{ N m}^{-1}$ and that the force, $\mathbf{F}$ (see Table 2.1), is given by $\mathbf{F} = k\ell$. When a collagen fiber is contracted reversibly, it absorbs heat $q_{rev} = 0.050 \text{ J}$. Calculate the change in the Helmholtz energy, ΔA, as the fiber contracts isothermally from $\ell = 0.20$ to 0.10 cm. Calculate also the reversible work performed w_{rev}, ΔS, and ΔU. Assume that the temperature is constant at $T = 310. \text{ K}$.

$$w_{rev} = \int_{0.20\,m}^{0.10\,m} k\ell \, d\ell = \left[\frac{1}{2}k\ell^2\right]_{0.20\,m}^{0.10\,m} = 10.0 \text{ N m}^{-1} \times \frac{(0.10 \text{ m})^2 - (0.20 \text{ m})^2}{2} = -0.015 \text{ J}$$

$$\Delta A = w_{rev} = -0.015 \text{ J}$$

$$\Delta S = \frac{q_{rev}}{T} = \frac{0.050 \text{ J}}{310. \text{ K}} = 1.61 \times 10^{-4} \text{ J K}^{-1}$$

$$\Delta U = \Delta A + T\Delta S = -0.015 \text{ J} + 0.050 \text{ J} = -0.035 \text{ J}$$

P6.17   Calculate $\mu_{O_2}^{mixture}(298.15 \text{ K}, 1 \text{ bar})$ for oxygen in air, assuming that the mole fraction of O_2 in air is 0.210. Use the conventional molar Gibbs energy defined in Section 6.17.

We calculate the conventional molar Gibbs energy as described in Example Problem 6.17.

$$\mu_{O_2}^{\circ}(T) = -TS^{\circ}(O_2, g, 298.15 \text{ K}) = -298.15 \text{ K} \times 205.2 \text{ J K}^{-1} \text{ mol}^{-1} = -61.2 \text{ kJ mol}^{-1}$$

$$\mu_{O_2}^{mixture}(T, P) = \mu_{O_2}^{\circ}(T) + RT \ln\frac{P}{P^{\circ}} + RT \ln x_{O_2}$$

$$= -61.2 \text{ kJ mol}^{-1} + RT \ln\frac{1 \text{ bar}}{1 \text{ bar}} + 8.314 \text{ J mol}^{-1} \text{ K}^{-1} \times 298.15 \text{ K} \times \ln 0.210$$

$$= -65.0 \times 10^3 \text{ J mol}^{-1}$$

P6.18   Calculate the maximum nonexpansion work that can be gained from the combustion of benzene(l) and of $H_2(g)$ on a per gram and a per mole basis under standard conditions. Is it apparent from this calculation why fuel cells based on H_2 oxidation are under development for mobile applications?

$$C_6H_6(l) + 15/2 O_2(g) \rightarrow 6CO_2(g) + 3H_2O(l)$$

$$w_{nonexpansion}^{max} = \Delta G_R^{\circ} = 3\Delta G_f^{\circ}(H_2O, l) + 6\Delta G_f^{\circ}(CO_2, g) - \frac{15}{2}\Delta G_f^{\circ}(O_2, g) - \Delta G_f^{\circ}(C_6H_6, l)$$

$$w_{nonexpansion}^{max} = 3 \times (-237.1 \text{ kJ mol}^{-1}) + 6 \times (-394.4 \text{ kJ mol}^{-1}) - \frac{15}{2} \times (0) - 124.5 \text{ kJ mol}^{-1}$$

$$= -3202 \text{ kJ mol}^{-1}$$

$$= -3202 \text{ kJ mol}^{-1} \times \frac{1 \text{ mol}}{78.18 \text{ g}} = -40.99 \text{ kJ g}^{-1}$$

$$H_2(g) + \frac{1}{2}O_2(g) \rightarrow H_2O(l)$$

$$w_{nonexpansion}^{max} = \Delta G_R^{\circ} = \Delta G_f^{\circ}(H_2O, l) = -237.1 \text{ kJ mol}^{-1}$$

$$w_{nonexpansion}^{max} = -237.1 \text{ kJ mol}^{-1} \times \frac{1 \text{ mol}}{2.016 \text{ g}} = -117.6 \text{ kJ g}^{-1}$$

On a per gram basis, nearly three times as much work can be extracted from the oxidation of hydrogen than benzene.

P6.20   Calculate ΔG for the isothermal expansion of 2.25 mol of an ideal gas at 325 K from an initial pressure of 12.0 bar to a final pressure of 2.5 bar.

$$dG = -SdT + VdP$$

At constant T, we consider the reversible process. Because G is a state function, any path between the same initial and final states will give the same result.

$$\Delta G = \int_{P_i}^{P_f} V dP = nRT \ln \frac{P_f}{P_i} = 2.25 \text{ mol} \times 8.314 \text{ J mol}^{-1} \text{ K}^{-1} \times 325 \text{ K} \times \ln \frac{2.50 \text{ bar}}{12.0 \text{ bar}}$$

$$= -9.54 \times 10^3 \text{ J}$$

P6.24 Consider the reaction $FeO(s) + CO(g) \rightleftharpoons Fe(s) + CO_2(g)$ for which K_P is found to have the following values:

T	700.°C	1200.°C
K_P	0.688	0.310

a. Calculate ΔG_R°, ΔS_R°, and ΔH_R° for this reaction at 700.°C. Assume that ΔH_R° is independent of temperature.

b. Calculate the mole fraction of $CO_2(g)$ present in the gas phase at 700.°C.

$$FeO(s) + CO(g) \rightleftharpoons Fe(s) + CO_2(g)$$

$$K_P = \frac{P_{CO_2}/P^\circ}{P_{CO}/P^\circ}$$

$$\ln \frac{K_P(1200.°C)}{K_P(700.°C)} = \frac{\Delta H_R^\circ}{R} \left(\frac{1}{1473.15 \text{ K}} - \frac{1}{973.15 \text{ K}} \right)$$

Assume that ΔH_R° is independent of temperature.

$$\Delta H_R^\circ = \frac{-R \ln \dfrac{K_P(1473.15 \text{ K})}{K_P(973.15 \text{ K})}}{\left(\dfrac{1}{1473.15 \text{ K}} - \dfrac{1}{973.15 \text{ K}} \right)}$$

$$= \frac{-8.314 \text{ J mol}^{-1} \text{ K}^{-1} \times \ln \dfrac{0.310}{0.688}}{\left(\dfrac{1}{1473.15 \text{ K}} - \dfrac{1}{973.15 \text{ K}} \right)} = -19.0 \text{ kJ mol}^{-1}$$

$$\Delta G_R^\circ(700.°C) = -RT \ln K_P(700.°C)$$

$$= -8.314 \text{ J mol}^{-1} \text{ K}^{-1} \times 973.15 \text{ K} \times \ln(0.688) = 3.03 \times 10^3 \text{ J mol}^{-1}$$

$$\Delta S_R^\circ(700.°C) = \frac{\Delta H_R^\circ - \Delta G_R^\circ(600.°C)}{T}$$

$$= \frac{-19.0 \times 10^3 \text{ J mol}^{-1} - 3.03 \times 10^3 \text{ J mol}^{-1}}{973.15 \text{ K}} = -22.6 \text{ J mol}^{-1} \text{ K}^{-1}$$

$$K_P = P_{CO_2}/P_{CO} = 0.688$$

$$K_p = K_x \text{ because } \Delta n = 0$$

$$\frac{x_{CO_2}}{x_{CO}} = 0.688 \text{ and } x_{CO_2} + x_{CO} = 1$$

$$x_{CO_2} = 0.408 \quad x_{CO} = 0.592$$

P6.27 A gas mixture with 4.50 mol of Ar, x moles of Ne, and y moles of Xe is prepared at a pressure of 1 bar and a temperature of 298 K. The total number of moles in the mixture is five times that of Ar. Write an expression for ΔG_{mixing} in terms of x. At what value of x does the magnitude of ΔG_{mixing} have its minimum value? Calculate ΔG_{mixing} for this value of x.

If the number of moles of Ne is x, the number of moles of Xe is $y = 18.0 - x$.

$$\Delta G_{mixing} = nRT \sum_i x_i \ln x_i = nRT\left(\frac{1}{5}\ln\frac{1}{5} + \frac{x}{22.5}\ln\frac{x}{22.5} + \frac{18.0-x}{22.5}\ln\frac{18.0-x}{22.5}\right)$$

$$\frac{d\Delta G_{mixing}}{dx} = nRT\left(\frac{1}{22.5}\ln\frac{x}{22.5} + \frac{x}{22.5}\frac{22.5}{x} - \frac{1}{22.5}\ln\frac{18.0-x}{22.5} + \frac{18.0-x}{22.5}\frac{22.5}{18.0-x}(-1)\right)$$

$$= nRT\left(\frac{1}{22.5}\ln\frac{x}{22.5} + 1 - \frac{1}{22.5}\ln\frac{18.0-x}{22.5} - 1\right) = \frac{nRT}{22.5}\ln\frac{x}{18.0-x} = 0$$

$$\frac{x}{18.0-x} = 1; \quad x = 9.00$$

$$\Delta G_{mixing} = nRT\left(\frac{1}{5}\ln\frac{1}{5} + \frac{2\times 9.00}{22.5}\ln\frac{9.00}{22.5}\right)$$

$$= 22.5\,\text{mol} \times 8.314\,\text{J mol}^{-1}\,\text{K}^{-1} \times 298.15\,\text{K} \times (-1.0549) = -58.8\times 10^3\,\text{J}$$

P6.34 You have containers of pure O_2 and N_2 at 298 K and 1 atm pressure. Calculate ΔG_{mixing} relative to the unmixed gases of

a. a mixture of 10. mol of O_2 and 10. mol of N_2

b. a mixture of 10. mol of O_2 and 20. mol of N_2

c. Calculate ΔG_{mixing} if 10 mol of pure N_2 is added to the mixture of 10 mol of O_2 and 10 mol of N_2.
$$\Delta G_{mixing} = nRT(x_1 \ln x_1 + x_2 \ln x_2)$$

(a) $$\Delta G_{mixing} = 20\,\text{mol} \times 8.314\,\text{J K}^{-1}\,\text{mol}^{-1} \times 298\,\text{K} \times \left(\frac{1}{2}\ln\frac{1}{2} + \frac{1}{2}\ln\frac{1}{2}\right) = -34.3\,\text{kJ}$$

(b) $$\Delta G_{mixing} = 30\,\text{mol} \times 8.314\,\text{J K}^{-1}\,\text{mol}^{-1} \times 298\,\text{K} \times \left(\frac{1}{3}\ln\frac{1}{3} + \frac{2}{3}\ln\frac{2}{3}\right) = -47.3\,\text{kJ}$$

(c) $$\Delta G_{mixing} = \Delta G_{mixing}\,(\text{separate gases}) - \Delta G_{mixing}\,(20\,\text{mol A} + 10\,\text{mol B})$$

$$= 30\,\text{mol} \times 8.314\,\text{J K}^{-1}\,\text{mol}^{-1} \times 298\,\text{K} \times \left(\frac{2}{3}\ln\frac{2}{3} + \frac{1}{3}\ln\frac{1}{3}\right) + 34.3\,\text{kJ}$$

$$= -47.3\,\text{kJ} + 34.3\,\text{kJ} = -13.0\,\text{kJ}$$

P6.36 Consider the equilibrium in the reaction $3O_2(g) \rightleftharpoons 2O_3(g)$. Assume that ΔH_R° is independent of temperature.

a. Without doing a calculation, predict whether the equilibrium position will shift toward reactants or products as the pressure is increased.

b. Without doing a calculation, predict whether the equilibrium position will shift toward reactants or products as the temperature is increased.

c. Calculate K_p at 600. and 700. K. Compare your results with your answer to part (b).

d. Calculate K_x at 600. K and pressures of 1.00 and 2.25 bar. Compare your results with your answer to part (a).

(a) The number of moles of products is fewer than the number of moles of reactants. Therefore, the equilibrium position will shift toward products as the pressure is increased.

(b) Using the data tables, $\Delta H_R^\circ > 0$, the equilibrium position will shift toward products as the temperature is increased.

(c) $$\Delta G_R^\circ = 2\Delta G_f^\circ(O_3, g) = 2 \times 163.2 \times 10^3\,\text{J mol}^{-1}$$

$$\Delta H_R^\circ = 2\Delta H_f^\circ(O_3, g) = 2 \times 142.7 \times 10^3\,\text{J mol}^{-1}$$

$$\ln K_P(T_f) = -\frac{\Delta G_R^\circ(298.15\,\text{K})}{R \times 298.15\,\text{K}} - \frac{\Delta H_R^\circ}{R}\left(\frac{1}{T_f} - \frac{1}{298.15\,\text{K}}\right)$$

$$\ln K_P(600.\,\mathrm{K}) = -\frac{2 \times 163.2 \times 10^3\,\mathrm{J\,mol^{-1}}}{8.314\,\mathrm{J\,K^{-1}\,mol^{-1}} \times 298.15\,\mathrm{K}} - \frac{2 \times 142.7 \times 10^3\,\mathrm{J\,mol^{-1}}}{8.314\,\mathrm{J\,K^{-1}\,mol^{-1}}} \times \left(\frac{1}{600.\,\mathrm{K}} - \frac{1}{298.15\,\mathrm{K}} \right)$$

$$\ln K_P(600.\,\mathrm{K}) = -73.7529$$

$$K_P(600.\,\mathrm{K}) = 9.32 \times 10^{-33}$$

$$\ln K_P(700.\,\mathrm{K}) = -\frac{2 \times 163.2 \times 10^3\,\mathrm{J\,mol^{-1}}}{8.314\,\mathrm{J\,K^{-1}\,mol^{-1}} \times 298.15\,\mathrm{K}} - \frac{2 \times 142.7 \times 10^3\,\mathrm{J\,mol^{-1}}}{8.314\,\mathrm{J\,K^{-1}\,mol^{-1}}} \times \left(\frac{1}{700.\,\mathrm{K}} - \frac{1}{298.15\,\mathrm{K}} \right)$$

$$\ln K_P(700.\,\mathrm{K}) = -65.5796$$

$$K_P(700.\,\mathrm{K}) = 3.30 \times 10^{-29}$$

(d)   Calculate K_x at 600. K and pressures of 1.00 and 2.25 bar.

$$K_x = K_P \left(\frac{P}{P^\circ} \right)^{-\Delta\nu} = 9.32 \times 10^{-33} \times \left(\frac{1.00\,\mathrm{bar}}{1\,\mathrm{bar}} \right)^{+1} = 9.32 \times 10^{-33} \text{ for } P = 1.00\,\mathrm{bar}$$

$$K_x = K_P \left(\frac{P}{P^\circ} \right)^{-\Delta\nu} = 9.32 \times 10^{-33} \times \left(\frac{2.25\,\mathrm{bar}}{1\,\mathrm{bar}} \right)^{+1} = 2.10 \times 10^{-32} \text{ for } P = 2.25\,\mathrm{bar}$$

P6.37 N_2O_3 dissociates according to the equilibrium $N_2O_3(g) \rightleftharpoons NO_2(g) + NO(g)$. At 298 K and one bar pressure, the degree of dissociation defined as the ratio of moles of $NO(g)$ or $NO_2(g)$ to the initial moles of the reactant assuming that no dissociatioin occurs is 3.5×10^{-3}. Calculate ΔG_R° for this reaction.

From Example Problem 6.10,

We set up the following table:

	$N_2O_3(g)$	$\rightleftharpoons$	$NO_2(g)$	+	$NO(g)$
Initial number of moles	n_0		0		0
Moles present at equilibrium	$n_0 - \delta_{eq}$		δ_{eq}		δ_{eq}
Mole fraction present at equilibrium, x_i	$\dfrac{n_0 - \delta_{eq}}{n_0 + \delta_{eq}}$		$\dfrac{\delta_{eq}}{n_0 + \delta_{eq}}$		$\dfrac{\delta_{eq}}{n_0 + \delta_{eq}}$
Partial pressure at equilibrium, $P_i = x_i P$	$\left(\dfrac{n_0 - \delta_{eq}}{n_0 + \delta_{eq}} \right) P$		$\dfrac{\delta_{eq}}{n_0 + \delta_{eq}} P$		$\dfrac{\delta_{eq}}{n_0 + \delta_{eq}} P$

We next express K_P in terms of n_0, δ_{eq}, and P:

$$K_P(T) = \frac{\left(\dfrac{P_{NO}^{eq}}{P^\circ} \right) \left(\dfrac{P_{NO_2}^{eq}}{P^\circ} \right)}{\left(\dfrac{P_{N_2O_3}^{eq}}{P^\circ} \right)} = \frac{\left[\left(\dfrac{\delta_{eq}}{n_0 + \delta_{eq}} \right) \dfrac{P}{P^\circ} \right]^2}{\left(\dfrac{n_0 - \delta_{eq}}{n_0 + \delta_{eq}} \right) \dfrac{P}{P^\circ}} = \frac{\delta_{eq}^2}{(n_0 + \delta_{eq})(n_0 - \delta_{eq})} \frac{P}{P^\circ} = \frac{\delta_{eq}^2}{(n_0)^2 - \delta_{eq}^2} \frac{P}{P^\circ}$$

This expression is converted into one in terms of δ_{eq}:

Because $\alpha = \delta_{eq}/n_0$

$$K_P(T) = \frac{\delta_{eq}^2}{(n_0)^2 - \delta_{eq}^2} \frac{P}{P^\circ} = \frac{\alpha^2}{1 - \alpha^2} \frac{P}{P^\circ}$$

$$\left(K_P(T) + \frac{P}{P^\circ} \right) \alpha^2 = K_P(T)$$

$$\alpha = \sqrt{\frac{K_P(T)}{K_P(T) + \dfrac{P}{P^\circ}}}$$

$$\Delta G_R^\circ = -RT \ln K_P = -RT \ln \frac{\alpha^2}{1-\alpha^2}\frac{P}{P^\circ} = -8.314 \text{ J K}^{-1} \text{ mol}^{-1} \times 298 \text{ K} \times \ln\frac{(3.5\times10^{-3})^2}{1-(3.5\times10^{-3})^2}$$

$$= 28.0\times10^3 \text{ J mol}^{-1}$$

P6.39 Assume the internal energy of an elastic fiber under tension (see Problem 6.16) is given by $dU = T\,dS - P\,dV - F\,d\ell$. Obtain an expression for $(\partial G/\partial \ell)_{P,T}$ and calculate the maximum nonexpansion work obtainable when a collagen fiber contracts from $\ell = 20.0$ to 10.0 cm at constant P and T. Assume other properties as described in Problem 6.16.

$$dU = T\,dS - P\,dV - \gamma\,d\ell$$

$$dG = d(U + PV - TS) = T\,dS - P\,dV - \gamma\,d\ell + P\,dV + V\,dP - T\,dS - S\,dT$$

$$= -\gamma\,d\ell + V\,dP - S\,dT$$

$$\left(\frac{\partial G}{\partial \ell}\right)_{P,T} = -\gamma = k\ell$$

$$\Delta G = w_{rev} = \int_{0.20\,cm}^{0.10\,cm} k\ell\,d\ell = \left[\frac{1}{2}k\ell^2\right]_{0.20\,cm}^{0.10\,cm} = -10 \text{ N m}^{-1} \times \frac{(0.20 \text{ cm})^2 - (0.10 \text{ cm})^2}{2} = 1.5\times10^{-5} \text{ J}$$

7 The Properties of Real Gases

Numerical Problems

P7.1 A van der Waals gas has a value of $z = 1.00061$ at 410. K and 1 bar and the Boyle temperature of the gas is 195 K. Because the density is low, you can calculate V_m from the ideal gas law. Use this information and the result of Problem 7.28, $z \approx 1 + (b - a/RT)(1/V_m)$, to estimate a and b.

$$z - 1 = \frac{1}{V_m}\left(b - \frac{a}{RT}\right); \quad T_B = \frac{a}{Rb}$$

$$z - 1 = \frac{b}{V_m}\left(1 - \frac{T_B}{T}\right)$$

$$b = \frac{z-1}{1 - \dfrac{T_B}{T}}\frac{RT}{P} = \frac{0.00061}{1 - \dfrac{195\,\text{K}}{410.\,\text{K}}} \times \frac{8.314 \times 10^{-2}\,\text{dm}^3\,\text{bar mol}^{-1}\,\text{K}^{-1} \times 410.\,\text{K}}{1\,\text{bar}}$$

$$= 0.0397\,\text{dm}^3\,\text{mol}^{-1} = 3.97 \times 10^{-5}\,\text{m}^3\,\text{mol}^{-1}$$

$$a = RbT_B = 8.314\,\text{J mol}^{-1}\,\text{K}^{-1} \times 3.97 \times 10^{-5}\,\text{m}^3\,\text{mol}^{-1} \times 195\,\text{K} = 6.43 \times 10^{-2}\,\text{m}^6\,\text{Pa mol}^{-2}$$

P7.3 Assume that the equation of state for a gas can be written in the form $P(V_m - b(T)) = RT$. Derive an expression for $\beta = 1/V(\partial V/\partial T)_P$ and $\kappa = -1/V(\partial V/\partial P)_T$ for such a gas in terms of $b(T)$, $db(T)/dT$, P, and V_m.

$$P\left(\frac{V}{n} - b(T)\right) = RT; \quad \frac{V}{n} = \frac{RT}{P} + b(T)$$

$$V = nb(T) + \frac{nRT}{P}$$

$$\beta = \frac{1}{V}\left(\frac{\partial V}{\partial T}\right)_P = \frac{1}{V}\left(\frac{ndb(T)}{dT} + \frac{nR}{P}\right) = \frac{1}{V_m}\left(\frac{db(T)}{dT} + \frac{R}{P}\right)$$

$$\kappa = -\frac{1}{V}\left(\frac{\partial V}{\partial P}\right)_T = -\frac{1}{V}\left(-\frac{nRT}{P^2}\right) = \frac{RT}{V_m P^2}$$

P7.6 For values of z near one, it is a good approximation to write $z(P) = 1 + (\partial z/\partial P)_T P$. If $z = 1.00104$ at 298 K and 1 bar, and the Boyle temperature of the gas is 155 K, calculate the values of a, b, and V_m for the van der Waals gas.

From Example Problem 7.2,

$$\left(\frac{\partial z}{\partial P}\right)_T = \frac{1}{RT}\left(b - \frac{a}{RT}\right)$$

We can write three equations in three unknowns:

$$z - 1 = \left(b - \frac{a}{RT}\right)\frac{P}{RT}$$

$$1.04 \times 10^{-3} = \left(b - \frac{a}{8.314 \times 10^{-2}\,\text{L bar mol}^{-1}\,\text{K}^{-1} \times 298\,\text{K}}\right)$$

$$\times \frac{1\,\text{bar}}{8.314 \times 10^{-2}\,\text{L bar mol}^{-1}\,\text{K}^{-1} \times 298\,\text{K}}$$

39

$$T_B = \frac{a}{Rb} = \frac{a}{8.314 \times 10^{-2} \text{ L bar mol}^{-1} \text{ K}^{-1} \times b} = 155 \text{ K}$$

$$P = \frac{RT}{V_m - b} - \frac{a}{V_m} = \frac{8.314 \times 10^{-2} \text{ L bar mol}^{-1} \text{ K}^{-1} \times 298 \text{ K}}{V_m - b} - \frac{a}{V_m} = 1 \text{ bar}$$

Using an equation solver, we obtain a and b by solving the first two equations simultaneously. We substitute these values into the third equation to obtain V_m.

The results are:

$$a = 0.692 \text{ L}^2 \text{ bar mol}^{-2}, \; b = 0.0537 \text{ L mol}^{-1}, \; V_m = 24.8 \text{ L mol}^{-1}$$

P7.8 The experimentally determined density of O_2 at 140. bar and 298 K is 192 g L^{-1}. Calculate z and V_m from this information. Compare this result with what you would have estimated from Figure 7.8. What is the relative error in using Figure 7.8 for this case?

$$V_m = \frac{M}{\rho} = \frac{32.0 \text{ g mol}^{-1}}{192 \text{ g L}^{-1}} = 0.167 \text{ L mol}^{-1}$$

$$z = \frac{PV_m}{RT} = \frac{140. \text{ bar} \times 1.67 \times 10^{-1} \text{ L mol}^{-1}}{8.314 \times 10^{-2} \text{ L bar mol}^{-1} \text{ K}^{-1} \times 298 \text{ K}} = 0.942$$

Because $P_r = \dfrac{140. \text{ bar}}{50.43 \text{ bar}} = 2.78$ and $T_r = \dfrac{298 \text{ K}}{154.58 \text{ K}} = 1.93$, Figure 7.8 predicts $z = 0.88$. The relative error in z is 6.6%.

P7.14 Use the law of corresponding states and Figure 7.8 to estimate the molar volume of propane at $T = 500.$ K and $P = 75.0$ bar. The experimentally determined value is 0.438 mol L^{-1}. What is the relative error of your estimate?

We use the values for the critical constants in Table 7.2.

$$T_r = \frac{500. \text{ K}}{369.83 \text{ K}} = 1.35 \quad P_r = \frac{75.0 \text{ bar}}{42.48 \text{ bar}} = 1.77$$

Therefore, $z \approx 0.72$.

$$\frac{PV_m}{RT} = 0.72; \quad V_m = 0.72 \frac{RT}{P} = 0.72 \times \frac{8.314 \times 10^{-2} \text{ dm}^3 \text{ bar K}^{-1} \text{ mol}^{-1} \times 500. \text{ K}}{75.0 \text{ bar}}$$

$$V_m = 0.399 \text{ L mol}^{-1}$$

$$\text{Relative error} = 100 \times \frac{V_m - V_m^{\text{exp}}}{V_m^{\text{exp}}} = \frac{0.399 \text{ L mol}^{-1} - 0.438 \text{ L mol}^{-1}}{0.438 \text{ L mol}^{-1}} = -8.9\%$$

P7.15 Another equation of state is the Berthelot equation, $V_m = (RT/P) + b - a/(RT^2)$. Derive expressions for $\beta = 1/V(\partial V/\partial T)_P$ and $\kappa = -1/V(\partial V/\partial P)_T$ from the Berthelot equation in terms of V, T, and P.

$$V = \frac{nRT}{P} + nb - \frac{na}{RT^2}$$

$$\beta = \frac{1}{V}\left(\frac{\partial V}{\partial T}\right)_P = \frac{1}{V}\left(\frac{nR}{P} + \frac{2na}{RT^3}\right) = \frac{1}{V_m}\left(\frac{R}{P} + \frac{2a}{RT^3}\right)$$

$$\kappa = -\frac{1}{V}\left(\frac{\partial V}{\partial P}\right)_T = -\frac{1}{V}\left(-\frac{nRT}{P^2}\right) = \frac{nRT}{P^2 V} = \frac{RT}{P^2 V_m}$$

P7.16 Show that $P\kappa = 1 - P\left(\dfrac{\partial \ln z}{\partial P}\right)_T$ for a real gas where κ is the isothermal compressibility.

$$\kappa = -\frac{1}{V}\left(\frac{\partial V}{\partial P}\right)_T; \quad P\kappa = -\frac{P}{V}\left(\frac{\partial V}{\partial P}\right)_T$$

$$z = \frac{V}{V_{ideal}} = \frac{PV}{nRT}; \quad \ln z = \ln(PV) - \ln(nRT)$$

$$\left(\frac{\partial \ln z}{\partial P}\right)_T = \frac{1}{PV} \times \left[V + P\left(\frac{\partial V}{\partial P}\right)_T\right] = \frac{1}{P} + \frac{1}{V}\left(\frac{\partial V}{\partial P}\right)_T$$

$$\frac{1}{V}\left(\frac{\partial V}{\partial P}\right)_T = \left(\frac{\partial \ln z}{\partial P}\right)_T - \frac{1}{P}$$

Therefore $P\kappa = 1 - P\left(\dfrac{\partial \ln z}{\partial P}\right)_T$.

P7.17 Calculate the van der Waals parameters of carbon dioxide from the values of the critical constants and compare your results with the values for a and b in Table 7.4.

We use the values for the critical constants in Table 7.2.

$$b = \frac{RT_c}{8P_c} = \frac{8.314 \times 10^{-2}\ dm^3\ bar\ K^{-1}\ mol^{-1} \times 304.13\ K}{8 \times 73.75\ bar} = 0.04286\ dm^3\ mol^{-1}$$

$$a = \frac{27R^2T_c^2}{64P_c} = \frac{27 \times (8.314 \times 10^{-2}\ dm^3\ bar\ K^{-1}\ mol^{-1})^2 \times (304.13\ K)^2}{64 \times 73.75\ bar}$$

$$= 3.657\ dm^6\ bar\ mol^{-2}$$

P7.21 At what temperature does the slope of the z versus P curve as $P \to 0$ have its maximum value for a van der Waals gas? What is the value of the maximum slope?

$$\left(\frac{\partial z}{\partial P}\right)_{T,P \to 0} = \frac{1}{RT}\left(b - \frac{a}{RT}\right)$$

For a van der Waals gas

$$\left(\frac{\partial}{\partial T}\left(\frac{\partial z}{\partial P}\right)_T\right)_{P \to 0} = -\frac{1}{RT^2}\left(b - \frac{a}{RT}\right) + \frac{1}{RT^3} = -\frac{1}{RT^2}\left(b - \frac{2a}{RT}\right)$$

Setting this derivative equalt to zero gives

$$b - \frac{2a}{RT_{max}} = 0 \quad T_{max} = \frac{2a}{Rb}$$

The maximum slope is $\dfrac{1}{RT_{max}}\left(b - \dfrac{a}{RT_{max}}\right) = \dfrac{b}{2a}\left[b - a\left(\dfrac{b}{2a}\right)\right] = \dfrac{b^2}{4a}$.

P7.22 Calculate the density of $O_2(g)$ at 480. K and 280. bar using the ideal gas and the van der Waals equations of state. Use a numerical equation solver to solve the van der Waals equation for V_m or use an iterative approach starting with V_m equal to the ideal gas result. Based on your result, does the attractive or repulsive contribution to the interaction potential dominate under these conditions? The experimentally determined result is 208 g/L. What is the relative error of each of your two calculations?

$$V_m = \frac{RT}{P} = \frac{8.314 \times 10^{-2}\ L\ bar\ K^{-1}\ mol^{-1} \times 480.\ K}{280.\ bar} = 0.1425\ L$$

$$P_{vdW} = \frac{RT}{V_m - b} - \frac{a}{V_m^2}$$

$$= \frac{8.314 \times 10^{-2} \text{ L bar K}^{-1} \text{ mol}^{-1} \times 480. \text{ K}}{V_m - 0.0319 \text{ L mol}^{-1}} - \frac{1.382 \text{ L}^2 \text{ bar mol}^{-2}}{(V_m)^2} = 280 \text{ bar}$$

The three solutions to this equation are

$$V_m = (0.01306 \pm 0.02985 \, i) \text{ L mol}^{-1} \text{ and } V_m = 0.1483 \text{ L mol}^{-1}$$

Only the real solution is of significance.

$$\rho_{ideal\ gas} = \frac{M}{V_m} = \frac{32.0 \text{ g mol}^{-1}}{0.1425 \text{ L mol}^{-1}} = 224 \text{ g L}^{-1}$$

$$\rho_{vdW} = \frac{M}{V_m} = \frac{32.0 \text{ g mol}^{-1}}{0.1483 \text{ L mol}^{-1}} = 216 \text{ g L}^{-1}$$

Because the van der Waals density is less than the ideal gas density, the repulsive part of the potential dominates. The relative error is 7.94% and 3.75% for the ideal gas and van der Waals equations of state, respectively.

P7.28   For a van der Waals gas, $z = V_m/(V_m - b) - a/RTV_m$. Expand the first term of this expression in a Taylor series in the limit $V_m \gg b$ to obtain $z \approx 1 + (b - a/RT)(1/V_m)$.

$$f(x) = f(0) + \left(\frac{df(x)}{dx}\right)_{x=0} x + \dots$$

In this case, $f(x) = \dfrac{1}{1-x}$ and $x = \dfrac{b}{V_m}$.

$$z = \frac{V_m}{V_m - b} - \frac{a}{RTV_m} = \frac{1}{1 - \dfrac{b}{V_m}} - \frac{a}{RTV_m}$$

Because $\dfrac{1}{1-x} \approx 1 + x + \dfrac{x^2}{2} + \dots$

$$\frac{1}{1 - \dfrac{b}{V_m}} \approx 1 + \frac{b}{V_m}$$

$$z \approx 1 + \frac{b}{V_m} - \frac{a}{RTV_m} = 1 + \frac{1}{V_m}\left(b - \frac{a}{RT}\right)$$

8 Phase Diagrams and the Relative Stability of Solids, Liquids, and Gases

Numerical Problems

P8.2 The vapor pressure of ethanol(l) is given by

$$\ln\left(\frac{P}{\text{Pa}}\right) = 23.58 - \frac{3.6745 \times 10^3}{\dfrac{T}{\text{K}} - 46.702}$$

    a.   Calculate the standard boiling temperature.

    b.   Calculate $\Delta H_{vaporization}$ at 298 K and at the standard boiling temperature.

    (a) $\ln\left(\dfrac{P}{\text{Pa}}\right) = 20.767 - \dfrac{2.7738 \times 10^3}{\dfrac{T}{\text{K}} - 53.08} = \ln 10^5 \text{ at } T_b$

            $T_b = \dfrac{2.7738 \times 10^3}{20.767 - \ln(10^5)} + 53.08 = 351\,\text{K}$

    (b) $\Delta H_{vaporization}(298\,\text{K}) = RT^2 \dfrac{d\ln P}{dT} = \dfrac{8.314\,\text{J mol}^{-1}\,\text{K}^{-1} \times (298\,\text{K})^2 \times 3.6745 \times 10^3}{(298 - 46.702)^2}$

                $= 43.0\,\text{kJ mol}^{-1} \text{ at } 298\,\text{K}$

          $\Delta H_{vaporization}(351\,\text{K}) = RT^2 \dfrac{d\ln P}{dT} = \dfrac{8.314\,\text{J mol}^{-1}\,\text{K}^{-1} \times (352.8\,\text{K})^2 \times 3.6745 \times 10^3}{(352.8 - 46.702)^2}$

                $= 40.6\,\text{kJ mol}^{-1} \text{ at } 351\,\text{K}$

P8.5 Within what range can you restrict the values of P and T if the following information is known about CO_2? Use Figure 8.12 to answer this question.

    a.   As the temperature is increased, the solid is first converted to the liquid and subsequently to the gaseous state.

    b.   An interface delineating liquid and gaseous phases is observed throughout the pressure range between 6 and 65 atm.

    c.   Solid, liquid, and gas phases coexist at equilibrium.

    d.   Only a liquid phase is observed in the pressure range from 10. to 50. atm.

    e.   An increase in temperature from $-80.°$ to $20.°C$ converts a solid to a gas with no intermediate liquid phase.

(a)   The temperature and pressure are greater than the values for the triple point, −56.6°C and 5.11 atm.

(b)   The temperature lies between the values for the triple point and the critical point.

(c)   The system is at the triple point, −56.6°C and 5.11 atm.

(d)   The temperature lies between the values for the triple point and the critical point.

(e)   The pressure is below the triple point pressure value of 5.11 atm.

P8.8   It has been suggested that the surface melting of ice plays a role in enabling speed skaters to achieve peak performance. Carry out the following calculation to test this hypothesis. At 1 atm pressure, ice melts at 273.15 K, $\Delta H_{fusion} = 6010. \, J \, mol^{-1}$, the density of ice is $920. \, kg \, m^{-3}$, and the density of liquid water is $997 \, kg \, m^{-3}$.

a.   What pressure is required to lower the melting temperature by 4.0°C?

b.   Assume that the width of the skate in contact with the ice has been reduced by sharpening to 19×10^{-3} cm, and that the length of the contact area is 18 cm. If a skater of mass 78 kg is balanced on one skate, what pressure is exerted at the interface of the skate and the ice?

c.   What is the melting point of ice under this pressure?

d.   If the temperature of the ice is −4.0°C, do you expect melting of the ice at the ice–skate interface to occur?

(a)

$$\left(\frac{dP}{dT}\right)_{fusion} = \frac{\Delta S_{fusion}}{\Delta V_{fusion}} \approx \frac{\Delta S_{fusion}}{\dfrac{M}{\rho_{H_2O,l}} - \dfrac{M}{\rho_{H_2O,s}}} = \frac{22.0 \, J \, mol^{-1} \, K^{-1}}{\dfrac{18.02 \times 10^{-3} \, kg}{997 \, kg \, m^{-3}} - \dfrac{18.02 \times 10^{-3} \, kg}{920 \, kg \, m^{-3}}}$$

$$= -1.45 \times 10^7 \, Pa \, K^{-1} = -145 \, bar \, K^{-1}$$

The pressure must be increased by 582 bar to lower the melting point by 4.0°C.

(b)

$$P = \frac{F}{A} = \frac{78 \, kg \times 9.81 \, m \, s^{-2}}{18 \times 10^{-2} \, m \times 19 \times 10^{-5} \, m} = 2.24 \times 10^7 \, Pa = 2.24 \times 10^2 \, bar$$

(c)

$$\Delta T = \left(\frac{dT}{dP}\right)_{fusion} \Delta P = \frac{1°C}{145 \, bar} \times 2.24 \times 10^2 \, bar = 1.5°C; \quad T_m = -1.5°C$$

(d)   No, because the lowering of the melting temperature is less than the temperature of the ice.

P8.9   Answer the following questions using the *P–T* phase diagram for carbon sketched below.

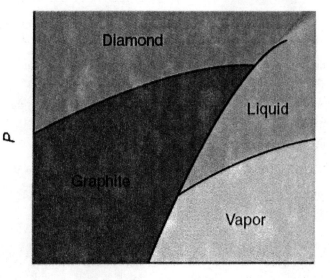

    a.   Which substance is denser, graphite or diamond? Explain your answer.

According to Le Chatelier's principle, for a given temperature, the phase with the higher density will be found at higher pressure. Therefore diamond is more dense than graphite.

    b.   Which phase is more dense, graphite or liquid carbon? Explain your answer.

Graphite is more dense because the slope of the graphite–liquid coexistence line is positive.

    c.   Why does the phase diagram have two triple points? Explain your answer.

Multiple triple points are possible only if there are several solid phases or in rare cases more than one liquid phase.

P8.10   You have a compound dissolved in chloroform and need to remove the solvent by distillation. Because the compound is heat sensitive, you hesitate to raise the temperature above 5.00°C and decide on vacuum distillation. What is the maximum pressure at which the distillation takes place?

We can calculate the vapor pressure of chloroform at the given temperature using the data in Table 8.3.

$$\ln P = a_1 - \frac{a_2}{T/K + a_3} = 20.907 - \frac{2696.1}{278.15 - 46.926} = \ln(10^5) = 9.2469$$

$$P = 10.4 \times 10^3 \text{ Pa}$$

P8.13   Autoclaves that are used to sterilize surgical tools require a temperature of 120.°C to kill some bacteria. If water is used for this purpose, at what pressure must the autoclave operate?

$$\ln \frac{P_f}{P_i} = -\frac{\Delta H_{vaporization}}{R}\left(\frac{1}{T_f} - \frac{1}{T_i}\right)$$

$$\ln \frac{P_f}{P_i} = -\frac{40.656 \times 10^3 \text{ J mol}^{-1}}{8.314 \text{ J mol}^{-1} \text{ K}^{-1}} \times \left(\frac{1}{393.15 \text{ K}} - \frac{1}{373.15 \text{ K}}\right) = 0.6667$$

$$\frac{P_f}{P_i} = 1.95; \quad P_f = 1.95 \text{ atm}$$

P8.20   The vapor pressure of liquid benzene is 20,170 Pa at 298.15 K, and $\Delta H_{vaporization} = 30.72 \text{ kJ mol}^{-1}$ at 1 atm pressure. Calculate the normal and standard boiling points. Does your result for the normal boiling point agree with that in Table 8.3? If not, suggest a possible cause.

$$\ln \frac{P_f}{P_i} = -\frac{\Delta H_{vaporization}}{R}\left(\frac{1}{T_f} - \frac{1}{T_i}\right)$$

$$T_f = \frac{\Delta H_{vaporization}}{R\left(\dfrac{\Delta H_{vaporization}}{RT_i} - \ln \dfrac{P_f}{P_i}\right)}$$

At the normal boiling point, $P = 101,325$ Pa.

$$T_{b, normal} = \frac{30.72 \times 10^3 \text{ J mol}^{-1}}{8.314 \text{ J mol}^{-1} \text{ K}^{-1} \times \left(\dfrac{30.72 \times 10^3 \text{ J mol}^{-1}}{8.314 \text{ J mol}^{-1} \text{ K}^{-1} \times 298.15 \text{ K}} - \ln \dfrac{101325}{20170}\right)} = 342.8 \text{ K}$$

At the standard boiling point, $P = 10^5$ Pa.

$$T_{b, standard} = \frac{30.72 \times 10^3 \text{ J mol}^{-1}}{8.314 \text{ J mol}^{-1} \text{ K}^{-1} \times \left(\dfrac{30.72 \times 10^3 \text{ J mol}^{-1}}{8.314 \text{ J mol}^{-1} \text{ K}^{-1} \times 298.15 \text{ K}} - \ln \dfrac{10^5}{7615}\right)} = 342.4 \text{ K}$$

The result for the normal boiling point is ~10 K lower than the value tabulated in Table 8.3. The most probable reason for this difference is that the calculation above has assumed that $\Delta H_m^{vaporization}$ is independent of T.

P8.21   Benzene(l) has a vapor pressure of 0.1269 bar at 298.15 K and an enthalpy of vaporization of 30.72 kJ mol^{-1}. The $C_{P,m}$ of the vapor and liquid phases at that temperature are 82.4 and 136.0 J K^{-1}mol^{-1}, respectively. Calculate the vapor pressure of $C_6H_6(l)$ at 340.0 K assuming

    a.   that the enthalpy of vaporization does not change with temperature.

    b.   that the enthalpy of sublimation at temperature T can be calculated from the equation $\Delta H_{vaporization}(T) = \Delta H_{vaporization}(T_0) + \Delta C_P(T - T_0)$ assuming that ΔC_P does not change with temperature.

    (a)   If the enthalpy of vaporization is constant

$$\ln \frac{P_2}{P_1} = -\frac{\Delta H_{vaporization}}{R}\left(\frac{1}{T_2} - \frac{1}{T_1}\right)$$

$$\ln P_2 = \ln(0.1269) - \frac{30.72 \times 10^3 \text{ J mol}^{-1}}{8.314 \text{ J mol}^{-1}\text{ K}^{-1}} \times \left(\frac{1}{340.\text{ K}} - \frac{1}{298.15\text{ K}}\right) = -0.5389$$

$$P_2 = 0.583 \text{ bar}$$

    (b)   If the enthalpy of vaporization is given by

$$\Delta H_{vaporization}(T) = \Delta H_{vaporization}(T_0) + \Delta C_P(T - T_0)$$

$$\int_{P_1}^{P_2} \frac{dP}{P} = \int_{T_1}^{T_2} \frac{\Delta H_{vaporization}}{RT^2}dT = \int_{T_1}^{T_2} \frac{\Delta H_{vaporization}(T_1) + \Delta C_P(T - T_1)}{RT^2}dT$$

$$\ln \frac{P_2}{P_1} = -\frac{\Delta H_{vaporization}(T_1)}{R}\left(\frac{1}{T_2} - \frac{1}{T_1}\right) + \frac{\Delta C_P T_1}{R}\left(\frac{1}{T_2} - \frac{1}{T_1}\right) + \frac{\Delta C_P}{R}\ln \frac{T_2}{T_1}$$

$$\ln P_2 = \ln(0.4741) - \frac{30.72 \times 10^3 \text{ J mol}^{-1}}{8.314 \text{ J mol}^{-1}\text{ K}^{-1}} \times \left(\frac{1}{340.\text{ K}} - \frac{1}{298.15\text{ K}}\right)$$

$$+ \frac{(82.4 - 136) \text{ J mol}^{-1}\text{ K}^{-1} \times 298.15\text{ K}}{8.314 \text{ J mol}^{-1}\text{ K}^{-1}}\left(\frac{1}{340.\text{ K}} - \frac{1}{298.15\text{ K}}\right)$$

$$+ \frac{(82.4 - 136) \text{ J mol}^{-1}\text{ K}^{-1}}{8.314 \text{ J mol}^{-1}\text{ K}^{-1}}\ln \frac{340.\text{ K}}{298.15\text{ K}}$$

$$\ln P_2 = -0.59218$$

$$P_2 = 0.553 \text{ bar}$$

P8.22   Use the values for $\Delta G_f^{\circ}(CCl_4, l)$ and $\Delta G_f^{\circ}(CCl_4, g)$ from Appendix B to calculate the vapor pressure of CCl_4 at 298.15 K.

    For the transformation $C_6H_6(l) \rightarrow C_6H_6(g)$

$$\ln K_P = \ln \frac{P_{C_6H_6(g)}}{P^{\circ}} = -\frac{\Delta G_f^{\circ}(C_6H_6, g) - \Delta G_f^{\circ}(C_6H_6, l)}{RT}$$

$$= -\frac{-62.500 \times 10^3 \text{ J mol}^{-1} + 66.800 \times 10^3 \text{ J mol}^{-1}}{8.314 \text{ J mol}^{-1}\text{ K}^{-1} \times 298.15\text{ K}} = -1.7347$$

$$K_P = \frac{P_{C_6H_6(g)}}{1 \text{ bar}} = 0.176 \quad P_{C_6H_6(g)} = 0.176 \text{ bar}$$

P8.25   For water, $\Delta H_{vaporization}$ is 40.656 kJ mol^{-1}, and the normal boiling point is 373.12 K. Calculate the boiling point for water on the top of Mt. Everest of (elevation 8848 m), where the normal barometric pressure is 253 Torr.

$$\ln \frac{P_f}{P_i} = -\frac{\Delta H_{vaporization}}{R}\left(\frac{1}{T_f} - \frac{1}{T_i}\right)$$

$$T_f = \frac{\Delta H_{vaporization}}{R\left(\dfrac{\Delta H_{vaporization}}{RT_i} - \ln\dfrac{P_f}{P_i}\right)}$$

At the normal boiling point, $P = 760.$ Torr.

$$T_{b,\,normal} = \frac{40.656 \times 10^3 \text{ J mol}^{-1}}{8.314 \text{ J mol}^{-1}\text{ K}^{-1} \times \left(\dfrac{40.656 \times 10^3 \text{ J mol}^{-1}}{8.314 \text{ J mol}^{-1}\text{ K}^{-1} \times 373.12 \text{ K}} - \ln\dfrac{253 \text{ Torr}}{760.\text{ Torr}}\right)} = 344 \text{ K}$$

P8.28   Use the vapor pressures of $SO_2(l)$ given in the following table to calculate the enthalpy of vaporization using a graphical method or a least squares fitting routine.

T (K)	P (Pa)	T (K)	P (Pa)
190.	824.1	230.	17950
200.	2050.	240.	32110
210.	4591	250.	54410
220.	9421	260.	87930

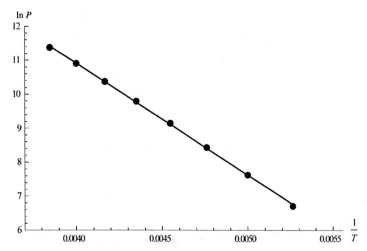

A least squares fit of ln P versus $1/T$ gives the result $\Delta H_{vaporization} = 27.4$ kJ mol^{-1}.

P8.36   The densities of a given solid and liquid of molar mass 122.5 g mol^{-1} at its normal melting temperature of 427.15 K are 1075 and 1012 kg m^{-3}, respectively. If the pressure is increased to 120. bar, the melting temperature increases to 429.35 K. Calculate $\Delta H^{\circ}_{fusion}$ and $\Delta S^{\circ}_{fusion}$ for this substance.

$$\frac{\Delta P}{\Delta T} \approx \frac{\Delta S}{\Delta V}; \quad \Delta S \approx \frac{\Delta P}{\Delta T}\Delta V$$

$$\Delta S_{fusion} = \frac{\Delta P}{\Delta T}M\left(\frac{1}{\rho_{liquid}} - \frac{1}{\rho_{solid}}\right)$$

$$\Delta S_{fusion} = \frac{119 \times 10^5 \text{ Pa}}{429.35 \text{ K} - 427.15 \text{ K}} \times 147.2 \times 10^{-3} \text{ kg mol}^{-1} \times \left(\frac{1}{1012 \text{ kg m}^{-3}} - \frac{1}{1075 \text{ kg m}^{-3}}\right)$$

$$= 38.4 \text{ J K}^{-1} \text{ mol}^{-1}$$

$$\Delta H_{fusion} = T_{fusion}\,\Delta S_{fusion} = 427.15 \text{ K} \times 38.4 \text{ J K}^{-1} \text{ mol}^{-1} = 16.4 \times 10^3 \text{ J mol}^{-1} \text{ at 1 bar}$$

P8.37   The variation of the vapor pressure of the liquid and solid forms of a pure substance near the triple point are given by $\ln\dfrac{P_{solid}}{Pa} = -8750\dfrac{K}{T} + 34.143$ and $\ln\dfrac{P_{liquid}}{Pa} = -4053\dfrac{K}{T} + 21.10$. Calculate the temperature and pressure at the triple point.

At the triple point, $P_{solid} = P_{liquid}$.

$$-8750.\frac{K}{T} + 34.143 = -4053\frac{K}{T} + 21.10$$

$$13.043 = 4697\frac{K}{T}$$

$$T_{tp} = 360.\,K$$

$$\ln\frac{P_{tp}}{Pa} = \frac{-8750.}{360.} + 34.143 = 9.84534$$

$$\frac{P_{tp}}{Pa} = 1.89 \times 10^4$$

P8.41   Calculate the vapor pressure of a droplet of benzene of radius 1.25×10^{-8} m at 38.0°C in equilibrium with its vapor. Use the tabulated value of the density and the surface tension at 298 K from Appendix B for this problem. (*Hint:* You need to calculate the vapor pressure of benzene at this temperature.)

$$\ln\frac{P(T)}{Pa} = A(1) - \frac{A(2)}{\dfrac{T}{K} + A(3)} = 20.767 - \frac{2.7738 \times 10^3}{311.15 - 53.08} = 10.0188$$

$$P = 2.24 \times 10^4 \text{ Pa}$$

Using Equation (8.29),

$$\Delta P = \frac{2\gamma}{r} = \frac{2 \times 28.22 \times 10^{-3} \text{ N m}^{-1}}{1.25 \times 10^{-8} \text{ m}} = 4.52 \times 10^6 \text{ Pa}$$

$$P_{inside} = P_{vapor} + \Delta P = 2.24 \times 10^4 \text{ Pa} + 4.52 \times 10^6 \text{ Pa} = 4.54 \times 10^6 \text{ Pa}$$

For a very large droplet, $\Delta P \rightarrow 0$, the vapor pressure is 2.24×10^4 Pa. For the small droplet, the vapor pressure is increased by the factor

$$\ln\left(\frac{P}{P_0}\right) = \frac{\dfrac{M}{\rho}(P - P_0)}{RT} = \frac{\dfrac{78.11 \times 10^{-3} \text{ kg mol}^{-1}}{876.5 \text{ kg m}^{-3}} \times (4.54 \times 10^6 - 2.24 \times 10^4)\text{Pa}}{8.314 \text{ J mol}^{-1}\text{ K}^{-1} \times 311.15 \text{ K}}$$

$$= 0.1555$$

$$P = 1.168\, P_0 = 2.62 \times 10^4 \text{ Pa}$$

P8.42   Solid iodine, $I_2(s)$, at 25.0°C has an enthalpy of sublimation of 56.30 kJ mol^{-1}. The $C_{P,m}$ of the vapor and solid phases at that temperature are 36.9 and 54.4 J K^{-1} mol^{-1}, respectively. The sublimation pressure at 25.0°C is 0.30844 Torr. Calculate the sublimation pressure of the solid at the melting point (113.6°C) assuming

a.   that the enthalpy of sublimation and the heat capacities do not change with temperature.

b.   that the enthalpy of sublimation at temperature T can be calculated from the equation $\Delta H_{sublimation}(T)$
$(T)\Delta H_{sublimation} = \Delta H_{sublimation}(T_0) + \Delta C_P(T - T_0)$ and ΔC_P does not change with T.

(a)   If the enthalpy of sublimation is constant

$$\ln\frac{P_2}{P_1} = -\frac{\Delta H_{sublimation}}{R}\left(\frac{1}{T_2} - \frac{1}{T_1}\right)$$

$$\ln P_2 = \ln 0.30844 - \frac{56.30 \times 10^3 \text{ J mol}^{-1}}{8.314 \text{ J mol}^{-1}\text{ K}^{-1}} \times \left(\frac{1}{386.8 \text{ K}} - \frac{1}{298.15 \text{ K}}\right)$$

$$P_2 = 56.1 \text{ Torr}$$

(b)   If the enthalpy of sublimation is given by

$$\Delta H_{sublimation}(T) = \Delta H_{sublimation}(T_0) + \Delta C_P(T - T_0)$$

$$\int_{P_1}^{P_2} \frac{dP}{P} = \int_{T_1}^{T_2} \frac{\Delta H_{vaporization}}{RT^2} dT = \int_{T_1}^{T_2} \frac{\Delta H_{vaporization}(T_1) + \Delta C_P(T - T_1)}{RT^2} dT$$

$$\ln \frac{P_2}{P_1} = -\frac{\Delta H_{vaporization}(T_1)}{R}\left(\frac{1}{T_2} - \frac{1}{T_1}\right) + \frac{\Delta C_P T_1}{R}\left(\frac{1}{T_2} - \frac{1}{T_1}\right) + \frac{\Delta C_P}{R}\ln\frac{T_2}{T_1}$$

$$\ln P_2 = \ln 0.30844 - \frac{56.30 \times 10^3 \text{ J mol}^{-1}}{8.314 \text{ J K}^{-1} \text{ mol}^{-1}} \times \left(\frac{1}{386.8 \text{ K}} - \frac{1}{298.15 \text{ K}}\right)$$

$$+ \frac{(36.9 - 54.4) \text{ J K}^{-1} \text{ mol}^{-1} \times 298.15 \text{ K}}{8.314 \text{ J K}^{-1} \text{ mol}^{-1}} \times \left(\frac{1}{386.8 \text{ K}} - \frac{1}{298.15 \text{ K}}\right)$$

$$+ \frac{(36.9 - 54.4) \text{ J K}^{-1} \text{ mol}^{-1}}{8.314 \text{ J K}^{-1} \text{ mol}^{-1}} \times \ln\frac{386.8 \text{ K}}{298.15 \text{ K}}$$

$$\ln P_2 = 3.964$$

$$P_2 = 52.5 \text{ Torr}$$

P8.43   Consider the transition between two forms of solid tin, $Sn(s, \text{gray}) \rightarrow Sn(s, \text{white})$. The two phases are in equilibrium at 1 bar and 18°C. The densities for gray and white tin are 5750 and 7280 kg m^{-3}, respectively, and the molar entropies for gray and white tin are 44.14 and 51.18 $\text{J K}^{-1} \text{mol}^{-1}$, respectively. Calculate the temperature at which the two phases are in equilibrium at 350. bar.

In going from 1 atm, 18°C to 200 atm, T

$$\Delta G^{gray} = V_m^{gray} \Delta P - S^{gray} \Delta T$$

$$\Delta G^{white} = V_m^{white} \Delta P - S^{white} \Delta T$$

At equilibrium

$$\Delta G^{gray} - \Delta G^{white} = 0 = (V_m^{gray} - V_m^{white}) \Delta P - (S^{gray} - S^{white}) \Delta T$$

$$\Delta T = \frac{(V_m^{gray} - V_m^{white}) \Delta P}{(S^{gray} - S^{white})} = \frac{M_{Sn}\left(\dfrac{1}{\rho_{gray}} - \dfrac{1}{\rho_{white}}\right) \Delta P}{\Delta S_{transition}}$$

$$= \frac{118.71 \times 10^{-3} \text{ kg mol}^{-1} \times \left(\dfrac{1}{5750 \text{ kg m}^{-3}} - \dfrac{1}{7280 \text{ kg m}^{-3}}\right) \times 349 \times 10^5 \text{ Pa}}{-7.04 \text{ J K}^{-1} \text{ mol}^{-1}} = -21.5°C$$

$$T_f = -3.5°C$$

9 Ideal and Real Solutions

Numerical Problems

P9.2 At a given temperature, a nonideal solution of the volatile components A and B has a vapor pressure of 795 Torr. For this solution, $y_A = 0.375$. In addition, $x_A = 0.310$, $P_A^* = 610.$ Torr, and $P_B^* = 495$ Torr. Calculate the activity and activity coefficient of A and B.

$$P_A = y_A P_{total} = 0.375 \times 795 \text{ Torr} = 298 \text{ Torr}$$

$$P_B = 795 \text{ Torr} - 298 \text{ Torr} = 497 \text{ Torr}$$

$$a_A = \frac{P_A}{P_A^*} = \frac{298 \text{ Torr}}{610. \text{ Torr}} = 0.489$$

$$\gamma_A = \frac{a_A}{x_A} = \frac{0.488}{0.310} = 1.58$$

$$a_B = \frac{P_B}{P_B^*} = \frac{497 \text{ Torr}}{495 \text{ Torr}} = 1.00$$

$$\gamma_B = \frac{a_B}{x_B} = \frac{1.00}{0.690} = 1.45$$

P9.4 At 350. K, pure toluene and hexane have vapor pressures of 3.57×10^4 Pa and 1.30×10^5 Pa, respectively.

    a.   Calculate the mole fraction of hexane in the liquid mixture that boils at 350. K at a pressure of 1 atm.

    b.   Calculate the mole fraction of hexane in the vapor that is in equilibrium with the liquid of part (a).

(a) $P_{total} = x_{hex} P_{hex}^* + (1 - x_{hex}) P_{tol}^*$

$$1.01325 \times 10^5 \text{ Pa} = 1.30 \times 10^5 \text{ Pa } x_{hex} + 3.57 \times 10^4 \text{ Pa } (1 - x_{hex})$$

$$x_{hex} = 0.697$$

(b) $y_B = \dfrac{x_{hex} P_{hex}^*}{P_{tol}^* + (P_{hex}^* - P_{tol}^*) x_{hex}}$

$$= \frac{0.697 \times 1.30 \times 10^5 \text{ Pa}}{3.57 \times 10^4 \text{ Pa} + 0.697 \times (1.30 \times 10^5 \text{ Pa} - 3.57 \times 10^4 \text{ Pa})}$$

$$= 0.893$$

P9.7 The osmotic pressure of an unknown substance is measured at 298 K. Determine the molecular weight if the concentration of this substance is 31.2 kg m^{-3} and the osmotic pressure is 5.30×10^4 Pa. The density of the solution is 997 kg m^{-3}.

$$\pi = \frac{n_{solute} RT}{V} = \frac{c_{solute} RT}{M_{solute}}; \quad M_{solute} = \frac{c_{solute} RT}{\pi}$$

$$M_{solute} = \frac{31.2 \text{ kg m}^{-3} \times 8.314 \text{ J mol}^{-1} \text{ K}^{-1} \times 298 \text{ K}}{5.34 \times 10^4 \text{ Pa}} = 1.45 \times 10^3 \text{ g mol}^{-1}$$

P9.9 The volatile liquids A and B, for which $P_A^* = 165$ Torr and $P_B^* = 85.1$ Torr are confined to a piston and cylinder assembly. Initially, only the liquid phase is present. As the pressure is reduced, the first vapor is observed at a total pressure of 110. Torr. Calculate x_A.

The first vapor is observed at a pressure of

$$P_{total} = x_A P_A^* + (1 - x_A)P_B^*$$

$$x_A = \frac{P_{total} - P_B^*}{P_A^* - P_B^*} = \frac{110.\,\text{Torr} - 85.1\,\text{Torr}}{165\,\text{Torr} - 85.1\,\text{Torr}} = 0.312$$

P9.15 At 39.9°C, a solution of ethanol ($x_1 = 0.9006$, $P_1^* = 130.4$ Torr) and isooctane ($P_2^* = 43.9$ Torr) forms a vapor phase with $y_1 = 0.6667$ at a total pressure of 185.9 Torr.

a. Calculate the activity and activity coefficient of each component.

b. Calculate the total pressure that the solution would have if it were ideal.

(a) The activity and activity coefficient for ethanol are given by

$$a_1 = \frac{y_1 P_{total}}{P_1^*} = \frac{0.6667 \times 185.9\,\text{Torr}}{130.4\,\text{Torr}} = 0.9504$$

$$\gamma_1 = \frac{a_1}{x_1} = \frac{0.9504}{0.9006} = 1.055$$

Similarly, the activity and activity coefficient for isooctane are given by

$$a_2 = \frac{(1 - y_1)P_{total}}{P_2^*} = \frac{0.3333 \times 185.9\,\text{Torr}}{43.9\,\text{Torr}} = 1.411$$

$$\gamma_2 = \frac{a_2}{x_2} = \frac{1.411}{1 - 0.9006} = 14.2$$

(b) If the solution were ideal, Raoult's law would apply.

$$P_{Total} = x_1 P_1^* + x_2 P_2^*$$
$$= 0.9006 \times 130.4\,\text{Torr} + (1 - 0.9006) \times 43.9\,\text{Torr}$$
$$= 121.8\,\text{Torr}$$

P9.16 Calculate the solubility of H_2S in 1 L of water if its pressure above the solution is 2.75 Pa. The density of water at this temperature is $997\,\text{kg m}^{-3}$.

$$x_{H_2S} = \frac{n_{H_2S}}{n_{H_2S} + n_{H_2O}} \approx \frac{n_{H_2S}}{n_{H_2O}} = \frac{P_{H_2S}}{k_{H_2S}^H} = \frac{2.75 \times 10^5\,\text{bar}}{568\,\text{bar}} = 4.84 \times 10^{-8}$$

$$n_{H_2O} = \frac{\rho_{H_2O}V}{M_{H_2O}} = \frac{10^{-3}\,\text{m}^3 \times 997\,\text{kg m}^{-3}}{18.02 \times 10^{-3}\,\text{kg mol}^{-1}} = 55.3$$

$$n_{H_2S} = x_{H_2S}n_{H_2O} = 4.84 \times 10^{-8} \times 55.3 = 2.68 \times 10^{-6}\,\text{mol}$$

P9.19 A and B form an ideal solution. At a total pressure of 0.720 bar, $y_A = 0.510$ and $x_A = 0.420$. Using this information, calculate the vapor pressure of pure A and of pure B.

$$P_{total} = x_A P_A^* + y_B P_{total}$$

$$P_A^* = \frac{P_{total} - y_B P_{total}}{x_A} = \frac{0.720\,\text{bar} \times (1 - 0.510)}{0.420} = 0.840\,\text{bar}$$

$$P_B^* = \frac{P_{total} - P_A^* x_A}{1 - x_A} = \frac{0.874\,\text{bar} \times (0.420 \times 0.510 - 0.420)}{(0.420 - 1) \times 0.510} = 0.608\,\text{bar}$$

P9.24   An ideal solution is formed by mixing liquids A and B at 298 K. The vapor pressure of pure A is 151 Torr and that of pure B is 84.3 Torr. If the mole fraction of A in the vapor is 0.610, what is the mole fraction of A in the solution?

$$\text{Using Equation (9.11), } x_A = \frac{y_A P_B^*}{P_A^* + (P_B^* - P_A^*)y_A} = \frac{0.610 \times 84.3\,\text{Torr}}{151\,\text{Torr} + (84.3\,\text{Torr} - 151\,\text{Torr}) \times 0.610} = 0.466$$

P9.25   A solution is prepared by dissolving 45.2 g of a nonvolatile solute in 119 g of water. The vapor pressure above the solution is 22.51 Torr and the vapor pressure of pure water is 23.76 Torr at this temperature. What is the molecular weight of the solute?

$$x_{H_2O} = \frac{P_{H_2O}}{P_{H_2O}^*} = \frac{22.51\,\text{Torr}}{23.76\,\text{Torr}} = 0.947$$

$$x_{solute} = 0.0526 = \frac{n_{solute}}{n_{solute} + n_{H_2O}};$$

$$n_{solute} = \frac{x_{solute}\dfrac{m_{H_2O}}{M_{H_2O}}}{x_{H_2O}} = \frac{0.0526 \times \dfrac{119\,\text{g}}{18.02\,\text{g mol}^{-1}}}{0.947} = 0.367\,\text{mol}$$

$$M = \frac{45.2\,\text{g}}{0.367\,\text{mol}} = 123\,\text{g mol}^{-1}$$

P9.28   The vapor pressures of 1-bromobutane and 1-chlorobutane can be expressed in the form

$$\ln\frac{P_{bromo}}{\text{Pa}} = 17.076 - \frac{1584.8}{\dfrac{T}{\text{K}} - 111.88}$$

and

$$\ln\frac{P_{chloro}}{\text{Pa}} = 20.612 - \frac{2688.1}{\dfrac{T}{\text{K}} - 55.725}$$

Assuming ideal solution behavior, calculate x_{bromo} and y_{bromo} at 305 K and a total pressure of 9750. Pa. At 305 K, $P_{bromo}^\circ = 7113\,\text{Pa}$ and $P_{chloro}^\circ = 18552\,\text{Pa}$.

$$P_{total} = x_{bromo}P_{bromo}^\circ + (1 - x_{bromo})P_{chloro}^\circ$$

$$x_{bromo} = \frac{P_{total} - P_{chloro}^\circ}{P_{bromo}^\circ - P_{chloro}^\circ} = \frac{9750\,\text{Pa} - 18552\,\text{Pa}}{7113\,\text{Pa} - 18552\,\text{Pa}} = 0.769$$

$$y_{bromo} = \frac{x_{bromo}P_{bromo}^\circ}{P_{total}} = \frac{0.769 \times 7113\,\text{Pa}}{9750\,\text{Pa}} = 0.561$$

P9.29   In an ideal solution of A and B, 3.00 mol are in the liquid phase and 5.00 mol are in the gaseous phase. The overall composition of the system is $Z_A = 0.375$ and $x_A = 0.250$. Calculate y_A.

$$n_{liq}^{tot}(Z_B - x_B) = n_{vapor}^{tot}(y_B - Z_B)$$

$$y_B = \frac{n_{liq}^{tot}(Z_B - x_B) + n_{vapor}^{tot}Z_B}{n_{vapor}^{tot}} = \frac{3.00\,\text{mol} \times (0.625 - 0.750) + 5.00\,\text{mol} \times 0.625}{5.00\,\text{mol}}$$

$$= 0.550$$

$$y_A = 1 - 0.550 = 0.450$$

P9.33   The dissolution of 7.75g of a substance in 825 g of benzene at 298 K raises the boiling point by 0.575°C. Note that $K_f = 5.12\,\text{K kg mol}^{-1}$, $K_b = 2.53\,\text{K kg mol}^{-1}$, and the density of benzene is $876.6\,\text{kg m}^{-3}$. Calculate the freezing point depression, the ratio of the vapor pressure above the solution to that of the pure solvent, the osmotic pressure, and the molecular weight of the solute. $P^*_{benzene} = 103$ Torr at 298 K.

$$\Delta T_b = K_b m_{solute}; \quad m_{solute} = \frac{\Delta T_b}{K_b} = \frac{0.575\,\text{K}}{2.53\,\text{K kg mol}^{-1}} = 0.227\,\text{mol kg}^{-1}$$

$$M = \frac{7.75\,\text{g}}{0.227\,\text{mol kg}^{-1} \times 0.825\,\text{kg}} = 41.3\,\text{g mol}^{-1}$$

$$\Delta T_f = -K_f m_{solute} = -5.12\,\text{K kg mol}^{-1} \times 0.227\,\text{mol kg}^{-1} = -1.16\,\text{K}$$

$$\frac{P_{benzene}}{P^*_{benzene}} = x_{benzene} = \frac{n_{benzene}}{n_{benzene} + n_{solute}}$$

$$= \frac{\dfrac{825\,\text{g}}{78.11\,\text{g mol}^{-1}}}{\dfrac{825\,\text{g}}{78.11\,\text{g mol}^{-1}} + \dfrac{7.75\,\text{g}}{41.3\,\text{g mol}^{-1}}} = 0.983$$

$$\pi = \frac{n_{solute}RT}{V} = \frac{\dfrac{7.75 \times 10^{-3}\,\text{kg}}{41.3 \times 10^{-3}\,\text{kg mol}^{-1}} \times 8.314\,\text{J mol}^{-1}\,\text{K}^{-1} \times 298\,\text{K}}{\dfrac{825 \times 10^{-3}\,\text{kg}}{876.6\,\text{kg m}^{-3}}} = 4.93 \times 10^5\,\text{Pa}$$

10 Electrolyte Solutions

Numerical Problems

P10.2 Calculate ΔS_R° for the reaction $AgNO_3(aq) + KCl(aq) \rightarrow AgCl(s) + KNO_3(aq)$.

$$\Delta S_R^\circ = S^\circ(AgCl, s) - S^\circ(Ag^+, aq) - S^\circ(Cl^-, aq)$$

$$\Delta S_R^\circ = 96.3 \ JK^{-1} \ mol^{-1} - 72.7 \ JK^{-1} \ mol^{-1} - 56.5 \ JK^{-1} \ mol^{-1} = -32.9 \ JK^{-1} \ mol^{-1}$$

P10.7 At 25°C, the equilibrium constant for the dissociation of acetic acid, K_a, is 1.75×10^{-5}. Using the Debye–Hückel limiting law, calculate the degree of dissociation in 0.150 m and 1.50 m solutions using an iterative calculation until the answer is constant in the second decimal place. Compare these values with what you would obtain if the ionic interactions had been ignored. Compare your results with the degree of dissociation of the acid assuming $\gamma_\pm = 1$.

A weak acid dissociates according to the following equilibrium reaction.

$$HA(aq) \rightleftharpoons H^+(aq) + A^-(aq)$$

$$K_a = \frac{a_{H^+} a_{A^-}}{a_{HA}}$$

Because the undissociated acid is a neutral species, it is a good approximation to set its activity equal to its concentration. The activity of the ionic species produced upon dissociation are given by $a_{H^+} = \gamma_\pm c_{H^+}$ and $a_{A^-} = \gamma_\pm c_{A^-}$. The equilibrium expression becomes

$$K_a = \frac{\gamma_\pm c_{H^+} \ \gamma_\pm c_{A^-}}{c_{HA}} = \frac{(\gamma_\pm)^2 \dfrac{m}{m^\circ}}{c^\circ - \dfrac{m}{m^\circ}}$$

This expression has two unknowns. We solve for m using an iterative method. We first assume that $\gamma_\pm = 1$ and solve for m. We use this value of m to calculate $\gamma_\pm$ and using this value, recalculate m. We repeat this procedure until x is sufficiently constant.

$$CH_3COOH(aq) \rightleftharpoons CH_3COO^-(aq) + H^+(aq)$$

For 0.150 m

$$\frac{m^2 \gamma_\pm^2}{0.150 \ mol \ kg^{-1} - m} = 1.75 \times 10^{-5}$$

When $\gamma_\pm = 1$

$$m = 1.611 \times 10^{-3} \ mol \ kg^{-1}$$

$$I = \frac{m}{2}(2) = m = 1.611 \times 10^{-3} \ mol \ kg^{-1}$$

$$\ln \gamma_\pm - 1.173 \times 1 \times \sqrt{1.611 \times 10^{-3}} = -0.04709$$

$$\gamma_\pm = 0.954$$

54

When $\gamma_\pm = 0.954$

$$\frac{\left(\dfrac{m}{m^\circ}\right)^2 \gamma_\pm^2}{0.150 - \dfrac{m}{m^\circ}} = 1.75 \times 10^{-5}$$

$$m = 1.689 \times 10^{-3} \text{ mol kg}^{-1}$$

We iterate several times.

When $m = 1.689 \times 10^{-3} \text{ mol kg}^{-1}$

$$I = \frac{m}{2}(2) = m = 1.689 \times 10^{-3} \text{ mol kg}^{-1}$$

$$\ln \gamma_\pm = -1.173 \times 1 \times \sqrt{1.689 \times 10^{-3}} = -0.04820$$

$$\gamma_\pm = 0.953$$

When $\gamma_\pm = 0.953$

$$\frac{\left(\dfrac{m}{m^\circ}\right)^2 \gamma_\pm^2}{0.150 - \dfrac{m}{m^\circ}} = 1.75 \times 10^{-5}$$

$$m = 1.691 \times 10^{-3} \text{ mol kg}^{-1}$$

This result has converged sufficiently to calculate the degree of dissociation.

$$m = 1.691 \times 10^{-3} \text{ mol kg}^{-1}$$

$$\frac{1.691 \times 10^{-3} \text{ mol kg}^{-1}}{0.150 \text{ mol kg}^{-1}} \times 100\% = 1.13\%$$

For 1.50 m

$$\frac{\left(\dfrac{m}{m^\circ}\right)^2 \gamma_\pm^2}{1.50 - \dfrac{m}{m^\circ}} = 1.75 \times 10^{-5}$$

When $\gamma_\pm = 1$

$$m = 5.115 \times 10^{-3} \text{ mol kg}^{-1}$$

$$I = m = 5.115 \times 10^{-3} \text{ mol kg}^{-1}$$

$$\ln \gamma_\pm = -0.08389$$

$$\gamma_\pm = 0.920$$

When $\gamma_\pm = 0.920$

$$\frac{\left(\dfrac{m}{m^\circ}\right)^2 \gamma_\pm^2}{1.50 - \dfrac{m}{m^\circ}} = 1.75 \times 10^{-5}$$

$$m = 5.561 \times 10^{-3} \text{ mol kg}^{-1}$$

We iterate several times.

When $m = 5.561 \times 10^{-3}$ mol kg^{-1}

$$I = \frac{m}{2}(2) = m = 5.561 \times 10^{-3} \text{ mol kg}^{-1}$$

$$\ln \gamma_\pm = -1.173 \times 1 \times \sqrt{5.561 \times 10^{-3}} = -0.08748$$

$$\gamma_\pm = 0.916$$

When $\gamma_\pm = 0.916$

$$\frac{\left(\dfrac{m}{m^\circ}\right)^2 \gamma_\pm^2}{1.50 - \dfrac{m}{m^\circ}} = 1.75 \times 10^{-5}$$

$$m = 5.581 \times 10^{-3} \text{ mol kg}^{-1} \rightarrow 5.58 \times 10^{-3} \text{ mol kg}^{-1}$$

This result has converged sufficiently to calculate the degree of dissociation.

$$m = 5.58 \times 10^{-3} \text{ mol kg}^{-1}$$

$$\frac{5.58 \times 10^{-3} \text{ mol kg}^{-1}}{1.50 \text{ mol kg}^{-1}} \times 100\% = 0.372\%$$

If ionic interactions are ignored:

For 0.150 m

$$\frac{\left(\dfrac{m}{m^\circ}\right)^2}{0.150 - \dfrac{m}{m^\circ}} = 1.75 \times 10^{-5}$$

$$m = 1.61 \times 10^{-3} \text{ mol kg}^{-1}$$

$$\frac{1.61 \times 10^{-3}}{0.150} \times 100\% = 1.07\%$$

For 1.50 m

$$\frac{\left(\dfrac{m}{m^\circ}\right)^2}{1.50 - \dfrac{m}{m^\circ}} = 1.75 \times 10^{-5}$$

$$m = 5.11 \times 10^{-3} \text{ mol kg}^{-1}$$

$$\frac{5.11 \times 10^{-3}}{1.50} \times 100\% = 0.341\%$$

P10.13   Calculate the ionic strength in a solution that is 0.0750 m in K_2SO_4, 0.0085 m in Na_3PO_4, and 0.0150 m in $MgCl_2$.

$$I_{K_2SO_4} = \frac{m}{2}(v_+ z_+^2 + v_- z_-^2)$$

$$= \frac{0.0750}{2}(2 + 4) = 0.225 \text{ mol kg}^{-1}$$

$$I_{Na_3PO_4} = \frac{m}{2}(v_+ z_+^2 + v_- z_-^2)$$

$$= \frac{0.0085}{2}(3 + 9) = 0.051 \text{ mol kg}^{-1}$$

$$I_{MgCl_2} = \frac{m}{2}(v_+ z_+^2 + v_- z_-^2)$$

$$= \frac{0.0150}{2}(4 + 2) = 0.0450 \text{ mol kg}^{-1}$$

Total ionic strength

$$I = (0.225 + 0.0510 + 0.0450) \text{ mol kg}^{-1}$$
$$= 0.321 \text{ mol kg}^{-1}$$

P10.14   Calculate I, $\gamma_\pm$, and $a_\pm$ for a 0.0120 m solution of Na_3PO_4 at 298 K. Assume complete dissociation.

$$Na_3PO_4 \Rightarrow v_+ = 3, v_- = 1, z_+ = 1, z_- = 3$$

$$I = \frac{m}{2}(v_+ z_+^2 + v_- z_-^2)$$

$$I = \frac{0.0120}{2}(3 + 9) = 0.0720 \text{ mol kg}^{-1}$$

$$\ln \gamma_\pm = -1.173 |z_+ z_-| \sqrt{\frac{I}{\text{mol kg}^{-1}}} = -1.173 \times 2 \times \sqrt{0.0720} = -0.9442$$

$$\gamma_\pm = 0.389$$

$$m_\pm^{(v_+ + v_-)} = m_+^{v_+} m_-^{v_-}$$

$$m_\pm^3 = (0.015)^2 (0.0075)^1 = 5.60 \times 10^{-7}$$

$$m_\pm = 0.0273 \text{ mol kg}^{-1}$$

$$a_\pm = \left(\frac{m_\pm}{m^\circ}\right)\gamma_\pm$$

$$a_\pm = 0.0273 \times 0.389 = 0.0106$$

P10.15   Express $\mu_\pm$ in terms of μ_+ and μ_- for (a) NaCl, (b) MgBr$_2$, (c) Li$_3$PO$_4$, and (d) Ca(NO$_3$)$_2$. Assume complete dissociation.

(a)   NaCl: $\mu_\pm = \dfrac{\mu_{solute}}{v} = \dfrac{v_+ \mu_+ + v_- \mu_-}{v} = \dfrac{\mu_+ + \mu_-}{2}$

(b)   MgBr$_2$: $\mu_\pm = \dfrac{\mu_{solute}}{v} = \dfrac{v_+ \mu_+ + v_- \mu_-}{v} = \dfrac{\mu_+ + 2\mu_-}{3}$

(c)   Li$_3$PO$_4$: $\mu_\pm = \dfrac{\mu_{solute}}{v} = \dfrac{v_+ \mu_+ + v_- \mu_-}{v} = \dfrac{3\mu_+ + \mu_-}{4}$

(d)   Ca(NO$_3$)$_2$: $\mu_\pm = \dfrac{\mu_{solute}}{v} = \dfrac{v_+ \mu_+ + v_- \mu_-}{v} = \dfrac{\mu_+ + 2\mu_-}{3}$

P10.17   Calculate the solubility of CaCO$_3$ ($K_{sp} = 3.4 \times 10^{-9}$) (a) in pure H$_2$O and (b) in an aqueous solution with $I = 0.0250 \text{ mol kg}^{-1}$. For part (a), do an iterative calculation of $\gamma_\pm$ and the solubility until the answer is constant in the second decimal place. Do you need to repeat this procedure in part (b)?

a. $CaCO_3(s) \rightleftharpoons Ca^{2+}(aq) + CO_3^{2-}(aq)$

$$v_+ = 1, \qquad v_- = 1$$
$$z_+ = 2, \qquad z_- = 2$$

$$K_{sp} = \left(\frac{c_{Ca^{2+}}}{c^\circ}\right)\left(\frac{c_{CO_3^{2-}}}{c^\circ}\right)\gamma_\pm^2 = 3.4 \times 10^{-9}$$

$$c_{Ca^{2+}} = c_{CO_3^{2-}}$$

$$K_{sp} = \left(\frac{c_{Ca^{2+}}}{c^\circ}\right)^2 \gamma_\pm^2 = 3.4 \times 10^{-9}$$

When $\gamma_\pm = 1$ $c_{Ca^{2+}} = 5.8 \times 10^{-5}$ mol L^{-1}

$$I = \frac{m}{2}\sum(v_+ z_+^2 + v_- z_-^2)$$

$$= \frac{5.8 \times 10^{-5}}{2} \times (4 + 4) = 2.33 \times 10^{-4} \text{ mol kg}^{-1}$$

$$\ln \gamma_\pm = -1.173 \times 4 \times \sqrt{2.33 \times 10^{-4}} = -0.07166$$

$$\gamma_\pm = 0.9309$$

When $\gamma_\pm = 0.9309$ $c_{Ca^{2+}} = 6.26 \times 10^{-5}$ mol L^{-1}

$$I = \frac{6.26 \times 10^{-5}}{2} \times (8) = 2.51 \times 10^{-4} \text{ mol kg}^{-1}$$

$$\ln \gamma_\pm = -1.173 \times 4 \times \sqrt{2.51 \times 10^{-4}} = -0.07427$$

$$\gamma_\pm = 0.928$$

When $\gamma_\pm = 0.928$ $c_{Ca^{2+}} = 6.28 \times 10^{-5}$ mol L^{-1}

b. $I = 0.0250$ mol kg^{-1}

$$\ln \gamma_\pm = 1.173 \times 4 \times \sqrt{0.0250} = -0.74187$$

$$\gamma_\pm = 0.476$$

$$K_{sp} = \left(\frac{c_{Ca^{2+}}}{c^\circ}\right)^2 \times (0.476)^2 = 3.4 \times 10^{-9}$$

$$c_{Ca^{2+}} = 1.22 \times 10^{-4} \text{ mol L}^{-1}$$

There is no need to repeat the calculation because the ionic strength in the solution is not influenced by the dissociation of the BaSO$_4$.

P10.21   The equilibrium constant for the hydrolysis of dimethylamine,

$$(CH_3)_2 NH(aq) + H_2O(aq) \rightarrow CH_3NH_3{}^+(aq) + OH^-(aq)$$

is 5.12×10^{-4}. Calculate the extent of hydrolysis for (a) a 0.210 m solution of $(CH_3)_2 NH$ in water using an iterative calculation until the answer is constant in the second decimal place. (b) Repeat the calculation for a solution that is also 0.500 m in NaNO$_3$. Do you need to use an iterative calculation in this case?

We use an iterative method as explained in the solution to P10.5.

(a) $(CH_3)_2 NH\ (aq) + H_2O(l) \rightleftharpoons CH_3NH_3^+(aq) + OH^-(aq)$

$$K = \frac{\left(\dfrac{m}{m^\circ}\right)^2 \gamma_\pm^2}{0.210 - \dfrac{m}{m^\circ}} = 5.12 \times 10^{-4}$$

If $\gamma_\pm = 1$, $m = 1.01 \times 10^{-2}$ mol kg^{-1}.

When $m = 1.01 \times 10^{-2}$ mol kg^{-1}

$$I = \frac{m}{2}(2) = m = 1.01 \times 10^{-2} \text{ mol kg}^{-1}$$

$$\ln \gamma_\pm = -1.173 \times 1 \times \sqrt{1.01 \times 10^{-2}} = -0.1180$$

$$\gamma_\pm = 0.889$$

When $\gamma_\pm = 0.889$

$$\frac{\left(\dfrac{m}{m^\circ}\right)^2 \gamma_\pm^2}{0.210 - \left(\dfrac{m}{m^\circ}\right)} = 5.12 \times 10^{-4}$$

$$m = 1.13 \times 10^{-2} \text{ mol kg}^{-1}$$

We iterate several times.

When $m = 1.13 \times 10^{-2}$ mol kg^{-1}

$$I = \frac{m}{2}(2) = m = 1.13 \times 10^{-2} \text{ mol kg}^{-1}$$

$$\ln \gamma_\pm = -1.173 \times 1 \times \sqrt{1.13 \times 10^{-2}} = -0.1250$$

$$\gamma_\pm = 0.883$$

When $\gamma_\pm = 0.883$

$$\frac{\left(\dfrac{m}{m^\circ}\right)^2 \gamma_\pm^2}{0.210 - \dfrac{m}{m^\circ}} = 5.12 \times 10^{-4}$$

$$m = 1.14 \times 10^{-2} \text{ mol kg}^{-1}$$

The degree of hydrolysis is

$$\frac{0.0114}{0.210} \times 100\% = 5.44\%$$

(b)   NaNO$_3$

$$\text{Na}^+ \quad \text{NO}_3^- \quad \nu_+ = 1 \quad \nu_- = 1$$

$$0.500 \quad 0.500 \quad z_+ = 1 \quad z_- = 1$$

$$I = \frac{500}{2}(1 + 1) = 0.500 \text{ mol kg}^{-1}$$

Add to this the ionic strength from the last iteration of part (a).

$$I_{total} = 0.500 + 0.0114 = 0.511 \text{ mol kg}^{-1}$$

$$\ln \gamma_\pm = -1.173 \times 1 \times \sqrt{0.511} = -0.8389$$

$$\gamma_\pm = 0.432$$

$$K = \frac{\left(\dfrac{m}{m^\circ}\right)^2 (\gamma_\pm)^2}{0.210 - \left(\dfrac{m}{m^\circ}\right)} = \frac{\left(\dfrac{m}{m^\circ}\right)^2 (0.432)^2}{0.210 - \left(\dfrac{m}{m^\circ}\right)} = 5.12 \times 10^{-4}$$

$$m = 0.0227 \text{ mol kg}^{-1}$$

Carrying out another iteration,

$$I_{total} = 0.500 + 0.0227 = 0.523 \text{ mol kg}^{-1}$$

$$\ln \gamma_\pm = -1.173 \times 1 \times \sqrt{0.523} = -0.8480$$

$$\gamma_\pm = 0.428$$

$$K = \frac{\left(\dfrac{m}{m^\circ}\right)^2 (\gamma_\pm)^2}{0.210 - \left(\dfrac{m}{m^\circ}\right)} = \frac{\left(\dfrac{m}{m^\circ}\right)^2 (0.428)^2}{0.125 - \left(\dfrac{m}{m^\circ}\right)} = 5.12 \times 10^{-4}$$

$$m = 0.0229 \text{ mol kg}^{-1}$$

The degree of hydrolysis is

$$\frac{0.0229}{0.210} \times 100\% = 10.9\%$$

P10.23   Calculate the Debye–Hückel screening length $1/\kappa$ at 298 K in a 0.0075 m solution of K_3PO_4.

$$\kappa = 3.29 \times 10^9 \sqrt{I} \ m^{-1}$$

$$I = \frac{m}{2}(v_+ z_+^2 + v_- z_-^2) = \frac{0.0075}{2}(3 \times 1^1 + 3^2) = 0.045 \ \text{mol kg}^{-1}$$

$$\kappa = 3.29 \times 10^9 \sqrt{0.045} \ m^{-1} = 6.98 \times 10^8 \ m^{-1}$$

$$\frac{1}{\kappa} = 1.4 \times 10^{-9} \ m = 1.4 \ nm$$

P10.26   Calculate the ionic strength of each of the solutions in Problem P10.4.

$$I = \frac{m}{2}(v_+ z_+^2 + v_- z_-^2)$$

For $Ca(NO_3)_2$

$$I = \frac{0.0750}{2}(1 \times 2^2 + 2 \times 1^2) = 0.225 \ \text{mol kg}^{-1}$$

For NaOH

$$I = \frac{0.0750}{2}(1 + 1) = 0.0750 \ \text{mol kg}^{-1}$$

For $MgSO_4$

$$I = \frac{0.0750}{2}(1 \times 2^2 + 1 \times 2^2) = 0.300 \ \text{mol kg}^{-1}$$

For $AlCl_3$

$$I = \frac{0.0750}{2}(1 \times 3^2 + 3 \times 1^2) = 0.450 \ \text{mol kg}^{-1}$$

P10.28   Calculate ΔH_R and ΔG_R for the reaction $AgNO_3(aq) + KCl(aq) \rightarrow AgCl(s) + KNO_3(aq)$.

$$\Delta G_R^\circ = \Delta G_f^\circ(AgCl, s) + \Delta G_f^\circ(K^+, aq) + \Delta G_f^\circ(NO_3^-, aq) - \Delta G_f^\circ(Ag^+, aq)$$

$$- \Delta G_f^\circ(NO_3^-, aq) - \Delta G_f^\circ(K^+, aq) - \Delta G_f^\circ(Cl^-, aq)$$

$$\Delta G_R^\circ = \Delta G_f^\circ(AgCl, s) - \Delta G_f^\circ(Ag^+, aq) - \Delta G_f^\circ(Cl^-, aq) = -109.8 \ \text{kJ mol}^{-1}$$

$$- 77.1 \ \text{kJ mol}^{-1} + 131.2 \ \text{kJ mol}^{-1} = -55.7 \ \text{kJ mol}^{-1}$$

$$\Delta H_R^\circ = \Delta H_f^\circ(AgCl, s) - \Delta H_f^\circ(Ag^+, aq) - \Delta H_f^\circ(Cl^-, aq)$$

$$\Delta H_R^\circ = -127.0 \ \text{kJ mol}^{-1} - 105.6 \ \text{kJ mol}^{-1} + 167.2 \ \text{kJ mol}^{-1} = -65.4 \ \text{kJ mol}^{-1}$$

11 Electrochemical Cells, Batteries, and Fuel Cells

Numerical Problems

P11.3   For the half-cell reaction $Hg_2Cl_2(s) + 2e^- \rightarrow 2Hg(l) + 2Cl^-(aq)$, $E° = +0.26808$ V.

Using this result and $\Delta G_f°(Hg_2Cl_2, s) = -210.7$ kJ mol^{-1}, determine $\Delta G_f°$ (Cl^-, aq).

$$\Delta G_R° = -nFE° = 2\Delta G_f°(Cl^-, aq) - \Delta G_f°(Hg_2Cl_2, s)$$

$$\Delta G_f°(Cl^-, aq) = \frac{-\Delta G_f°(Hg_2Cl_2, s) - nFE°}{2}$$

$$\Delta G_f°(Cl^-, aq) = \frac{-210.7 \text{ kJ mol}^{-1} - 2 \text{ mol} \times 96485 \text{ C mol}^{-1} \times 0.26808 \text{ V}}{2} = -131.2 \text{ kJ mol}^{-1}$$

P11.9   Consider the half-cell reaction $AgCl(s) + e^- \rightarrow Ag(s) + Cl^-(aq)$. If μ $(AgCl, s) = -109.71$ kJ mol^{-1}, and if $E° = +0.222$ V for this half-cell, calculate the standard Gibbs energy of formation of $Cl^-(aq)$.

$$\Delta G_f°(AgCl, s) = -109.71 \text{ kJ mol}^{-1}$$

$$\Delta G_R° = -1 \times 96,485 \text{ C mol}^{-1} \times 0.222 \text{ V} = 21.4 \text{ kJ mol}^{-1}$$

$$\Delta G_f°(Cl^-(aq)) = \Delta G_f° + \Delta G_R°(AgCl, s) = -131.1 \text{ kJ mol}^{-1}$$

P11.11   Consider the cell $Fe(s)|FeSO_4(aq, a = 0.0250)|Hg_2SO_4(s)|Hg(l)$.

a.   Write the cell reaction.

b.   Calculate the cell potential, the equilibrium constant for the cell reaction, and $\Delta G_R°$ at 25°C.

(a) $Hg_2SO_4(s) + 2e^- \rightarrow 2Hg(l) + SO_4^{2-}(aq)$ $E° = 0.6125$ V

$Fe(s) \rightarrow Fe^{2+}(aq) + 2e^-$ $E° = 0.447$ V

Cell reaction:

$$Hg_2SO_4(s) + Fe(s) \rightleftharpoons 2Hg(l) + Fe^{2+}(aq) + SO_4^{2-}(aq)$$

$$E_{cell}° = 0.6125 + 0.447 = 1.0595 \text{ V}$$

$$E_{cell} = E_{cell}° - \frac{RT}{nF} \ln a_{Fe^{2+}} a_{SO_4^{2-}}$$

$$= 1.0595 - \frac{8.3145 \text{ J mol}^{-1} \text{ K}^{-1} \times 298.15 \text{ K}}{2 \times 96,485 \text{ C mol}^{-1}} \ln ((0.0250)^2)$$

$$= 1.154 \text{ V}$$

$$\Delta G_{reaction}° = -nFE° = -2 \times 96,485 \text{ C mol}^{-1} \times (1.0595 \text{ V})$$

$$= -204.5 \text{ kJ mol}^{-1}$$

$$K = e^{-\Delta G_R°/RT} = 6.61 \times 10^{35}$$

P11.13   a. Calculate ΔG_R and the equilibrium constant, K, at 298.15 K for the reaction $2Hg(l) + Cl_2(g) \rightleftharpoons Hg_2Cl_2(s)$.
b. Calculate K using Table 4.1. What value of ΔG_R would make the value of K the same as calculated from the half-cell potentials?

(a) $2Hg(l) + 2Cl^-(aq) \rightarrow Hg_2Cl_2(s) + 2e^-$ $E° = -0.26808$ V

$Cl_2(g) + 2e^- \rightarrow 2Cl^-(aq)$ $E° = 1.35827$ V

$2Hg(l) + Cl_2(g) \rightarrow Hg_2Cl_2(s)$ $E° = 1.09019$ V

$$\Delta G_R° = -nFE° = -2 \times 96485 \text{ C mol}^{-1} \times 1.09019 \text{ V} = -210.4 \text{ kJ mol}^{-1}$$

$$\ln K = \frac{nF}{RT} E° = \frac{2 \times 96485 \text{ C mol}^{-1} \times 1.09019 \text{ V}}{8.314 \text{ J K}^{-1} \text{ mol}^{-1} \times 298.15 \text{ K}}$$

$$= 84.8686$$

$$K = 7.21 \times 10^{36}$$

(b) $$\ln K = -\frac{\Delta G_R°}{RT} = \frac{210700 \text{ J mol}^{-1}}{8.314 \text{ J K}^{-1} \text{ mol}^{-1} \times 298.15 \text{ K}}$$

$$= 85.0002$$

$$K = 8.22 \times 10^{36}$$

The results would agree exactly if $\Delta G_R° = -210374$ J mol^{-1}.

P11.20   Determine $E°$ for the reaction $Cr^{2+}(aq) + 2e^- \rightarrow Cr(s)$ from the one-electron reduction potential for $Cr^{3+}(aq)$ and the three-electron reduction potential for $Cr^{3+}(aq)$ given in Table 11.1 (see Appendix B).

$Cr^{3+}(aq) + 3e^- \rightarrow Cr(s)$ $\Delta G° = -nFE° = -3 \times 96485 \text{ C mol}^{-1} \times (-0.744 \text{ V}) = 215.4 \text{ kJ mol}^{-1}$

$Cr^{2+}(aq) \rightarrow Cr^{3+}(aq) + e^-$ $\Delta G° = -nFE° = -1 \times 96485 \text{ C mol}^{-1} \times 0.407 \text{ V} = -39.27 \text{ kJ mol}^{-1}$

$Cr^{2+}(aq) + 2e^- \rightarrow Cr(s)$ $\Delta G = 215.4 \text{ kJ mol}^{-1} - 39.27 \text{ kJ mol}^{-1} = 176.1 \text{ kJ mol}^{-1}$

$$E°_{Cr^{2+}/Cr} = -\frac{\Delta G°}{nF} = \frac{-176.1 \times 10^3 \text{ J mol}^{-1}}{2 \times 96485 \text{ C mol}^{-1}} = -0.913 \text{ V}$$

P11.23   Consider the half-cell reaction $O_2(g) + 4H^+(aq) + 4e^- \rightarrow 2H_2O(l)$. By what factor are n, Q, E, and $E°$ changed if all the stoichiometric coefficients are multiplied by the factor two? Justify your answers.

n is proportional to the number of electrons transferred, and increases by the factor two.

Q is squared if all stoichiometric factors are doubled. The factor by which it is increased depends on the activities of O_2 and H^+.

$E° = \dfrac{\Delta G°}{nF}$ is unchanged because both $\Delta G°$ and n are doubled.

$E = E° - \dfrac{RT}{nF} \ln Q$ is unchanged because the squaring of Q is offset exactly by the doubling of n.

P11.25   The half-cell potential for the reaction $O_2(g) + 4H^+(aq) + 4e^- \rightarrow 2H_2O(l)$ is +1.03 V at 298.15 K when $a_{O_2} = 1$. Determine a_{H^+}.

$$E = E° - \frac{RT}{nF} \ln \frac{1}{a_{O_2} a_{H^+}^4}$$

$$1.03 \text{ V} = 1.23 \text{ V} - \frac{0.05916 \text{ V}}{4} \log_{10} \frac{1}{a_{H^+}^4}$$

$$\log_{10} a_{H^+} = \frac{1.03 \text{ V} - 1.23 \text{ V}}{0.05916 \text{ V}} = -3.381$$

$$a_{H^+} = 4.16 \times 10^{-4}$$

P11.26 Using half-cell potentials, calculate the equilibrium constant at 298.15 K for the reaction $2H_2O(l) \rightleftharpoons 2H_2(g) + O_2(g)$. Compare your answer with that calculated using $\Delta G_f°$ values from Table 4.1 (see Appendix B). What is the value of $E°$ for the overall reaction that makes the two methods agree exactly?

$$2H_2O(l) + 2e^- \rightarrow H_2(g) + 2OH^-(aq) \qquad E° = -0.8277 \text{ V}$$

$$4OH^-(aq) \rightarrow O_2(g) + 4e^- + 2H_2O(l) \qquad E° = -0.401 \text{ V}$$

$$2H_2O(l) \rightleftharpoons 2H_2(g) + O_2(g) \qquad E° = -1.2287 \text{ V}$$

$$\ln K = \frac{nF}{RT} E° = -\frac{4 \times 96485 \text{ C mol}^{-1} \times 1.2287 \text{ V}}{8.314 \text{ J K}^{-1} \text{ mol}^{-1} \times 298.15 \text{ K}}$$

$$= -191.303$$

$$K = 8.28 \times 10^{-84}$$

$$\ln K = -\frac{2\Delta G_f°(H_2O, l)}{RT} = -\frac{2 \text{ mol} \times 237.1 \times 10^3 \text{ J mol}^{-1}}{8.3145 \text{ J mol}^{-1} \text{ K}^{-1} \times 298.15 \text{ K}} = -191.301$$

$$K = 8.30 \times 10^{-84}$$

For the two results to agree, $E°$ must be given by

$$E° = -\frac{191.325 \times 8.3145 \text{ J mol}^{-1} \text{ K}^{-1} \times 298.15 \text{ K}}{4 \times 96485 \text{ C mol}^{-1}} = -1.22869 \text{ V}$$

This value lies within the error limits of the determination of $E°$.

P11.29 Determine K_{sp} for AgBr at 298.15 K using the electrochemical cell described by

$$Ag(s) \left| Ag^+(aq, a_{Ag^+}) \right\| Br^-(aq, a_{Br^-}) \left| AgBr(s) \right| Ag(s)$$

The half-cell and overall reactions are

$$Ag(s) + Br^-(aq) \rightarrow AgBr(s) + e^- \qquad E° = -0.07133 \text{ V}$$

$$Ag^+(aq) + e^- \rightarrow Ag(s) \qquad E° = 0.7996 \text{ V}$$

$$Ag^+(aq) + Br^-(aq) \rightleftharpoons AgBr(s) \qquad E° = 0.7283 \text{ V}$$

$$\log_{10} K_{sp} = \frac{nE°}{0.05916 \text{ V}} = -\frac{0.72827 \text{ V}}{0.05916 \text{ V}} = 12.31$$

$$K = 2.04 \times 10^{12}$$

$$K_{sp} = 1/K = 4.89 \times 10^{-13}$$

12 From Classical to Quantum Mechanics

Numerical Problems

P12.5   Calculate the highest possible energy of a photon that can be observed in the emission spectrum of H.

The highest energy photon corresponds to a transition from $n = \infty$ to $n = 1$.

$$\tilde{\nu} = 109678\left(\frac{1}{1} - \frac{1}{\infty^2}\right) = 109678 \text{ cm}^{-1}$$

$$E = hc\tilde{\nu} = 2.17871 \times 10^{-18} \text{ J}$$

P12.8   What speed does a F_2 molecule have if it has the same momentum as a photon of wavelength 225 nm?

$$p = \frac{h}{\lambda} = m_{F_2}\nu_{F_2}$$

$$\nu_{F_2} = \frac{h}{m_{F_2}\lambda} = \frac{6.626 \times 10^{-34} \text{ J s}}{37.9968 \text{ amu} \times 1.661 \times 10^{-27} \text{ kg (amu)}^{-1} \times 225. \times 10^{-9} \text{ m}} = 0.0467 \text{ m s}^{-1}$$

P12.12   Electrons have been used to determine molecular structure by diffraction. Calculate the speed and kinetic energy of an electron for which the wavelength is equal to a typical bond length, namely, 0.125 nm.

$$v = \frac{p}{m} = \frac{h}{m\lambda} = \frac{6.626 \times 10^{-34} \text{ J s}}{9.109 \times 10^{-31} \text{ kg} \times 0.125 \times 10^{-9} \text{ m}} = 5.82 \times 10^{6} \text{ m s}^{-1}$$

$$E_{kinetic} = \frac{1}{2}mv^2 = \frac{1}{2} \times 9.109 \times 10^{-31} \text{ kg} \times (5.82 \times 10^{6} \text{ m s}^{-1})^2 = 1.54 \times 10^{-17} \text{ J}$$

P12.13   For a monatomic gas, one measure of the "average speed" of the atoms is the root mean square speed, $\nu_{rms} = \langle v^2 \rangle^{1/2} = \sqrt{3k_B T/m}$, in which m is the molecular mass and k_B is the Boltzmann constant. Using this formula, calculate the de Broglie wavelength for H_2 and Ar at 200. and at 900. K.

$$\lambda = \frac{h}{m\nu_{rms}} = \frac{h}{\sqrt{3k_B Tm}} = \frac{6.626 \times 10^{-34} \text{ J s}}{\sqrt{3 \times 1.381 \times 10^{-23} \text{ J K}^{-1} \times 200. \text{ K} \times 2.016 \text{ amu} \times 1.661 \times 10^{-27} \text{ kg amu}^{-1}}}$$

$$= 1.26 \times 10^{-10} \text{ m}$$

For H_2 at 900. K, $\lambda = 5.93 \times 10^{-11}$ m. For Ar, $\lambda = 2.83 \times 10^{-11}$ m and 1.33×10^{-11} m at 200. K and 900. K, respectively.

P12.16   If an electron passes through an electrical potential difference of 1 V, it has an energy of 1 electron-volt. What potential difference must it pass through in order to have a wavelength of 0.300 nm?

$$E = \frac{1}{2}m_e v^2 = \frac{1}{2}m_e \times \left(\frac{h}{m_e\lambda}\right)^2 = \frac{h^2}{2m_e\lambda^2}$$

$$= \frac{(6.626 \times 10^{-34} \text{ J s})^2}{2 \times 9.109 \times 10^{-31} \text{ kg} \times (3.00 \times 10^{-10} \text{ m})^2} \times \frac{1 \text{ eV}}{1.602 \times 10^{-19} \text{ J}} = 16.7 \text{ eV}$$

The electron must pass through an electrical potential of 16.7 V.

P12.20   The power (energy per unit time) radiated by a blackbody per unit area of surface expressed in units of W m^{-2} is given by $P = \sigma T^4$ with $\sigma = 5.67 \times 10^{-8}$ W m^{-2} K^{-4}. The radius of the sun is 7.00×10^5 km and the surface temperature is 5800. K. Calculate the total energy radiated per second by the sun. Assume ideal blackbody behavior.

$$E = PA = \sigma T^4 \times 4\pi r^2$$
$$= 5.67 \times 10^{-8} \text{ W m}^{-2} \text{ K}^{-4} \times (5800. \text{ K})^4 \times 4\pi \times (7.00 \times 10^8 \text{ m})^2$$
$$= 3.95 \times 10^{26} \text{ W}$$

P12.21   The work function of palladium is 5.22 eV. What is the minimum frequency of light required to observe the photoelectric effect on Pd? If light with a 200. nm wavelength is absorbed by the surface, what is the velocity of the emitted electrons?

a.   For electrons to be emitted, the photon energy must be greater than the work function of the surface.

$$E = h\nu \geq 5.22 \text{ eV} \times \frac{1.602 \times 10^{-19} \text{ J}}{\text{eV}} = 8.36 \times 10^{-19} \text{ J}$$

$$\nu \geq \frac{E}{h} \geq \frac{8.36 \times 10^{-19} \text{ J}}{6.626 \times 10^{-34} \text{ J s}} \geq 1.26 \times 10^{15} \text{ s}^{-1}$$

b.   The outgoing electron must first surmount the barrier arising from the work function, so not all the photon energy is converted to kinetic energy.

$$E_e = h\nu - \phi = \frac{hc}{\lambda} - \phi$$
$$= \frac{6.626 \times 10^{-34} \text{ J s} \times 2.998 \times 10^8 \text{ m s}^{-1}}{200. \times 10^{-9} \text{ m}} - 8.36 \times 10^{-19} \text{ J} = 1.57 \times 10^{-19} \text{ J}$$

$$\nu = \sqrt{\frac{2E_e}{m_e}} = \sqrt{\frac{2 \times 1.62 \times 10^{-19} \text{ J}}{9.11 \times 10^{-31} \text{ kg}}} = 5.87 \times 10^5 \text{ m s}^{-1}$$

P12.23   Calculate the longest and the shortest wavelength observed in the Lyman series.

For the longest wavelength, the transition is from $n = 2$ to $n = 1$.

$$\tilde{\nu} = 109678\left(\frac{1}{1} - \frac{1}{2^2}\right) = 82257.8 \text{ cm}^{-1}$$

$$\lambda = \frac{1}{\tilde{\nu}} = 1.21569 \times 10^{-5} \text{ cm}$$

For the shortest wavelength, the transition is from $n = \infty$ to $n = 1$.

$$\tilde{\nu} = 109678\left(\frac{1}{1} - \frac{1}{\infty^2}\right) = 109678 \text{ cm}^{-1}$$

$$\lambda = \frac{1}{\tilde{\nu}} = 9.11768 \times 10^{-6} \text{ cm}$$

P12.24   A 1000. W gas discharge lamp emits 4.50 W of ultraviolet radiation in a narrow range centered near 275 nm. How many photons of this wavelength are emitted per second?

$$n' = \frac{E_{total}}{E_{photon}} = \frac{4.50 \text{ W} \times 1 \text{ J s}^{-1} \text{ W}^{-1}}{\dfrac{hc}{\lambda}} = \frac{5.25 \text{ W} \times 1 \text{ J s}^{-1} \text{ W}^{-1}}{\dfrac{6.626 \times 10^{-34} \text{ J s} \times 2.998 \times 10^8 \text{ m s}^{-1}}{275 \times 10^{-9} \text{ m}}} = 6.23 \times 10^{18} \text{ s}^{-1}$$

P12.25   The power per unit area emitted by a blackbody is given by $P = \sigma T^4$ with $\sigma = 5.67 \times 10^{-8}$ W m^{-2} K^{-4}. Calculate the power radiated by a spherical blackbody of radius 0.500 m at 925 K. What would the radius of a blackbody at 3000. K be if it emitted the same energy as the spherical blackbody of radius 0.500 m at 925 K?

$E = A\sigma T^4 = 4\pi(0.500 \text{ m})^2 \times 5.67 \times 10^{-8} \text{ J s}^{-1} \text{ m}^{-2} \text{K}^{-4} \times (925 \text{ K})^4 = 1.30 \times 10^5 \text{ J s}^{-1}$. Because the total energy radiated by the spheres must be equal,

$$4\pi r_1^2 \sigma T_1^4 = 4\pi r_2^2 \sigma T_2^4$$

$$r_2 = \sqrt{\frac{r_1^2 T_1^4}{T_2^4}} = \sqrt{\frac{(0.500 \text{ m})^2 (925 \text{ K})^4}{(3000. \text{ K})^4}} = 0.0475 \text{ m}$$

P12.27   Pulsed lasers are powerful sources of nearly monochromatic radiation. Lasers that emit photons in a pulse of 5.00 ns duration with a total energy in the pulse of 0.175 J at 875 nm are commercially available.

    a.   What is the average power (energy per unit time) in units of watts (1 W = 1 J/s) associated with such a pulse?

    b.   How many 1000-nm photons are emitted in such a pulse?

(a) $P = \dfrac{\Delta E}{\Delta t} = \dfrac{0.175 \text{ J}}{5.00 \times 10^{-9} \text{ s}} = 3.50 \times 10^7 \text{ J s}^{-1}$

(b) $N = \dfrac{E_{pulse}}{E_{photon}} = \dfrac{E_{pulse}}{h\dfrac{c}{\lambda}} = \dfrac{0.175 \text{ J}}{6.626 \times 10^{-34} \text{ J s}^{-1} \times \dfrac{2.998 \times 10^8 \text{ m s}^{-1}}{875 \times 10^{-9} \text{ m}}} = 7.71 \times 10^{17}$

13 The Schrödinger Equation

Numerical Problems

P13.1 A wave traveling in the z direction is described by the wave function $\Psi(z,t) = A_1\,\mathbf{x}\sin(kz - \omega t + \phi_1) + A_2\,\mathbf{y}\sin(kz - \omega t + \phi_2)$, where $\mathbf{x}$ and $\mathbf{y}$ are vectors of unit length along the x and y axes, respectively. Because the amplitude is perpendicular to the propagation direction, $\Psi(z,t)$ represents a transverse wave.

    a.   What requirements must A_1 and A_2 satisfy for a plane polarized wave in the x-z plane? The amplitude of a plane polarized wave is nonzero only in one plane.

    b.   What requirements must A_1 and A_2 satisfy for a plane polarized wave in the y-z plane?

    c.   What requirements must A_1 and A_2 and ϕ_1 and ϕ_2 satisfy for a plane polarized wave in a plane oriented at $45°$ to the x-z plane?

    d.   What requirements must A_1 and A_2 and ϕ_1 and ϕ_2 satisfy for a circularly polarized wave? The phases of the two components of a circularly polarized wave differ by $\pi/2$.

    (a)   The amplitude along the x axis must oscillate, and the amplitude along the y axis must vanish. Therefore $A_1 \neq 0$ and $A_2 = 0$.

    (b)   The amplitude along the y axis must oscillate, and the amplitude along the x axis must vanish. Therefore $A_1 = 0$ and $A_2 \neq 0$.

    (c)   The amplitude along both the x and y axes must oscillate in phase. Therefore $A_2 = A_1 \neq 0$. Because they must oscillate in phase, $\phi_1 = \phi_2$.

    (d)   The amplitude along both the x and y axes must oscillate with the same amplitude. Therefore $A_1 = A_2 \neq 0$. For a circularly polarized wave, the x and y components must be out of phase by $\pi/2$. Therefore $\phi_1 = \phi_2 \pm \dfrac{\pi}{2}$. This can be seen by comparing the x and y amplitudes for the positive sign.

$$\Psi(z,t) = A_1\,\mathbf{x}\sin(kz - \omega t + \phi_1) + A_1\,\mathbf{y}\sin\left(kz - \omega t + \phi_1 + \frac{\pi}{2}\right)$$

Let $kz + \phi = kz'$

$$\Psi(z,t) = A_1\,\mathbf{x}\sin(kz' - \omega t) + A_1\,\mathbf{y}\sin\left(kz' - \omega t + \frac{\pi}{2}\right)$$
$$= A_1\,\mathbf{x}\sin(kz' - \omega t) + A_1\,\mathbf{y}\left[\sin(kz' - \omega t)\cos\frac{\pi}{2} + \cos(kz' - \omega t)\sin\frac{\pi}{2}\right]$$
$$= A_1\,\mathbf{x}\sin(kz' - \omega t) + A_1\,\mathbf{y}\cos(kz' - \omega t)$$

The x and y amplitudes are $\pi/2$ out of phase and the sum of the squares of their amplitudes is a constant as required for a circle.

P13.2 Because $\int_{-d}^{d}\cos(n\pi x/d)\cos(m\pi x/d)dx = 0$, $m \neq n$, the functions $\cos(n\pi x/d)$ for $n = 1, 2, 3, \ldots$ form an orthogonal set in the interval $(-d, d)$. What constant must these functions be multiplied by to form an orthonormal set?

$$1 = N^2 \int_{-d}^{d} \cos\left(\frac{m\pi x}{d}\right)\cos\left(\frac{m\pi x}{d}\right) dx = N^2 \left[\frac{x}{2} + \frac{d}{4m\pi}\sin\left(\frac{2m\pi x}{d}\right)\right]_{-d}^{d}$$

where we have used the standard integral $\int (\cos^2 ax)\, dx = \frac{1}{2}x + \frac{1}{4a}\sin 2ax.$

$$1 = N^2\left[\frac{d}{2} + \frac{d}{4m\pi}\sin(2m\pi) + \frac{d}{2} + \frac{d}{4m\pi}\sin(-2m\pi)\right] = dN^2 = 1$$

$$N = \sqrt{\frac{1}{d}}$$

P13.6   Carry out the following coordinate transformations:

a.   Express the point $x = 3, y = 1,$ and $z = 1$ in spherical coordinates.

b.   Express the point $r = 5, \theta = \frac{\pi}{4},$ and $\phi = \frac{3\pi}{4}$ in Cartesian coordinates.

(a) $r = \sqrt{x^2 + y^2 + z^2} = \sqrt{3^2 + 1^2 + 1^2} = \sqrt{11}$

$\theta = \cos^{-1}\dfrac{z}{\sqrt{x^2 + y^2 + z^2}} = \cos^{-1}\dfrac{1}{\sqrt{11}} = 1.26$ radians

$\phi = \tan^{-1}\dfrac{y}{x} = \tan^{-1}\dfrac{1}{3} = 0.322$ radians

(b) $x = r\sin\theta\cos\phi = 5\sin\dfrac{\pi}{4}\cos\dfrac{\pi}{4} = 5/2$

$y = r\sin\theta\sin\phi = 5\sin\dfrac{\pi}{4}\sin\dfrac{3\pi}{4} = 5/2$

$z = r\cos\theta = 5\cos\dfrac{\pi}{4} = \dfrac{5}{\sqrt{2}}$

P13.8   Show that

$$\frac{a + ib}{c + id} = \frac{ac + bd + i(bc - ad)}{c^2 + d^2}$$

$$\frac{a + ib}{c + id} = \left(\frac{a + ib}{c + id}\right)\left(\frac{c - id}{c - id}\right) = \frac{ac + bd + ibc - iad}{c^2 + d^2} = \frac{ac + bd + i(bc - ad)}{c^2 + d^2}$$

P13.9   Express the following complex numbers in the form $re^{i\theta}$.

a. $5 + 6i$ c. 4 e. $\dfrac{2 - i}{1 + i}$

b. $2i$ d. $\dfrac{5 + i}{3 - 4i}$

In the notation $re^{i\theta}$, $r = |z| = \sqrt{a^2 + b^2}$ and $\theta = \sin^{-1}\left(\dfrac{\operatorname{Im} z}{|z|}\right).$

(a) $5 + 6i = \sqrt{61}\exp\left(i\sin^{-1}\dfrac{6}{\sqrt{61}}\right) = \sqrt{61}\exp(-0.876i)$

(b) $2i = 2\exp\left(i\sin^{-1}\dfrac{2}{\sqrt{2}}\right) = 2\exp i\left(\dfrac{\pi}{2}\right)$

(c) $4 = 4 \exp\left(i\sin^{-1}\dfrac{0}{4} \right) = 4\exp(0i)$

(d) $\dfrac{5+i}{3-4i} = \dfrac{5+i}{3-4i} \times \dfrac{3+4i}{3+4i} = \dfrac{15+23i-4}{25} = \dfrac{11}{25} + \dfrac{23i}{25}$

We next calculate the magnitude of the complex number

$$\sqrt{\left(\dfrac{11}{25}\right)^2 + \left(\dfrac{23}{25}\right)^2} = \sqrt{\dfrac{650}{625}} = \dfrac{\sqrt{26}}{5}$$

$$\dfrac{5+i}{3-4i} = \dfrac{\sqrt{26}}{5} \exp\left(i\sin^{-1}\dfrac{23/25}{\sqrt{26}/5} \right) = \dfrac{\sqrt{26}}{5}\exp(1.125i)$$

(e) $\dfrac{2-i}{1+i} = \dfrac{2-i}{1+i} \times \dfrac{1-i}{1-i} = \dfrac{2-3i-1}{2} = \dfrac{1}{2} - \dfrac{3i}{2}$

We next calculate the magnitude of the complex number

$$\sqrt{\left(\dfrac{1}{2}\right)^2 + \left(\dfrac{3}{2}\right)^2} = \sqrt{\dfrac{5}{2}}$$

$$\dfrac{2-i}{1+i} = \sqrt{\dfrac{5}{2}} \exp\left(i\sin^{-1}\dfrac{3/2}{\sqrt{5/2}} \right) = \sqrt{\dfrac{5}{2}}\exp(-1.125i)$$

P13.15   Show by carrying out the integration that $\sin(m\pi x/a)$ and $\cos(m\pi x/a)$, where m is an integer, are orthogonal over the interval $0 \le x \le a$. Would you get the same result if you used the interval $0 \le x \le 3a/4$? Explain your result.

$$\int_0^a \cos\left(\dfrac{m\pi x}{a}\right)\sin\left(\dfrac{m\pi x}{a}\right) dx = \left[\dfrac{a}{2m\pi}\sin^2\left(\dfrac{m\pi x}{a}\right)\right]_0^a = \dfrac{a}{2m\pi}\left[\sin^2(m\pi) - 0\right] = 0$$

$$\int_0^{\frac{3a}{4}} \cos\left(\dfrac{m\pi x}{a}\right)\sin\left(\dfrac{m\pi x}{a}\right) dx = \int_0^{\frac{3a}{4}} \cos\left(\dfrac{m\pi x}{a}\right)\sin\left(\dfrac{m\pi x}{a}\right) dx$$

$$= \left[\dfrac{a}{2m\pi}\sin^2\left(\dfrac{m\pi x}{a}\right)\right]_0^{\frac{3a}{4}} = \dfrac{a}{2m\pi}\left[\sin^2\left(\dfrac{3m\pi}{4}\right) - 0\right] \ne 0$$

except for the special case $\dfrac{3m}{4} = n$ where n is an integer. The length of the integration interval must be n periods (for n an integer) to make the integral zero.

P13.19   Is the function $2x^2 - 1$ an eigenfunction of the operator $-(3/2 - x^2)(d^2/dx^2) + 2x(d/dx)$? If so, what is the eigenvalue?

$$-(3/2 - x^2)(d^2[2x^2 - 1]/dx^2) + 2x(d[2x^2 - 1]/dx)$$
$$= -4(3/2 - x^2) + 8x^2 = 12x^2 - 6 = 6[2x^2 - 1]$$

Eigenfunction with eigenvalue 6

P13.20   Find the result of operating with $d^2/dx^2 - 2x^2$ on the function e^{-ax^2}. What must the value of a be to make this function an eigenfunction of the operator? What is the eigenvalue?

$$\dfrac{d^2 e^{-ax^2}}{dx^2} - 2x^2 e^{-ax^2} = \dfrac{d(-2axe^{-ax^2})}{dx} - 2x^2 e^{-ax^2}$$
$$= -2ae^{-ax^2} + 4a^2x^2 e^{-ax^2} - 2x^2 e^{-ax^2}$$

For the function to be an eigenfunction of the operator, the terms containing $x^2 e^{-ax^2}$ must vanish. This is the case if $4a^2 = 2$ or $a = \pm\dfrac{1}{\sqrt{2}}$.

P13.21 Determine in each of the following cases if the function in the first column is an eigenfunction of the operator in the second column. If so, what is the eigenvalue?

a. $\sin\theta\cos\phi$ $\partial^4/\partial\phi^4$

b. $e^{\frac{3x^2}{\sqrt{2}}}$ $(1/x)\,d/dx$

c. $\cos\dfrac{2\pi x}{a}$ $\left(1\Big/\tan\dfrac{2\pi x}{a}\right)d/dx$

(a) $\dfrac{\partial^4}{\partial\phi^4}\sin\theta\cos\phi = \sin\theta\cos\phi$ Eigenfunction with eigenvalue 1

(b) $\dfrac{1}{x}\dfrac{d}{dx}e^{-\left(3x^2\big/\sqrt{2}\right)} = -3\sqrt{2}e^{-\left(3x^2\big/\sqrt{2}\right)}$ Eigenfunction with eigenvalue $-3\sqrt{2}$

(c) $\dfrac{1}{\tan\dfrac{2\pi x}{a}}\dfrac{d}{dx}\cos\dfrac{2\pi x}{a} = \dfrac{-1}{\tan\dfrac{2\pi x}{a}}\dfrac{2\pi}{a}\sin\dfrac{2\pi x}{a}$

$$= \dfrac{-\cos\dfrac{2\pi x}{a}}{\sin\dfrac{2\pi x}{a}}\dfrac{2\pi}{a}\sin\dfrac{2\pi x}{a} = -\dfrac{2\pi}{a}\cos\dfrac{2\pi x}{a}$$

Eigenfunction with eigenvalue $-\dfrac{2\pi}{a}$

P13.24 If two operators act on a wave function as indicated by $\hat{A}\hat{B}f(x)$, it is important to carry out the operations in succession, with the first operation being that nearest to the function. Mathematically, $\hat{A}\hat{B}f(x) = \hat{A}(\hat{B}f(x))$ and $\hat{A}^2 f(x) = \hat{A}(\hat{A}f(x))$. Evaluate the following successive operations $\hat{A}\hat{B}f(x)$. The operators $\hat{A}$ and $\hat{B}$ are listed in the first and second columns and $f(x)$ is listed in the third column. Compare your answers to parts (a) and (b), and to (c) and (d).

a. $\dfrac{d}{dx}$ x $x^2 + e^{ax^2}$

b. x $\dfrac{d}{dx}$ $x^2 + e^{ax^2}$

c. $\dfrac{\partial^2}{\partial y^2}$ y^2 $(\cos 3y)\sin^2 x$

d. y^2 $\dfrac{\partial^2}{\partial y^2}$ $(\cos 3y)\sin^2 x$

(a) $\dfrac{d}{dx}\left[x(x^2 + e^{ax^2})\right] = 3x^2 + e^{ax^2}(1 + 2ax^2)$

(b) $x\dfrac{d}{dx}\left[(x^2 + e^{ax^2})\right] = 2x^2 + 2ax^2 e^{ax^2}$

(c) $\dfrac{\partial^2}{\partial y^2}[y^2(\cos 3y)\sin^2 x] = \sin^2 x\dfrac{\partial}{\partial y}[2y(\cos 3y) - 3y^2\sin 3y]$

$$= \sin^2 x(2\cos 3y - 6y\sin 3y - 6y\sin 3y - 9y^2\cos 3y)$$

$$= \sin^2 x([2 - 9y^2]\cos 3y - 12y\sin 3y)$$

(d) $y^2 \dfrac{\partial^2}{\partial y^2}\left[(\cos 3y)\sin^2 x\right] = -9y^2 \sin^2 x(\cos 3y)$

Changing the order of operators changes the outcome in both cases.

P13.26   Consider a two-level system with $\varepsilon_1 = 2.25 \times 10^{-22}$ J and $\varepsilon_2 = 4.50 \times 10^{-21}$ J. If $g_2 = 2g_1$, what value of T is required to obtain $n_2/n_1 = 0.175$? What value of T is required to obtain $n_2/n_1 = 0.750$?

$$\frac{n_2}{n_1} = \frac{g_2}{g_1}\exp\left[\frac{-(\varepsilon_2 - \varepsilon_1)}{k_B T}\right]$$

$$\ln\left(\frac{n_2}{n_1}\right) = \ln\left(\frac{g_2}{g_1}\right) - \frac{(\varepsilon_2 - \varepsilon_1)}{k_B T}$$

$$\frac{1}{T} = \frac{k_B}{(\varepsilon_2 - \varepsilon_1)}\left[\ln\left(\frac{g_2}{g_1}\right) - \ln\left(\frac{n_2}{n_1}\right)\right]$$

$$T = \frac{(\varepsilon_2 - \varepsilon_1)}{k_B\left[\ln\left(\dfrac{g_2}{g_1}\right) - \ln\left(\dfrac{n_2}{n_1}\right)\right]}$$

for $n_2/n_1 = 0.175$ $T = \dfrac{4.275 \times 10^{-21}\ \text{J}}{1.381 \times 10^{-23}\ \text{J K}^{-1} \times \left[\ln(1) - \ln(0.175)\right]} = 127\ \text{K}$

for $n_2/n_1 = 0.750$ $T = \dfrac{4.275 \times 10^{-21}\ \text{J}}{1.381 \times 10^{-23}\ \text{JK}^{-1} \times \left[\ln(1) - \ln(0.750)\right]} = 316\ \text{K}$

P13.28   Normalize the set of functions $\phi_n(\theta) = e^{in\theta}$, $0 \le \theta \le 2\pi$. To do so, you need to multiply the functions by a so-called normalization constant N so that the integral

$$NN^* \int_0^{2\pi} \phi_m^*(\theta)\,\phi_n(\theta)\,d\theta = 1 \quad \text{for } m = n$$

$NN^* \displaystyle\int_0^{2\pi} e^{-in\theta}\,e^{in\theta}\,d\theta = NN^* \int_0^{2\pi} d\theta = 2\pi\,NN^* = 1$. This equation is satisfied for $N = \dfrac{1}{\sqrt{2\pi}}$ and the normalized

functions are $\phi_n(\theta) = \dfrac{1}{\sqrt{2\pi}}\,e^{in\theta}$, $\quad 0 \le \theta \le 2\pi$.

P13.30   Operate with (a) $\dfrac{\partial}{\partial x} + \dfrac{\partial}{\partial y} + \dfrac{\partial}{\partial z}$ and (b) $\dfrac{\partial^2}{\partial x^2} + \dfrac{\partial^2}{\partial y^2} + \dfrac{\partial^2}{\partial z^2}$ on the function $A\,e^{-ik_1 x}e^{-ik_1 y}e^{-ik_1 z}$. Under what conditions is the function an eigenfunction of one or both operators? What is the eigenvalue?

$$\left[\frac{\partial}{\partial x} + \frac{\partial}{\partial y} + \frac{\partial}{\partial z}\right]A\,e^{-ik_1 x}e^{-ik_1 y}e^{-ik_1 z}$$

$$= -Aik_1 e^{-ik_1 x}e^{-ik_1 y}e^{-ik_1 z} - Aik_1 e^{-ik_1 x}e^{-ik_1 y}e^{-ik_1 z} - Aik_1 e^{-ik_1 x}e^{-ik_1 y}e^{-ik_1 z}$$

$$= -3A\,ik_1 e^{-ik_1 x}e^{-ik_1 y}e^{-ik_1 z}$$

Eigenfunction with eigenvalue $-3ik_1$

$$\left[\frac{\partial^2}{\partial x^2} + \frac{\partial^2}{\partial y^2} + \frac{\partial^2}{\partial z^2}\right]A e^{-ik_1 x}e^{-ik_1 y}e^{-ik_1 z}$$

$$= 3A\,e^{-ik_1 x}e^{-ik_1 y}e^{-ik_1 z}$$

Eigenfunction with eigenvalue $3k_1^2$

P13.36   Which of the following wave functions are eigenfunctions of the operator d/dx? If they are eigenfunctions, what is the eigenvalue?

a. $ae^{-3x} + be^{-3ix}$ d. $\cos ax$

b. $\sin^2 x$ e. e^{-ix^2}

c. e^{-ix}

(a) $\dfrac{d(ae^{-3x} + be^{-3ix})}{dx} = -3ae^{-3x} - 3ibe^{-3ix}$ Not an eigenfunction

(b) $\dfrac{d\sin^2 x}{dx} = 2\sin x \cos x$ Not an eigenfunction

(c) $\dfrac{de^{-ix}}{dx} = -ie^{-ix}$ Eigenfunction with eigenvalue $-i$

(d) $\dfrac{d\cos ax}{dx} = -a\sin ax$ Not an eigenfunction

(e) $\dfrac{de^{-ix^2}}{dx} = -2ixe^{-ix^2}$ Not an eigenfunction

14 The Quantum Mechanical Postulates

Numerical Problems

P14.2 Which of the following functions are single-valued functions of the variable x?

a. $\sin\dfrac{2\pi x}{a}$ b. $e^{3\sqrt{x}}$ c. $1-3\sin^2 x$ d. $e^{2\pi i x}$

(a) Yes. The function $\sin\dfrac{2\pi x}{a}$ has only a single value for a given value of x.

(b) No. The function $e^{3\sqrt{x}}$ has two values for a given value of x.

(c) Yes. The function $1-3\sin^2 x$ has only a single value for a given value of x.

(d) Yes. The function $e^{2\pi i x}$ has only a single value for a given value of x.

P14.5 Is the function $(x^2-1)/(x-1)$ continuous at $x=1$? Answer this question by evaluating $f(1)$ and $\lim\limits_{x\to 1} f(x)$.

It is not continuous because $f(1)$ is not defined. The expression cannot be simplified to

$$(x^2-1)/(x-1) = \frac{(x+1)(x-1)}{x-1} = x+1$$ because it is not permissible to divide by zero. To calculate the limit, we use L'Hôpital's rule

$$\lim_{x\to 1}\left[\frac{f(x)}{g(x)}\right] = \lim_{x\to 1}\left[\frac{df(x)/dx}{dg(x)/dx}\right] = \lim_{x\to 1}\left[\frac{2x}{1}\right] = 2$$

P14.8 Which of the following functions are acceptable wave functions over the indicated interval?

a. e^{-x} $0 < x < \infty$

c. $e^{-2\pi i x}$ $-100 < x < 100$

b. e^{-x} $-\infty < x < \infty$

d. $\dfrac{1}{x}$ $1 < x < \infty$

Explain your answers.

(a) e^{-x} $0 < x < \infty$ is an acceptable wave function because it is finite over the interval, is defined at every point in the interval, and the first and second derivatives exist.

(b) e^{-x} $-\infty < x < \infty$ is not an acceptable wave function because it becomes infinite as $x \to -\infty$.

(c) $e^{-2\pi i x}$ $-100 < x < 100$ is an acceptable wave function because it is finite over the interval, is defined at every point in the interval, and the first and second derivatives exist.

(d) $\dfrac{1}{x}$ $1 < x < \infty$ is an acceptable wave function because it is finite over the interval, is defined at every point in the interval and the first and second derivatives exist.

P14.11 For a Hermetian operator $\hat{A}$, $\int \psi^*(x)[\hat{A}\psi(x)]\,dx = \int \psi(x)[\hat{A}\psi(x)]^*\,dx$. Assume that $\hat{A}f(x) = (a + ib)f(x)$, where a and b are constants. Show that if $\hat{A}$ is a Hermetian operator, $b = 0$ so that the eigenvalues of $f(x)$ are real.

$$\int \psi^*(x)\left[\hat{A}\psi(x)\right]dx = \int \psi(x)\left[\hat{A}\psi(x)\right]^*dx$$

$$\int \psi^*(x)[(a+ib)\psi(x)]\,dx = \int \psi(x)[(a+ib)\psi(x)]^*\,dx = \int \psi(x)\left[(a-ib)\psi^*(x)\right]dx$$

$$a\int \psi^*(x)\psi(x)\,dx + ib\int \psi^*(x)\psi(x)\,dx = a\int \psi^*(x)\psi(x)\,dx - ib\int \psi^*(x)\psi(x)\,dx$$

The last statement is only true if $-b = b$, which is only satisfied if $b = 0$.

P14.13 Is the relation $(\hat{A}f(x))/f(x) = \hat{A}$ always obeyed? If not, give an example to support your conclusion.

No.

Assume $f(x) = x$ and $\hat{A} = d/dx$.

$$\frac{df(x)}{dx} = 1; \frac{\hat{A}f(x)}{f(x)} = \frac{1}{x} \neq \frac{d}{dx}$$

15 Using Quantum Mechanics on Simple Systems

Numerical Problems

P15.3 Normalize the total energy eigenfunctions for the three-dimensional box in the interval $0 \le x \le a$, $0 \le y \le b$, $0 \le z \le c$.

$$1 = \int_0^a \int_0^b \int_0^c \psi^*(x, y, z)\psi(x, y, z)\, dx\, dy\, dz$$

$$= N^2 \int_0^a \sin^2\left(\frac{n_x \pi x}{a}\right) dx \int_0^b \sin^2\left(\frac{n_y \pi y}{b}\right) dy \int_0^c \sin^2\left(\frac{n_z \pi z}{c}\right) dz$$

Using the standard integral,

$$\int \sin^2 \alpha x\, dx = \frac{x}{2} - \frac{\sin(2\alpha x)}{4\alpha}$$

$$1 = N^2 \int_0^a \sin^2\left(\frac{n_x \pi x}{a}\right) dx \int_0^b \sin^2\left(\frac{n_y \pi y}{b}\right) dy \int_0^c \sin^2\left(\frac{n_z \pi z}{c}\right) dz$$

$$1 = N^2 \left[\frac{a}{2} - \frac{a}{4 n_x \pi}(\sin n_x \pi - \sin 0)\right] \times \left[\frac{b}{2} - \frac{b}{4 n_y \pi}(\sin n_y \pi - \sin 0)\right] \times \left[\frac{c}{2} - \frac{c}{4 n_z \pi}(\sin n_z \pi - \sin 0)\right]$$

$$= N^2 \frac{abc}{8}$$

$$N = \sqrt{\frac{8}{abc}} \text{ and } \psi(x,y) = \sqrt{\frac{8}{abc}} \sin\left(\frac{n_x \pi x}{a}\right)\sin\left(\frac{n_y \pi y}{b}\right)\sin\left(\frac{n_z \pi z}{c}\right)$$

P15.4 Is the superposition wave function for the free particle $\psi(x) = A_+ e^{+i\sqrt{\frac{2mE}{\hbar^2}}x} + A_- e^{-i\sqrt{\frac{2mE}{\hbar^2}}x}$ an eigenfunction of the momentum operator? Is it an eigenfunction of the total energy operator? Explain your result.

$$-i\hbar\frac{d}{dx}\left(A_+ e^{+i\sqrt{\frac{2mE}{\hbar^2}}x} + A_- e^{-i\sqrt{\frac{2mE}{\hbar^2}}x}\right) = -(i)^2\hbar\sqrt{\frac{2mE}{\hbar^2}}A_+ e^{+i\sqrt{\frac{2mE}{\hbar^2}}x} + (i)^2\hbar\sqrt{\frac{2mE}{\hbar^2}}A_- e^{-i\sqrt{\frac{2mE}{\hbar^2}}x}$$

Using the relation $k = \sqrt{\frac{2mE}{\hbar^2}}$ the previous equation becomes

$$= \hbar k\, A_+ e^{+i\sqrt{\frac{2mE}{\hbar^2}}x} - \hbar k\, A_- e^{-i\sqrt{\frac{2mE}{\hbar^2}}x}$$

This function is not an eigenfunction of the momentum operator, because the operation does not return the original function multiplied by a constant.

$$-\frac{\hbar^2}{2m}\frac{d^2}{dx^2}\left(A_+e^{+i\sqrt{\frac{2mE}{\hbar^2}}x} + A_-e^{-i\sqrt{\frac{2mE}{\hbar^2}}x}\right) = -(i)^2\frac{\hbar^2}{2m}\frac{2mE}{\hbar^2}A_+e^{+i\sqrt{\frac{2mE}{\hbar^2}}x} - (-i)^2\frac{\hbar^2}{2m}\frac{2mE}{\hbar^2}A_-e^{-i\sqrt{\frac{2mE}{\hbar^2}}x}$$

$$= E\left(A_+e^{+i\sqrt{\frac{2mE}{\hbar^2}}x} + A_-e^{-i\sqrt{\frac{2mE}{\hbar^2}}x}\right)$$

This function is an eigenfunction of the total energy operator. Because the energy is proportional to p^2, the difference in sign of the momentum of these two components does not affect the energy.

P15.8 Evaluate the normalization integral for the eigenfunctions of $\hat{H}$ for the particle in the box $\psi_n(x) = A\sin(n\pi x/a)$ using the trigonometric identity $\sin^2 y = (1 - \cos 2y)/2$.

$$1 = \int_0^a A^2\sin^2\left(\frac{n\pi x}{a}\right)dx$$

$$\text{Let } y = \frac{n\pi x}{a}; dx = \frac{a}{n\pi}dy$$

$$1 = A^2\frac{a}{n\pi}\int_0^{n\pi}\sin^2 y\,dy = A^2\frac{a}{n\pi}\int_0^{n\pi}\frac{1-\cos 2y}{2}dy = A^2\frac{a}{n\pi}\left[\frac{y}{2} - \frac{\sin 2y}{4}\right]_0^{n\pi}$$

$$= \frac{A^2}{2}\frac{a}{n\pi}n\pi - \frac{A^2}{4}\frac{a}{n\pi}(\sin 2n\pi - \sin 0) = \frac{A^2 a}{2}$$

$$A = \sqrt{\frac{2}{a}}$$

P15.10 What is the solution of the time-dependent Schrödinger equation $\psi(x,t)$ for the total energy eigenfunction $\psi_4(x) = \sqrt{2/a}\sin(3\pi x/a)$ for an electron in a one-dimensional box of length 1.00×10^{-10} m? Write explicitly in terms of the parameters of the problem. Give numerical values for the angular frequency ω and the wavelength of the particle.

$$\psi(x,t) = \psi(x)e^{-i\omega t} = \psi(x)e^{-i\frac{Et}{\hbar}}$$

$$\text{Because } E = \frac{n^2 h^2}{8ma^2} = \frac{9h^2}{8ma^2},$$

$$\psi(x,t) = \sqrt{\frac{2}{a}}\sin\left(\frac{3\pi x}{a}\right)e^{-i\frac{9\pi ht}{4ma^2}}$$

$$\omega = \frac{9\pi h}{4ma^2} = \frac{9\pi \times 6.626 \times 10^{-34}\text{ J s}}{4 \times 9.109 \times 10^{-31}\text{ kg} \times (1.00 \times 10^{-10}\text{ m})^2} = 5.14 \times 10^{17}\text{ s}^{-1}$$

$$\lambda = \frac{2a}{3} = 6.67 \times 10^{-11}\text{ m}$$

P15.13 Show that the energy eigenvalues for the free particle, $E = \hbar^2 k^2/2m$, are consistent with the classical result $E = (1/2)mv^2$.

$$E = \frac{1}{2}mv^2 = \frac{p^2}{2m}$$

From the de Broglie relation, $p = \frac{h}{\lambda}$, and using the relation $k = \frac{2\pi}{\lambda}$

$$E = \frac{1}{2m}\left(\frac{h}{\lambda}\right)^2 = \frac{\hbar^2 k^2}{2m}, \text{ showing consistency between the classical and quantum result.}$$

P15.15 Calculate the wavelength of the light emitted when an electron in a one-dimensional box of length 5.0 nm makes a transition from the $n = 7$ state to the $n = 6$ state.

$$E = h\nu = \frac{h^2}{8\,ma^2}(n_2^2 - n_1^2); \quad \nu = \frac{h}{8\,ma^2}(n_2^2 - n_1^2)$$

$$= \frac{6.26 \times 10^{-34} \text{ J s}}{8 \times 9.109 \times 10^{-31} \text{ kg} \times 25.0 \times 10^{-18} \text{ m}^2}(7^2 - 6^2) = 4.73 \times 10^{13} \text{ s}^{-1}$$

$$\lambda = \frac{c}{\nu} = \frac{2.998 \times 10^8 \text{ m s}^{-1}}{4.73 \times 10^{13} \text{ s}^{-1}} = 6.34 \times 10^{-6} \text{ m}$$

P15.20 Calculate (a) the zero point energy of a He atom in a one-dimensional box of length 1.00 cm and (b) the ratio of the zero point energy to $k_B T$ at 300. K.

$$E_1 = \frac{h^2}{8ma^2} = \frac{(6.26 \times 10^{-34} \text{ J s})^2}{8 \times 4.003 \text{ amu} \times 1.661 \times 10^{-27} \text{ kg amu}^{-1} \times 1.00 \times 10^{-4} \text{ m}^2}$$

$$= 8.25 \times 10^{-38} \text{ J}$$

$$\frac{E_1}{k_B T} = \frac{8.25 \times 10^{-38} \text{ J}}{1.381 \times 10^{-23} \text{ J K}^{-1} \times 300. \text{ K}} = 1.99 \times 10^{-17}$$

P15.21 Normalize the total energy eigenfunction for the rectangular two-dimensional box,

$$\psi_{n_x, n_y}(x, y) = N \sin\left(\frac{n_x \pi x}{a}\right) \sin\left(\frac{n_y \pi y}{b}\right)$$

in the interval $0 \le x \le a$, $0 \le y \le b$.

$$1 = \int_0^a \int_0^b \psi^*(x, y)\psi(x, y)\, dx\, dy = N^2 \int_0^a \sin^2\left(\frac{n_x \pi x}{a}\right) dx \int_0^b \sin^2\left(\frac{n_y \pi y}{b}\right) dy$$

Using the standard integral $\int \sin^2(\alpha x)\, dx = \frac{x}{2} - \frac{\sin(2\alpha x)}{4\alpha}$,

$$N^2 \int_0^a \sin^2\left(\frac{n_x \pi x}{a}\right) dx \int_0^b \sin^2\left(\frac{n_y \pi y}{b}\right) dy$$

$$= N^2\left[\frac{a}{2} - \frac{a}{4n_x\pi}(\sin n_x\pi - \sin 0)\right] \times \left[\frac{b}{2} - \frac{b}{4n_y\pi}(\sin n_y\pi - \sin 0)\right] = N^2\frac{ab}{4}$$

$$N = \sqrt{\frac{4}{ab}} \text{ and } \psi(x, y) = \sqrt{\frac{4}{ab}}\sin\left(\frac{n_x\pi x}{a}\right)\sin\left(\frac{n_y\pi y}{b}\right)$$

P15.24 What is the zero point energy and what are the energies of the lowest seven energy levels in a three-dimensional box with $a = b = c$? What is the degeneracy of each level?

$$n_x^2 + n_y^2 + n_z^2 = 3: \text{ (111) degeneracy 1; } E = \frac{3h^2}{8ma^2}$$

$$n_x^2 + n_y^2 + n_z^2 = 6: \text{ (112), (121), (211) degeneracy 3; } E = \frac{6h^2}{8ma^2}$$

$$n_x^2 + n_y^2 + n_z^2 = 9: \text{ (212), (122), (221) degeneracy 3; } E = \frac{9h^2}{8ma^2}$$

$$n_x^2 + n_y^2 + n_z^2 = 11: (311), (131), (113) \text{ degeneracy 3; } E = \frac{11h^2}{8ma^2}$$

$$n_x^2 + n_y^2 + n_z^2 = 12: (222) \text{ degeneracy 1; } E = \frac{12h^2}{8ma^2}$$

$$n_x^2 + n_y^2 + n_z^2 = 14: (123), (132), (213), (231), (312), (321) \text{ degeneracy 6; } E = \frac{14h^2}{8ma^2}$$

$$n_x^2 + n_y^2 + n_z^2 = 17: (223), (232), (322) \text{ degeneracy 3; } E = \frac{17h^2}{8ma^2}$$

P15.25   In discussing the Boltzmann distribution in Chapter 13, we used the symbols g_i and g_j to indicate the degeneracies of the energy levels i and j. By degeneracy, we mean the number of distinct quantum states (different quantum numbers) all of which have the same energy.

a.   Using your answer to Problem P15.17a, what is the degeneracy of the energy level $9h^2/8\,ma^2$ for the square two-dimensional box of edge length a?

b.   Using your answer to Problem P15.14b, what is the degeneracy of the energy level $17h^2/8\,ma^2$ for a three-dimensional cubic box of edge length a?

(a)   The only pair n_x, n_y that satisfy the equation $n_x^2 + n_y^2 = 18$ is 3,3. Therefore the degeneracy of this energy level is 1.

(b)   The only trios of nonzero numbers n, q, r that satisfy the equation $n_x^2 + n_y^2 + n_z^2 = 17$ are 3, 2, 2; 2, 2, 3; and 2, 3, 2. Therefore the degeneracy of this energy level is 3.

P15.28   Is the superposition wave function $\psi(x) = \sqrt{2/a}[\sin(n\pi x/a) + \sin(m\pi x/a)]$ an eigenfunction of the total energy operator for the particle in the box?

$$\hat{H}\psi(x) = -\frac{\hbar^2}{2m}\frac{d^2}{dx^2}\left[\sqrt{\frac{2}{a}}\sin\left(\frac{n\pi x}{a}\right) + \sqrt{\frac{2}{a}}\sin\left(\frac{m\pi x}{a}\right)\right]$$

$$= \frac{h^2 n^2}{8\,ma^2}\sqrt{\frac{2}{a}}\sin\left(\frac{n\pi x}{a}\right) + \frac{h^2 m^2}{8\,ma^2}\sqrt{\frac{2}{a}}\sin\left(\frac{m\pi x}{a}\right)$$

Because the result is not the wave function multiplied by a constant, the superposition wave function is not an eigenfunction of the total energy operator.

P15.34   Calculate the probability that a particle in a one-dimensional box of length a is found between $0.32a$ and $0.35a$ when it is described by the following wave functions:

a. $\sqrt{\dfrac{2}{a}}\sin\left(\dfrac{\pi x}{a}\right)$

b. $\sqrt{\dfrac{2}{a}}\sin\left(\dfrac{3\pi x}{a}\right)$

What would you expect for a classical particle? Compare your results for (a) and (b) with the classical result.

(a)   Using the standard integral

$$\int \sin^2(by)\,dy = \frac{y}{2} - \frac{1}{4b}\sin(2by)$$

$$P = \frac{2}{a}\int_{0.32a}^{0.35a} \sin^2\left(\frac{\pi x}{a}\right) dx = \frac{2}{a}\left[\frac{x}{2} - \frac{a}{4\pi}\sin\left(\frac{2\pi x}{a}\right)\right]_{0.32a}^{0.35a}$$

$$= \frac{2}{a}\left[\frac{0.35a}{2} - \frac{a}{4\pi}\sin(0.70\pi) - \frac{0.32a}{2} + \frac{a}{4\pi}\sin(0.64\pi)\right]$$

$$= 0.03 + \frac{1}{2\pi}[-\sin(0.70\pi) + \sin(0.64\pi)] = 0.045$$

(b)   Using the standard integral

$$\int \sin^2(by)\,dy = \frac{y}{2} - \frac{1}{4b}\sin(2by)$$

$$P = \frac{2}{a}\int_{0.32a}^{0.35a} \sin^2\left(\frac{3\pi x}{a}\right) dx = \frac{2}{a}\left[\frac{0.35a}{2} - \frac{a}{12\pi}\sin(2.10\pi) - \frac{0.32a}{2} + \frac{a}{12\pi}\sin(1.92\pi)\right]$$

$$= 0.03 + \frac{1}{6\pi}\left[-\sin(2.10\pi) + \sin(1.92\pi)\right] = 0.00041$$

Because a classical particle is equally likely to be in any given interval, the probability will be 0.04 independent of the energy. In the ground state, the interval chosen is near the maximum of the wave function so that the quantum mechanical probability is greater than the classical probability. For the $n = 3$ state, the interval chosen is near a node of the wave function so that the quantum mechanical probability is much less than the classical probability.

$$E = \frac{n^2 h^2}{8ma^2} = \frac{(1)^2(6.626\times10^{-34})}{8}$$

$$\left(\left(\frac{2}{a}\right)^{1/2}\right)^2 \sin^2\left(\frac{\pi x}{a}\right)$$

17 Commuting and Noncommuting Operators and the Surprising Consequences of Entanglement

Numerical Problems

P17.4 (a) Show that $\psi(x) = e^{-x^2/2}$ is an eigenfunction of $\hat{A} = x^2 - \partial^2/\partial x^2$; and (b) show that $\hat{B}\psi(x)$, (where $\hat{B} = x - \partial/\partial x$) is another eigenfunction of $\hat{A}$.

  a. $\hat{A}\psi(x) = x^2 e^{-x^2/2} - \partial^2 e^{-x^2/2}/\partial x^2 = x^2 e^{-x^2/2} - \partial(-xe^{-x^2/2})/\partial x$

  $= e^{-x^2/2} + x^2 e^{-x^2/2} - x^2 e^{-x^2/2} = e^{-x^2/2}$

  b. $\hat{B}\psi(x) = xe^{-x^2/2} - \partial e^{-x^2/2}/\partial x = xe^{-x^2/2} + xe^{-x^2/2} = 2xe^{-x^2/2}$

  $\hat{A}(\hat{B}\psi(x)) = x^2(2xe^{-x^2/2}) - \partial^2(2xe^{-x^2/2})/\partial x^2$

  $= 2x^3 e^{-x^2/2} - \partial(2e^{-x^2/2} - 2x^2 e^{-x^2/2})/\partial x$

  $= 2x^3 e^{-x^2/2} + 2xe^{-x^2/2} + 4xe^{-x^2/2} - 2x^3 e^{-x^2/2}$

  $= 6xe^{-x^2/2}$

P17.5 Another important uncertainty principle is encountered in time-dependent systems. It relates the lifetime of a state Δt with the measured spread in the photon energy ΔE associated with the decay of this state to a stationary state of the system. "Derive" the relation $\Delta E \Delta t \geq \hbar/2$ in the following steps.

  a. Starting from $E = p_x^2/2m$ and $\Delta E = (dE/dp_x)\Delta p_x$, show that $\Delta E = v_x \Delta p_x$.

  b. Using $v_x = \Delta x/\Delta t$, show that $\Delta E \Delta t = \Delta p_x \Delta x \geq \hbar/2$.

  c. Estimate the width of a spectral line originating from the decay of a state of lifetime 1.0×10^{-9} s and 1.0×10^{-11} s in inverse seconds and inverse centimeters.

  (a) $\dfrac{dE}{dp_x}\Delta p_x = \dfrac{p_x}{m}\Delta p_x = v_x \Delta p_x$

  (b) $\Delta E \, \Delta t = \dfrac{\Delta x}{\Delta t}\Delta p_x \Delta t = \Delta x \, \Delta p_x \geq \dfrac{\hbar}{2}$

  (c) $\Delta E = h\Delta \nu \geq \dfrac{\hbar}{2\Delta t}$

  $\Delta \nu \geq \dfrac{1}{4\pi \Delta t} = \dfrac{1}{4\pi(1.0 \times 10^{-9}\,\text{s})} = 8.0 \times 10^7\,\text{s}^{-1}$

$$\Delta v(\text{cm}^{-1}) = \frac{\Delta v(\text{s}^{-1})}{c} \geq \frac{8.0 \times 10^7 \text{ s}^{-1}}{2.998 \times 10^{10} \text{ cm s}^{-1}} = 0.00265 \text{ cm}^{-1}$$

The corresponding answers for 1.0×10^{-11} s are 8.0×10^9 s^{-1} and 0.265 cm^{-1}, respectively.

P17.6 Evaluate the commutator $[x(\partial/\partial y), y(\partial/\partial x)]$ by applying the operators to an arbitrary function $f(x,y)$.

$$\left[x\frac{\partial}{\partial y}, y\frac{\partial}{\partial x}\right]f(x,y) = x\frac{\partial}{\partial y}\left[y\frac{\partial f(x,y)}{\partial x}\right] - y\frac{\partial}{\partial x}\left[x\frac{\partial f(x,y)}{\partial y}\right]$$

$$= x\left(\frac{\partial f(x,y)}{\partial x} + y\frac{\partial^2 f(x,y)}{\partial y \partial x}\right) - y\left(\frac{\partial f(x,y)}{\partial y} + x\frac{\partial^2 f(x,y)}{\partial x \partial y}\right)$$

$$= x\frac{\partial f(x,y)}{\partial x} - y\frac{\partial f(x,y)}{\partial y}$$

Therefore, $\left[x\dfrac{\partial}{\partial y}, y\dfrac{\partial}{\partial x}\right] = x\dfrac{\partial}{\partial x} - y\dfrac{\partial}{\partial y}$.

P17.8 Consider the entangled wave function for two photons,

$$\psi_{12} = \frac{1}{\sqrt{2}}(\psi_1(H)\psi_2(V) + \psi_1(V)\psi_2(H))$$

Assume that the polarization operator $\hat{P}_i$ has the properties $\hat{P}_i\psi_i(H) = -\psi_i(H)$ and $\hat{P}_i\psi_i(V) = +\psi_i(V)$, where $i = 1$ or $i = 2$. H and V designate horizontal and vertical polarization, respectively.

a. Show that ψ_{12} is not an eigenfunction of $\hat{P}_1$ or $\hat{P}_2$.

b. Show that each of the two terms in ψ_{12} is an eigenfunction of the polarization operator $\hat{P}_1$.

c. What is the average value of the polarization P_1 that you will measure on identically prepared systems?

(a) $\hat{P}_1\psi_{12} = \dfrac{1}{\sqrt{2}}(-\psi_1(H)\psi_2(V) + \psi_1(V)\psi_2(H))$

$\hat{P}_2\psi_{12} = \dfrac{1}{\sqrt{2}}(\psi_1(H)\psi_2(V) - \psi_1(V)\psi_2(H))$

In neither case does the operation return the original function multiplied by a constant. Therefore, the function is not an eigenfunction of either operator.

(b) Each of the two terms of the above expression is the original term multiplied by a constant. Therefore each individual term is an eigenfunction of the operators.

(c) A measurement will project the system into the wave function $\psi_1(H)\psi_2(V)$ or $-\psi_1(V)\psi_2(H)$ with equal probability. Therefore it is equally likely to measure the eigenvalue +1 as −1, and the average of the measured values will be zero.

P17.10 Revisit the double-slit experiment of Example Problem 17.2. Using the same geometry and relative uncertainty in the momentum, what electron momentum would give a position uncertainty of 2.50×10^{-10} m? What is the ratio of the wavelength and the slit spacing for this momentum? Would you expect a pronounced diffraction effect for this wavelength?

$$\Delta p = \frac{1}{2}\frac{\hbar}{\Delta x} = \frac{1.055 \times 10^{-34} \text{ J s}}{2.50 \times 10^{-10} \text{ m}} = 2.11 \times 10^{-25} \text{ kg m s}^{-1}$$

$$p = \frac{\Delta p}{0.01} = 2.11 \times 10^{-23} \text{ kg m s}^{-1}$$

$$\lambda = \frac{h}{p} = \frac{6.626 \times 10^{-34} \text{ J s}}{2.11 \times 10^{-23} \text{ kg m s}^{-1}} = 3.14 \times 10^{-11} \text{ m}$$

Because $\dfrac{\lambda}{b} = \dfrac{3.14 \times 10^{-11} \text{ m}}{1.00 \times 10^{-8} \text{ m}} = 0.00314$, where b is the slit spacing, the diffraction will hardly be noticeable.

P17.16 Evaluate the commutator $[d/dx, x^2]$ by applying the operators to an arbitrary function $f(x)$.

$$\left[\frac{d}{dx}, x^2\right] f(x) = \frac{d}{dx}(x^2 f(x)) - x^2 \frac{d}{dx} f(x)$$

$$= 2xf(x) + x^2 \frac{d}{dx} f(x) - x^2 \frac{d}{dx} f(x) = 2xf(x)$$

$$\left[\frac{d}{dx}, x^2\right] = 2x$$

P17.17 Evaluate the commutator $[\hat{x}, \hat{p}_x]$ by applying the operators to an arbitrary function $f(x)$. What value does the commutator $[\hat{p}_x, \hat{x}]$ have?

$$[\hat{x}, \hat{p}_x] f(x) = \hat{x} \hat{p}_x f(x) - \hat{p}_x \hat{x} f(x)$$

$$= -i\hbar x \frac{df(x)}{dx} + i\hbar \frac{d}{dx}(x f(x))$$

$$= -i\hbar x \frac{df(x)}{dx} + i\hbar f(x) + i\hbar x \frac{df(x)}{dx}$$

$$[\hat{x}, \hat{p}_x] = i\hbar$$

Because the order of operation is interchanged, the commutator would have the same value, but an opposite sign.

P17.19 Evaluate the commutator $[(d^2/dy^2) - y, (d^2/dy^2) + y]$ by applying the operators to an arbitrary function $f(y)$.

$$\left[\frac{d^2}{dy^2} - y, \frac{d^2}{dy^2} + y\right] f(y) = \left(\frac{d^2}{dy^2} - y\right)\left(\frac{d^2}{dy^2} + y\right) f(y) - \left(\frac{d^2}{dy^2} + y\right)\left(\frac{d^2}{dy^2} - y\right) f(y)$$

$$= \left(\frac{d^2}{dy^2} - y\right)\left(\frac{d^2 f(y)}{dy^2} + yf(y)\right) - \left(\frac{d^2}{dy^2} + y\right)\left(\frac{d^2 f(y)}{dy^2} - yf(y)\right)$$

$$= \frac{d^4 f(y)}{dy^4} + 2\frac{df(y)}{dy} + y\frac{d^2 f(y)}{dy^2} - y\frac{d^2 f(y)}{dy^2} - y^2 f(y)$$

$$- \left[\frac{d^4 f(y)}{dy^4} - 2\frac{df(y)}{dy} - y\frac{d^2 f(y)}{dy^2} + y\frac{d^2 f(y)}{dy^2} - y^2 f(y)\right]$$

$$= 4\frac{df(y)}{dy}$$

Therefore,

$$\left[\frac{d^2}{dy^2} - y, \frac{d^2}{dy^2} + y\right] = 4\frac{d}{dy}$$

P17.23 The muzzle velocity of a rifle bullet is $890. \text{ m s}^{-1}$ along the direction of motion. If the bullet weighs 35 g, and the uncertainty in its momentum is 0.20%, how accurately can the position of the bullet be measured along the direction of motion?

$$p = mv = 35 \times 10^{-3} \text{ kg} \times 890. \text{ m s}^{-1} = 31.1 \text{ kg m s}^{-1}$$

$$\Delta p = 2 \times 10^{-3} \, p = 6.23 \times 10^{-2} \text{ kg m s}^{-1}$$

$$\Delta x = \frac{\hbar}{2 \, \Delta p} = \frac{1.055 \times 10^{-34} \text{ J s}}{2 \times 6.23 \times 10^{-2} \text{ kg m s}^{-1}} = 8.5 \times 10^{-34} \text{ m}$$

18 A Quantum Mechanical Model for the Vibration and Rotation of Molecules

Numerical Problems

P18.1   A gas-phase $^1H^{127}I$ molecule, with a bond length of 160.92 pm, rotates in three-dimensional space.

    a.   Calculate the zero point energy associated with this rotation.

    b.   What is the smallest quantum of energy that can be absorbed by this molecule in a rotational excitation?

    (a)   There is no zero point energy because the rotation is not constrained.

    (b)   The smallest energy that can be absorbed is

$$E = \frac{\hbar^2}{2I} J(J+1) = \frac{\hbar^2}{2I} 1(1+1)$$

$$= \frac{2 \times (1.055 \times 10^{-34} \text{ J s})^2}{2 \times \dfrac{1.0078 \text{ amu} \times 126.9045 \text{ amu}}{1.0078 \text{ amu} + 126.9045 \text{ amu}} \times 1.66 \times 10^{-27} \text{ kg amu}^{-1} \times (160.92 \times 10^{-12} \text{ m})^2}$$

$$E = 2.59 \times 10^{-22} \text{ J}$$

P18.3   In discussing molecular rotation, the quantum number J is used rather than l. Using the Boltzmann distribution, calculate n_J/n_0 for $^1H^{35}Cl$ for $J = 0$, 5, 10, and 20 at $T = 1025$ K. Does n_J/n_0 go through a maximum as J increases? If so, what can you say about the value of J corresponding to the maximum?

$$\frac{n_J}{n_0} = (2J+1)e^{-(E_J - E_0)/k_B T} = (2J+1)\exp[-E_J/k_B T]$$

$$I = \mu r_0^2 = \frac{1.0078 \text{ amu} \times 34.6698 \text{ amu}}{1.0078 \text{ amu} + 34.6698 \text{ amu}} \times 1.66 \times 10^{-27} \text{ kg amu}^{-1} \times (127.5 \times 10^{-12} \text{ m})^2$$

$$= 2.64 \times 10^{-47} \text{ kg m}^2$$

$$E = \frac{\hbar^2}{2I} J(J+1) = \frac{(1.055 \times 10^{-34} \text{ J s})^2}{2 \times 2.64 \times 10^{-47} \text{ kg m}^2} J(J+1) = 2.104 \times 10^{-22} J(J+1)$$

$$\frac{n_0}{n_0} = 1$$

$$\frac{n_5}{n_0} = (2 \times 5 + 1)\exp\left[-(30 \times 2.104 \times 10^{-22} \text{ J})/1.381 \times 10^{-23} \text{ J K}^{-1} \times 650.\,\text{K}\right] = 7.04$$

$$\frac{n_{10}}{n_0} = (2 \times 10 + 1)\exp\left[-(110 \times 2.104 \times 10^{-22} \text{ J})/1.381 \times 10^{-23} \text{ J K}^{-1} \times 650.\,\text{K}\right] = 4.09$$

$$\frac{n_{20}}{n_0} = (2 \times 20 + 1)\exp\left[-(420 \times 2.104 \times 10^{-22} \text{ J})/1.381 \times 10^{-23} \text{ J K}^{-1} \times 650.\,\text{K}\right] = 0.0795$$

$\dfrac{n_J}{n_0}$ goes through a maximum because it has a value greater than one for $J = 5$. You can only conclude that $J_{max} \leq 5$.

P18.8   The vibrational frequency for D_2 expressed in wave numbers is 3115 cm^{-1}. What is the force constant associated with the D—D bond? How much would a classical spring with this force constant be elongated if a mass of 1.50 kg were attached to it? Use the gravitational acceleration on Earth at sea level for this problem.

$$v = c\tilde{v} = \frac{1}{2\pi}\sqrt{\frac{k}{\mu}}$$

so $k = (2\pi c\tilde{v})^2 \mu$

$$k = (2\pi \times 2.998 \times 10^{10} \text{ cm s}^{-1} \times 3115 \text{ cm}^{-1})^2 \times \frac{2.0142 \text{ amu} \times 2.0142 \text{ amu}}{2 \times 2.0142 \text{ amu}} \times \frac{1.661 \times 10^{-27} \text{ kg}}{\text{amu}}$$

$$k = 575 \text{ N m}^{-1}$$

$$x = \frac{F}{k} = \frac{mg}{k} = \frac{1.50 \text{ kg} \times 9.81 \text{ m s}^{-2}}{575 \text{ N m}^{-1}} = 2.55 \times 10^{-2} \text{ m}$$

P18.9   In discussing molecular rotation, the quantum number J is used rather than l. Calculate $E_{rot}/k_B T$ for $^1\text{H}^{81}\text{Br}$ for $J = 0, 5, 10,$ and 20 at 298 K. The bond length is 141.4 pm. For which of these values of J is $E_{rot}/k_B T \geq 10$.?

$$I = \mu r_0^2 = \frac{1.0078 \times 80.9163}{1.0078 + 80.9163} \times 1.66 \times 10^{-27} \text{ kg amu}^{-1} \times (141.4 \times 10^{-12} \text{ m})^2 = 3.30 \times 10^{-47} \text{ kg m}^2$$

$$E = \frac{\hbar^2}{2I}J(J+1) = \frac{(1.055 \times 10^{-34} \text{ J s})^2}{2 \times 3.31 \times 10^{-47} \text{ kg m}^2}J(J+1) = 1.68 \times 10^{-22}J(J+1)$$

$$E_{J=0} = 0$$

$$E_{J=5} = 30 \times 1.68 \times 10^{-22} \text{ J} = 5.05 \times 10^{-21} \text{ J}$$

$$\frac{E_{J=5}}{k_B T} = \frac{5.05 \times 10^{-21} \text{ J}}{1.381 \times 10^{-23} \text{ J K}^{-1} \times 298 \text{ K}} = 1.23$$

$$E_{J=10} = 110 \times 1.68 \times 10^{-22} \text{ J} = 1.85 \times 10^{-20} \text{ J}$$

$$\frac{E_{J=10}}{k_B T} = \frac{1.85 \times 10^{-20} \text{ J}}{1.381 \times 10^{-23} \text{ J K}^{-1} \times 298 \text{ K}} = 4.50$$

$$E_{J=20} = 20 \times 21 \times 1.85 \times 10^{-22} \text{ J} = 7.07 \times 10^{-19} \text{ J}$$

$$\frac{E_{J=20}}{k_B T} = \frac{7.07 \times 10^{-20} \text{ J}}{1.381 \times 10^{-23} \text{ J K}^{-1} \times 298 \text{ K}} = 17.2$$

P18.12   Show by carrying out the appropriate integration that the total energy eigenfunctions for the harmonic oscillator $\psi_0(x) = (\alpha/\pi)^{1/4}e^{-(1/2)\alpha x^2}$ and $\psi_2(x) = (\alpha/4\pi)^{1/4}(2\alpha x^2 - 1)e^{-(1/2)\alpha x^2}$ are orthogonal over the interval $-\infty < x < \infty$ and that $\psi_2(x)$ is normalized over the same interval. In evaluating integrals of this type, $\int_{-\infty}^{\infty} f(x)\,dx = 0$ if $f(x)$ is an odd function of x and $\int_{-\infty}^{\infty} f(x)\,dx = 2\int_0^{\infty} f(x)\,dx$ if $f(x)$ is an even function of x.

We use the standard integrals $\int_0^{\infty} x^{2n}e^{-ax^2}\,dx = \frac{1 \cdot 3 \cdot 5 \cdots (2n-1)}{2^{n+1}a^n}\sqrt{\frac{\pi}{a}}$ and $\int_0^{\infty} e^{-ax^2}\,dx = \left(\frac{\pi}{4a}\right)^{1/2}$.

$$\int_{-\infty}^{\infty} \psi_2^*(x)\psi_0(x)\,dx = \int_{-\infty}^{\infty} \left(\frac{\alpha}{4\pi}\right)^{1/4}(2\alpha x^2 - 1)e^{-\frac{1}{2}\alpha x^2}\left(\frac{\alpha}{\pi}\right)^{1/4}e^{-\frac{1}{2}\alpha x^2}\,dx$$

$$= \left(\frac{\alpha^2}{4\pi^2}\right)^{1/4}\int_{-\infty}^{\infty}(2\alpha x^2 - 1)e^{-\alpha x^2}\,dx = 2\left(\frac{\alpha^2}{4\pi^2}\right)^{1/4}\int_0^{\infty}(2\alpha x^2 - 1)e^{-\alpha x^2}\,dx$$

$$= \left(\frac{\alpha^2}{4\pi^2}\right)^{1/4}\left(2\alpha\frac{1}{4\alpha}\sqrt{\frac{\pi}{\alpha}} - \frac{1}{2}\sqrt{\frac{\pi}{\alpha}}\right) = 0$$

$$\int_{-\infty}^{\infty} \psi_2^*(x)\psi_2(x)\,dx = \int_{-\infty}^{\infty} \left(\frac{\alpha}{4\pi}\right)^{1/4}(2\alpha x^2 - 1)e^{-\frac{1}{2}\alpha x^2}\left(\frac{\alpha}{4\pi}\right)^{1/4}(2\alpha x^2 - 1)e^{-\frac{1}{2}\alpha x^2}\,dx$$

$$= 2\left(\frac{\alpha}{4\pi}\right)^{1/2}\int_0^{\infty}(4\alpha^2 x^4 - 4\alpha x^2 + 1)e^{-\alpha x^2}\,dx$$

$$= 2\left(\frac{\alpha}{4\pi}\right)^{1/2}\left(4\alpha^2\frac{3}{2^3\alpha^2}\sqrt{\frac{\pi}{\alpha}} - 4\alpha\frac{1}{2^2\alpha}\sqrt{\frac{\pi}{\alpha}} + \frac{1}{2}\sqrt{\frac{\pi}{\alpha}}\right) = 2\left(\frac{\alpha}{4\pi}\right)^{1/2}\sqrt{\frac{\pi}{\alpha}}\left(\frac{3}{2} - 1 + \frac{1}{2}\right) = 1$$

P18.14   Calculate the frequency and wavelength of the radiation absorbed when a quantum harmonic oscillator with a frequency of 3.15×10^{13} s^{-1} makes a transition from the $n = 2$ to the $n = 3$ state.

$$\Delta E = (3 + 1/2)h\nu - (2 + 1/2)h\nu = h\nu$$

$$= 6.626 \times 10^{-34}\text{ J s} \times 3.15 \times 10^{13}\text{ s}^{-1} = 2.09 \times 10^{-20}\text{ J}$$

$$\nu = \frac{E}{h} = 3.15 \times 10^{13}\text{ s}^{-1}$$

$$\lambda = \frac{c}{\nu} = \frac{2.998 \times 10^8\text{ m s}^{-1}}{3.15 \times 10^{13}\text{ s}^{-1}} = 9.52 \times 10^{-6}\text{ m}$$

P18.16   The vibrational frequency of ^{35}Cl$_2$ is 1.68×10^{13} s^{-1}. Calculate the force constant of the molecule. How large a mass would be required to stretch a classical spring with this force constant by 2.25 cm? Use the gravitational acceleration on Earth at sea level for this problem.

$$\nu = \frac{1}{2\pi}\sqrt{\frac{k}{\mu}}; k = 4\pi^2\mu\nu^2$$

$$k = 4 \times \pi^2 \times \frac{34.9648\text{ amu} \times 34.9648\text{ amu}}{34.9648\text{ amu} + 34.9648\text{ amu}} \times \frac{1.661 \times 10^{-27}\text{ kg}}{\text{amu}} \times (1.68 \times 10^{13}\text{ s}^{-1})^2$$

$$k = 324\text{ kg s}^{-2}$$

$$F = kx = mg$$

$$m = \frac{kx}{g} = \frac{324\text{ kg s}^{-2} \times 2.25 \times 10^{-2}\text{ m}}{9.81\text{ m s}^{-2}} = 0.742\text{ kg}$$

P18.17   Evaluate $\langle x^2 \rangle$ for the ground state ($n = 0$) and first two excited states ($n = 1$ and $n = 2$) of the quantum harmonic oscillator. Use the hint about evaluating integrals in Problem P18.12.

We use the standard integrals $\int_0^{\infty} x^{2n}e^{-ax^2}\,dx = \frac{1 \cdot 3 \cdot 5 \cdots (2n-1)}{2^{n+1}a^n}\sqrt{\frac{\pi}{a}}$ and $\int_0^{\infty} e^{-ax^2}\,dx = \left(\frac{\pi}{4a}\right)^{1/2}$.

$$\langle x^2 \rangle = \int_{-\infty}^{\infty} \psi_n^*(x)(x^2)\psi_n\,dx$$

For $n = 0$,

$$\langle x^2 \rangle = \int_{-\infty}^{\infty}\left(\frac{\alpha}{\pi}\right)^{1/4}e^{-\frac{1}{2}\alpha x^2}(x^2)\left(\frac{\alpha}{\pi}\right)^{1/4}e^{-\frac{1}{2}\alpha x^2}\,dx$$

$$\langle x^2 \rangle = \left(\frac{\alpha}{\pi}\right)^{1/2}\int_{-\infty}^{\infty} x^2 e^{-\alpha x^2}\,dx = 2\left(\frac{\alpha}{\pi}\right)^{1/2}\int_0^{\infty} x^2 e^{-\alpha x^2}\,dx$$

$$\langle x^2 \rangle = 2\left(\frac{\alpha}{\pi}\right)^{1/2}\frac{1}{2^2\alpha}\sqrt{\frac{\pi}{\alpha}} = \frac{1}{2\alpha} = \frac{\hbar}{2\sqrt{k\mu}}$$

For $n = 1$,

$$\langle x^2 \rangle = \int_{-\infty}^{\infty} \left(\frac{4\alpha^3}{\pi} \right)^{1/4} xe^{-\frac{1}{2}\alpha x^2} (x^2) \left(\frac{4\alpha^3}{\pi} \right)^{1/4} xe^{-\frac{1}{2}\alpha x^2} dx$$

$$\langle x^2 \rangle = \left(\frac{4\alpha^3}{\pi} \right)^{1/2} \int_{-\infty}^{\infty} x^4 e^{-\alpha x^2} dx = 2\left(\frac{4\alpha^3}{\pi} \right)^{1/2} \int_{0}^{\infty} x^4 e^{-\alpha x^2} dx$$

$$\langle x^2 \rangle = 2\left(\frac{4\alpha^3}{\pi} \right)^{1/2} \frac{3}{2^3\alpha^2} \sqrt{\frac{\pi}{\alpha}} = \frac{3}{2\alpha} = \frac{3\hbar}{2\sqrt{k\mu}}$$

For $n = 2$,

$$\langle x^2 \rangle = \int_{-\infty}^{\infty} \left(\frac{\alpha}{4\pi} \right)^{1/4} (2\alpha x^2 - 1)e^{-\frac{1}{2}\alpha x^2} (x^2) \left(\frac{\alpha}{4\pi} \right)^{1/4} (2\alpha x^2 - 1)e^{-\frac{1}{2}\alpha x^2} dx$$

$$\langle x^2 \rangle = \left(\frac{\alpha}{4\pi} \right)^{1/2} \int_{-\infty}^{\infty} (4\alpha^2 x^6 - 4\alpha x^4 + x^2)e^{-\alpha x^2} dx = 2\left(\frac{\alpha}{4\pi} \right)^{1/2} \int_{0}^{\infty} (4\alpha^2 x^6 - 4\alpha x^4 + x^2)e^{-\alpha x^2} dx$$

$$\langle x^2 \rangle = 2\left(\frac{\alpha}{4\pi} \right)^{1/2} \left(4\alpha^2 \frac{15}{2^4\alpha^3} \sqrt{\frac{\pi}{\alpha}} - 4\alpha \frac{3}{2^3\alpha^2} \sqrt{\frac{\pi}{\alpha}} + \frac{1}{2^2\alpha} \sqrt{\frac{\pi}{\alpha}} \right) = \frac{5}{2\alpha} = \frac{5\hbar}{2\sqrt{k\mu}}$$

P18.18 A coin with a mass of 8.31 g suspended on a rubber band has a vibrational frequency of $7.50\ \text{s}^{-1}$. Calculate (a) the force constant of the rubber band, (b) the zero point energy, (c) the total vibrational energy if the maximum displacement is 0.725 cm, and (d) the vibrational quantum number corresponding to the energy in part (c).

a. $$\nu = \frac{1}{2\pi} \sqrt{\frac{k}{m}}; \quad k = 4\pi^2 \nu^2 m$$

$$k = 4\pi^2 \times 7.50^2\ \text{s}^{-2} \times 8.31 \times 10^{-3}\ \text{kg} = 18.5\ \text{N m}^{-1}$$

b. $$E_0 = \frac{1}{2}h\nu = 6.626 \times 10^{-34}\ \text{J s} \times 7.50\ \text{s}^{-1} = 2.48 \times 10^{-33}\ \text{J}$$

c. $$E_n = \frac{1}{2}kx_{max}^2 = \frac{1}{2} \times 18.5\ \text{N m}^{-1} \times (0.725 \times 10^{-2}\ \text{m})^2 = 4.86 \times 10^{-4}\ \text{J}$$

$$E_n = \left(n + \frac{1}{2} \right)h\nu; \quad n = \frac{E_n}{h\nu} - \frac{1}{2}$$

d. $$n = \frac{4.86 \times 10^{-4}\ \text{J}}{6.626 \times 10^{-34}\ \text{J s} \times 7.50\ \text{s}^{-1}} - \frac{1}{2} = 9.78 \times 10^{28}$$

P18.21 Is it possible to simultaneously know the angular orientation of a molecule rotating in a two-dimensional space and its angular momentum? Answer this question by evaluating the commutator $[\phi, -i\hbar(\partial/\partial\phi)]$.

$$\left[\phi, -i\hbar\frac{\partial}{\partial\phi} \right] f(\phi) = -i\hbar\phi\frac{df(\phi)}{d\phi} + i\hbar\frac{d[\phi f(\phi)]}{d\phi}\phi = i\hbar f(\phi)$$

$$\left[\phi, -i\hbar\frac{\partial}{\partial\phi} \right] = i\hbar$$

Because the commutator is not equal to zero, it is not possible to simultaneously know the angular orientation of a molecule rotating in a two-dimensional space and its angular momentum.

P18.23 The force constant for a $^1\text{H}^{127}\text{I}$ molecule is $314\ \text{N m}^{-1}$.

a. Calculate the zero point vibrational energy for this molecule for a harmonic potential.

b. Calculate the light frequency needed to excite this molecule from the ground state to the first excited state.

(a) $E_1 = \hbar\sqrt{\dfrac{k}{\mu}}\left(1 + \dfrac{1}{2}\right) = \dfrac{3}{2} \times 1.055 \times 10^{-34} \text{ J s} \times \sqrt{\dfrac{314 \text{ N m}^{-1}}{\dfrac{1.0078 \text{ amu} \times 126.9045 \text{ amu}}{1.0078 \text{ amu} + 126.9045 \text{ amu}} \times 1.66 \times 10^{-27} \text{ kg amu}^{-1}}}$

$E_1 = 6.88 \times 10^{-20} \text{ J}$

$E_0 = \hbar\sqrt{\dfrac{k}{\mu}}\left(\dfrac{1}{2}\right) = \dfrac{1}{3}E_1 = 2.29 \times 10^{-20} \text{ J}$

(b) $\nu = \dfrac{E_1 - E_0}{h} = \dfrac{6.88 \times 10^{-20} \text{ J} - 2.29 \times 10^{-20} \text{ J}}{6.626 \times 10^{-34} \text{ J s}} = 6.92 \times 10^{13} \text{ s}^{-1}$

P18.25 A ^{1}H^{19}F molecule, with a bond length of 91.68 pm, absorbed on a surface rotates in two dimensions.

    a.   Calculate the zero point energy associated with this rotation.

    b.   What is the smallest quantum of energy that can be absorbed by this molecule in a rotational excitation?

    (a)   There is no zero point energy because the rotation is not constrained.

    (b)   The smallest energy that can be absorbed is

$$E = \frac{\hbar^2 m_l^2}{2I} = \frac{\hbar^2}{2I} = \frac{(1.055 \times 10^{-34} \text{ J s})^2}{2 \times \dfrac{1.0078 \text{ amu} \times 18.9984 \text{ amu}}{1.0078 \text{ amu} + 18.9984 \text{ amu}} \times 1.66 \times 10^{-27} \text{ kg amu}^{-1} \times (91.68 \times 10^{-12} \text{ m})^2}$$

$$E = 4.17 \times 10^{-22} \text{ J}$$

P18.29 Evaluate the average linear momentum of the quantum harmonic oscillator, $\langle p_x \rangle$, for the ground state ($n = 0$) and first two excited states ($n = 1$ and $n = 2$). Use the hint about evaluating integrals in Problem P18.12.

We use the standard integrals $\displaystyle\int_0^\infty x^{2n}e^{-ax^2}\,dx = \dfrac{1 \cdot 3 \cdot 5 \cdots (2n-1)}{2^{n+1}a^n}\sqrt{\dfrac{\pi}{a}}$ and $\displaystyle\int_0^\infty e^{-ax^2}\,dx = \left(\dfrac{\pi}{4a}\right)^{1/2}$.

$$\langle p_x \rangle = \int_{-\infty}^\infty \psi_n^*(x)\left(-i\hbar\frac{d}{dx}\right)\psi_n\,dx$$

For $n = 0$,

$$\langle p_x \rangle = \int_{-\infty}^\infty \left(\frac{\alpha}{\pi}\right)^{1/4} e^{-\frac{1}{2}\alpha x^2}\left(-i\hbar\frac{d}{dx}\right)\left(\frac{\alpha}{\pi}\right)^{1/4} e^{-\frac{1}{2}\alpha x^2}\,dx$$

$$\langle p_x \rangle = \left(\frac{\alpha}{\pi}\right)^{1/2}(-i\hbar)\int_{-\infty}^\infty -\alpha x e^{-\alpha x^2}\,dx$$

Because the integrand is an odd function of x, $\langle p_x \rangle = 0$ for $n = 0$.

For $n = 1$,

$$\langle p_x \rangle = \int_{-\infty}^\infty \left(\frac{4\alpha^3}{\pi}\right)^{1/4} x e^{-\frac{1}{2}\alpha x^2}\left(-i\hbar\frac{d}{dx}\right)\left(\frac{4\alpha^3}{\pi}\right)^{1/4} x e^{-\frac{1}{2}\alpha x^2}\,dx$$

$$\langle p_x \rangle = \left(\frac{4\alpha^3}{\pi}\right)^{1/2}(-i\hbar)\int_{-\infty}^\infty x(1 - \alpha x^2)e^{-\alpha x^2}\,dx$$

Because the integrand is an odd function of x, $\langle p_x \rangle = 0$ for $n = 1$.

For $n = 2$,

$$\langle p_x \rangle = \int_{-\infty}^{\infty} \left(\frac{\alpha}{4\pi}\right)^{1/4} (2\alpha x^2 - 1) e^{-\frac{1}{2}\alpha x^2} \left(-i\hbar \frac{d}{dx}\right) \left(\frac{\alpha}{4\pi}\right)^{1/4} (2\alpha x^2 - 1) e^{-\frac{1}{2}\alpha x^2} dx$$

$$\langle p_x \rangle = \left(\frac{\alpha}{4\pi}\right)^{1/2} (-i\hbar) \int_{-\infty}^{\infty} (2\alpha x^2 - 1) e^{-\alpha x^2} (-2\alpha^2 x^3 + 4\alpha x + \alpha x) dx$$

$$\langle p_x \rangle = \left(\frac{\alpha}{4\pi}\right)^{1/2} (-i\hbar) \int_{-\infty}^{\infty} e^{-\alpha x^2} (-4\alpha^3 x^5 + 12\alpha^2 x^3 - 5\alpha x) dx$$

Because the integrand is an odd function of x, $\langle p_x \rangle = 0$ for $n = 3$.

The result is general. $\langle p_x \rangle = 0$ for all values of n.

P18.33 Using your results for Problems P18.11, 17, 29, and 32, calculate the uncertainties in the position and momentum $\sigma_p^2 = \langle p^2 \rangle - \langle p \rangle^2$ and $\sigma_x^2 = \langle x^2 \rangle - \langle x \rangle^2$ for the ground state ($n = 0$) and first two excited states ($n = 1$ and $n = 2$) of the quantum harmonic oscillator. Compare your results with the predictions of the Heisenberg uncertainty principle.

$$\sigma_p^2 = \langle p^2 \rangle - \langle p \rangle^2 \quad \text{and} \quad \sigma_x^2 = \langle x^2 \rangle - \langle x \rangle^2$$

For $n = 0$,

$$\sigma_p^2 = \frac{1}{2}\hbar\sqrt{k\mu} - 0 = \frac{1}{2}\hbar\sqrt{k\mu}$$

$$\sigma_x^2 = \frac{\hbar}{2\sqrt{k\mu}} - 0 = \frac{\hbar}{2\sqrt{k\mu}}$$

$$\Delta p \, \Delta x = \sqrt{\sigma_p^2 \sigma_x^2} = \frac{1}{2}\hbar \geq \frac{1}{2}\hbar$$

For $n = 1$,

$$\sigma_p^2 = \frac{3}{2}\hbar\sqrt{k\mu} - 0 = \frac{3}{2}\hbar\sqrt{k\mu}$$

$$\sigma_x^2 = \frac{3\hbar}{2\sqrt{k\mu}} - 0 = \frac{3\hbar}{2\sqrt{k\mu}}$$

$$\Delta p \, \Delta x = \sqrt{\sigma_p^2 \sigma_x^2} = \frac{3}{2}\hbar \geq \frac{1}{2}\hbar$$

For $n = 2$,

$$\sigma_p^2 = \frac{5}{2}\hbar\sqrt{k\mu} - 0 = \frac{5}{2}\hbar\sqrt{k\mu}$$

$$\sigma_x^2 = \frac{5\hbar}{2\sqrt{k\mu}} - 0 = \frac{5\hbar}{2\sqrt{k\mu}}$$

$$\Delta p \, \Delta x = \sqrt{\sigma_p^2 \sigma_x^2} = \frac{5}{2}\hbar \geq \frac{1}{2}\hbar$$

19 The Vibrational and Rotational Spectroscopy of Diatomic Molecules

Problems

P19.2 The infrared spectrum of $^7Li^{19}F$ has an intense line at 910.57 cm^{-1}. Calculate the force constant and period of vibration of this molecule.

$$k = 4\pi^2 v^2 \mu$$
$$= 4\pi^2 (2.998 \times 10^{10} \text{ cm s}^{-1} \times 910.57 \text{ cm}^{-1})^2$$
$$\times \frac{18.9984 \text{ amu} \times 7.0160 \text{ amu}}{18.9984 \text{ amu} + 7.0160 \text{ amu}} \times 1.6605 \times 10^{-27} \text{ kg amu}^{-1}$$
$$= 250. \text{ N m}^{-1}$$
$$T = \frac{1}{v} = 3.66 \times 10^{-14} \text{ s}$$

P19.3 Purification of water for drinking using UV light is a viable way to provide potable water in many areas of the world. Experimentally, the decrease in UV light of wavelength 250 nm follows the empirical relation $I/I_0 = e^{-\varepsilon' l}$, where l is the distance that the light passed through the water and ε' is an effective absorption coefficient. $\varepsilon' = 0.070 \text{ cm}^{-1}$ for pure water and 0.30 cm^{-1} for water exiting a waste water treatment plant. What distance corresponds to a decrease in I of 15% from its incident value for (a) pure water and (b) waste water?

$$\ln\left[\frac{I(\lambda)}{I_0(\lambda)}\right] = -\varepsilon'(\lambda)l; \quad l = -\frac{\ln\left[\dfrac{I(\lambda)}{I_0(\lambda)}\right]}{\varepsilon'}$$

$$l = -\frac{\ln[0.85]}{0.070 \text{ cm}^{-1}} = 2.3 \text{ cm for pure water}$$

$$l = -\frac{\ln[0.85]}{0.30 \text{ cm}^{-1}} = 0.54 \text{ cm for treatment plant water}$$

P19.8 An infrared absorption spectrum of an organic compound is shown in the following figure. Use the characteristic group frequencies listed in Section 19.5 to decide whether this compound is more likely to be ethyl amine, pentanol, or acetone.

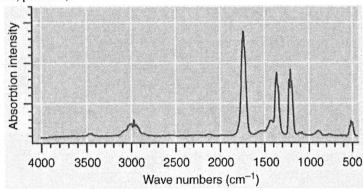

The major peak near 1700 cm^{-1} is the C=O stretch, and the peak near 1200 cm^{-1} is a C—C—C stretch. These peaks are consistent with the compound being acetone. Ethyl amine should show a strong peak near 3350 cm^{-1} and pentanol should show a strong peak near 3400 cm^{-1}. Because these peaks are absent, these compounds can be ruled out.

P19.10   Write an expression for the moment of inertia of the acetylene molecule in terms of the bond distances. Does this molecule have a pure rotational spectrum?

The two mutually perpendicular axes of rotation are perpendicular to the molecular axis and go through the center of the molecule. The two moments of inertia are equal.

$$I = \sum_i m_i x_i^2 = 2\left(m_C \cdot \left[\frac{x_{C\equiv C}}{2}\right]^2 + m_H \left[\frac{x_{C\equiv C}}{2} + x_{C-H}\right]^2 \right)$$

Because the molecule has no dipole moment, it does not have a pure rotational spectrum.

P19.15   Calculating the motion of individual atoms in the vibrational modes of molecules (called normal modes) is an advanced topic. Given the normal modes shown in the following figure, decide which of the normal modes of CO_2 and H_2O have a nonzero dynamical dipole moment and are therefore infrared active. The motion of the atoms in the second of the two doubly degenerate bend modes for CO_2 is identical to the first, but is perpendicular to the plane of the page.

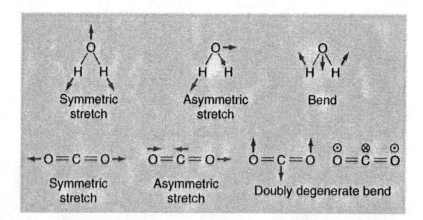

All three vibrational modes of water will lead to a change in the dipole moment and are therefore infrared active. The symmetric stretch of carbon dioxide will not lead to a change in the dipole moment and is infrared inactive. The other two modes will lead to a change in the dipole moment and are infrared active.

P19.16   The force constants for F_2 and I_2 are 470. and 172 N m^{-1}, respectively. Calculate the ratio of the vibrational state populations n_1/n_0 and n_2/n_0 at $T = 300.$ and at $1000.$ K.

The vibrational energy is given by $E_n = \left(n + \dfrac{1}{2}\right)h\nu = \left(n + \dfrac{1}{2}\right)h\sqrt{\dfrac{k}{\mu}}$.

For F_2,

$$E_0 = \frac{1}{2}\hbar\sqrt{\frac{k}{\mu}} = \frac{1}{2} \times 1.055 \times 10^{-34}\ \text{J s} \times \sqrt{\frac{470.\ \text{N m}^{-1}}{\dfrac{18.994\ \text{amu} \times 18.994\ \text{amu}}{18.994\ \text{amu} + 18.994\ \text{amu}} \times 1.661 \times 10^{-27}\ \text{kg amu}^{-1}}}$$

$$= 9.11 \times 10^{-21}\ \text{J}$$

$$E_1 = \frac{3}{2}\hbar\sqrt{\frac{k}{\mu}} = \frac{3}{2} \times 1.055 \times 10^{-34}\ \text{J s} \times \sqrt{\frac{470.\ \text{N m}^{-1}}{\dfrac{18.994\ \text{amu} \times 18.994\ \text{amu}}{18.994\ \text{amu} + 18.994\ \text{amu}} \times 1.661 \times 10^{-27}\ \text{kg amu}^{-1}}}$$

$$= 2.73 \times 10^{-20}\ \text{J}$$

$$E_2 = \frac{5}{2}\hbar\sqrt{\frac{k}{\mu}} = \frac{5}{2}\times 1.055\times 10^{-34}\text{ J s}\times\sqrt{\frac{470.\text{ N m}^{-1}}{\dfrac{18.994\text{ amu}\times 18.994\text{ amu}}{18.994\text{ amu}+18.994\text{ amu}}\times 1.661\times 10^{-27}\text{ kg amu}^{-1}}}$$

$$= 4.55\times 10^{-20}\text{ J}$$

For I_2: $E_0 = \frac{1}{2}\hbar\sqrt{\frac{k}{\mu}} = \frac{1}{2}\times 1.055\times 10^{-34}\text{ J s}\times\sqrt{\dfrac{172\text{ N m}^{-1}}{\dfrac{126.9045\text{ amu}\times 126.9045\text{ amu}}{126.9045\text{ amu}+126.9045\text{ amu}}\times 1.661\times 10^{-27}\text{ kg amu}^{-1}}}$

$$= 2.13\times 10^{-21}\text{ J}$$

$$E_1 = \frac{3}{2}\hbar\sqrt{\frac{k}{\mu}} = \frac{3}{2}\times 1.055\times 10^{-34}\text{ J s}\times\sqrt{\dfrac{172\text{ N m}^{-1}}{\dfrac{126.9045\text{ amu}\times 126.9045\text{ amu}}{126.9045\text{ amu}+126.9045\text{ amu}}\times 1.661\times 10^{-27}\text{ kg amu}^{-1}}}$$

$$= 6.39\times 10^{-21}\text{ J}$$

$$E_2 = \frac{5}{2}\hbar\sqrt{\frac{k}{\mu}} = \frac{5}{2}\times 1.055\times 10^{-34}\text{ J s}\times\sqrt{\dfrac{172\text{ N m}^{-1}}{\dfrac{126.9045\text{ amu}\times 126.9045\text{ amu}}{126.9045\text{ amu}+126.9045\text{ amu}}\times 1.661\times 10^{-27}\text{ kg amu}^{-1}}}$$

$$= 1.07\times 10^{-20}\text{ J}$$

For F_2 at 300. K, $\dfrac{n_1}{n_0} = e^{-\frac{E_1-E_0}{k_B T}} = e^{-\frac{(2.73-0.911)\times 10^{-20}\text{ J}}{1.381\times 10^{-23}\text{ J K}^{-1}\times 300.\text{ K}}} = 0.0123.$

For F_2 at 1000. K, $\dfrac{n_1}{n_0} = e^{-\frac{E_1-E_0}{k_B T}} = e^{-\frac{(1.31-0.437)\times 10^{-19}\text{ J}}{1.381\times 10^{-23}\text{ J K}^{-1}\times 1000.\text{ K}}} = 0.267.$

For F_2 at 300. K, $\dfrac{n_2}{n_0} = e^{-\frac{E_2-E_0}{k_B T}} = e^{-\frac{(4.55-0.915)\times 10^{-20}\text{ J}}{1.381\times 10^{-23}\text{ J K}^{-1}\times 300.\text{ K}}} = 1.52\times 10^{-4}.$

For F_2 at 1000. K, $\dfrac{n_2}{n_0} = e^{-\frac{E_2-E_0}{k_B T}} = e^{-\frac{(4.55-0.915)\times 10^{-20}\text{ J}}{1.381\times 10^{-23}\text{ J K}^{-1}\times 1000.\text{ K}}} = 0.0715.$

For I_2 at 300. K, $\dfrac{n_1}{n_0} = e^{-\frac{E_1-E_0}{k_B T}} = e^{-\frac{(6.39-2.13)\times 10^{-21}\text{ J}}{1.381\times 10^{-23}\text{ J K}^{-1}\times 300.\text{ K}}} = 0.357.$

For I_2 at 1000. K, $\dfrac{n_1}{n_0} = e^{-\frac{E_1-E_0}{k_B T}} = e^{-\frac{(6.39-2.13)\times 10^{-21}\text{ J}}{1.381\times 10^{-23}\text{ J K}^{-1}\times 1000.\text{ K}}} = 0.734.$

For I_2 at 300. K, $\dfrac{n_2}{n_0} = e^{-\frac{E_2-E_0}{k_B T}} = e^{-\frac{(1.07-0.213)\times 10^{-20}\text{ J}}{1.381\times 10^{-23}\text{ J K}^{-1}\times 300.\text{ K}}} = 0.127.$

For I_2 at 1000. K, $\dfrac{n_2}{n_0} = e^{-\frac{E_2-E_0}{k_B T}} = e^{-\frac{(1.07-0.213)\times 10^{-20}\text{ J}}{1.381\times 10^{-23}\text{ J K}^{-1}\times 1000.\text{ K}}} = 0.539.$

P19.19   Show that the Morse potential approaches the harmonic potential for small values of the vibrational amplitude. (*Hint:* Expand the Morse potential in a Taylor–Maclaurin series.)

$$V(R) = D_e \left[1 - e^{-\alpha(R-R_e)} \right]^2$$

Expanding in a Taylor–Maclaurin series and keeping only the first term,

$$V(R) = D_e \left[1 - \left[e^{-\alpha(R-R_e)} \right]_{R=R_e} - \left[\frac{\partial e^{-\alpha(R-R_e)}}{\partial(R-R_e)} \right]_{R=R_e} (R-R_e) \right]^2$$

$$= D_e \left[1 - 1 - \alpha(R-R_e) \right]^2 = D_e \alpha^2 (R-R_e)^2$$

P19.20   The rotational constant for $^7\text{Li}^{19}\text{F}$ determined from microwave spectroscopy is 1.342583 cm^{-1}. The atomic masses of $_7\text{Li}$ and $_{19}\text{F}$ are 7.00160041 and 18.9984032 amu, respectively. Calculate the bond length in $^7\text{Li}^{19}\text{F}$ to the maximum number of significant figures consistent with this information.

$$B = \frac{h}{8\pi^2 \mu r_0^2}; \quad r_0 = \sqrt{\frac{h}{8\pi^2 \mu B}}$$

$$r_0 = \sqrt{\frac{6.62606957 \times 10^{-34} \text{ J s}}{8\pi^2 \times \dfrac{7.00160 \text{ amu} \times 18.9984 \text{ amu}}{7.00160 \text{ amu} + 18.9984 \text{ amu}} \times 1.66053892 \times 10^{-27} \text{ kg amu}^{-1} \times 1.342583 \text{ cm}^{-1}}}$$
$$\times 2.99792458 \times 10^{10} \text{ cm s}^{-1}$$

$$r_0 = 1.56660 \times 10^{-10} \text{ m}$$

P19.23   The fundamental vibrational frequencies for $^1\text{H}_2$ and $^2\text{D}_2$ are 4401 and 3115 cm^{-1}, respectively, and D_e for both molecules is 7.667×10^{-19} J. Using this information, calculate the bond energy of both molecules.

$$\left(D_e - \frac{1}{2} hc\tilde{\nu} \right)_{\text{H}_2} = \frac{1}{2} \times 6.626 \times 10^{-34} \text{ J s} \times 2.998 \times 10^{10} \text{ cm s}^{-1} \times 4401 \text{ cm}^{-1} = 7.240 \times 10^{-19} \text{ J}$$

$$\left(D_e - \frac{1}{2} hc\tilde{\nu} \right)_{\text{D}_2} = \frac{1}{2} \times 6.626 \times 10^{-34} \text{ J s} \times 2.998 \times 10^{10} \text{ cm s}^{-1} \times 3115 \text{ cm}^{-1} = 7.368 \times 10^{-19} \text{ J}$$

P19.27   Fill in the missing step in the derivation that led to the calculation of the spectral line shape in Figure 19.24. Starting from

$$a_2(t) = \mu_x^{21} \frac{E_0}{2} \left(\frac{1 - e^{\frac{i}{\hbar}(E_2 - E_1 + h\nu)t}}{E_2 - E_1 + h\nu} + \frac{1 - e^{-\frac{i}{\hbar}(E_2 - E_1 - h\nu)t}}{E_2 - E_1 - h\nu} \right)$$

and neglecting the first term in the parentheses, show that

$$a_2^*(t) a_2(t) = E_0^2 \left[\mu_x^{21} \right]^2 \frac{\sin^2[(E_2 - E_1 - h\nu)t/2\hbar]}{(E_2 - E_1 - h\nu)^2}$$

We start with

$$a_2^*(t) a_2(t) = \left[\mu_z^{21} \right]^2 E_0^2 \left(\frac{1 - e^{+\frac{i}{\hbar}(E_2 - E_1 - h\nu)t}}{E_2 - E_1 - h\nu} \right) \left(\frac{1 - e^{-\frac{i}{\hbar}(E_2 - E_1 - h\nu)t}}{E_2 - E_1 - h\nu} \right)$$

$$= \left[\mu_z^{21} \right]^2 E_0^2 \frac{\left(2 - 2\cos\left[(E_2 - E_1 - h\nu)\dfrac{t}{\hbar} \right] \right)}{(E_2 - E_1 - h\nu)^2}$$

Using the identity $1 - \cos x = 1 - \cos\left(\dfrac{x}{2} + \dfrac{x}{2}\right) = \cos^2\dfrac{x}{2} + \sin^2\dfrac{x}{2} - \left(\cos^2\dfrac{x}{2} - \sin^2\dfrac{x}{2}\right) = 2\sin^2\dfrac{x}{2}$,

$$a_2^*(t)a_2(t) = \left[\mu_z^{21}\right]^2 \frac{E_0^2}{4}\left(\frac{2 - 2\cos\left[(E_2 - E_1 - h\nu)\dfrac{t}{\hbar}\right]}{(E_2 - E_1 - h\nu)^2}\right) = \left[\mu_z^{21}\right]^2 E_0^2 \frac{\sin^2\left[(E_2 - E_1 - h\nu)\dfrac{t}{2\hbar}\right]}{(E_2 - E_1 - h\nu)^2}$$

P19.30   A strong absorption band in the infrared region of the electromagnetic spectrum is observed at $\tilde{\nu} = 1298\ \text{cm}^{-1}$ for $^{40}\text{Ca}^1\text{H}$. Assuming that the harmonic potential applies, calculate the fundamental frequency ν in units of inverse seconds, the vibrational period in seconds, and the zero point energy for the molecule in joules and electron-volts.

$$\nu = \tilde{\nu}c = 1298\ \text{cm}^{-1} \times 2.998 \times 10^{10}\ \text{cm s}^{-1} = 3.89 \times 10^{13}\ \text{s}^{-1}$$

$$T = \frac{1}{\nu} = \frac{1}{3.89 \times 10^{13}\ \text{s}^{-1}} = 2.57 \times 10^{-14}\ \text{s}$$

$$E = \frac{1}{2}h\nu = \frac{1}{2} \times 6.626 \times 10^{-34}\ \text{J s} \times 3.89 \times 10^{13}\ \text{s}^{-1} = 1.29 \times 10^{-20}\ \text{J} \times \frac{6.241 \times 10^{18}\ \text{eV}}{\text{J}} = 0.0805\ \text{eV}$$

P19.31   The spacing between lines in the pure rotational spectrum of $^{11}\text{B}^2\text{D}$ is $3.9214 \times 10^{11}\ \text{s}^{-1}$. Calculate the bond length of this molecule.

$$B = \frac{\nu}{2c} = \frac{3.9214 \times 10^{11}\ \text{s}^{-1}}{2 \times 2.998 \times 10^{10}\ \text{cm s}^{-1}} = 6.540\ \text{cm}^{-1}$$

$$r_0 = \sqrt{\frac{h}{8\pi^2 \mu c B}}$$

$$= \sqrt{\frac{6.626 \times 10^{-34}\ \text{J s}}{8\pi^2 \times \dfrac{2.014\ \text{amu} \times 11.0093\ \text{amu}}{2.014\ \text{amu} + 11.0093\ \text{amu}} \times 1.6605 \times 10^{-27}\ \text{kg amu}^{-1} \times 2.998 \times 10^{10}\ \text{cm s}^{-1} \times 6.540\ \text{cm}^{-1}}}$$

$$r_0 = 1.230 \times 10^{-10}\ \text{m}$$

P19.33   Calculate the moment of inertia, the magnitude of the rotational angular momentum, and the energy in the $J = 4$ rotational state for $^{14}\text{N}_2$.

$$I = \mu r_0^2 = \frac{14.0031 \times 14.0031\ \text{amu}}{2 \times (14.0031 + 14.0031)} \times 1.6605402 \times 10^{-27}\ \text{kg amu}^{-1} \times (109.8 \times 10^{-12}\ \text{m})^2$$

$$= 1.401 \times 10^{-46}\ \text{kg m}^2$$

$$|J| = \sqrt{J(J+1)}\hbar = \sqrt{4 \times 5} \times 1.0554 \times 10^{-34}\ \text{J s} = 4.718 \times 10^{-34}\ \text{J s}$$

$$E_J = \frac{J(J+1)\hbar^2}{2I} = \frac{4 \times 5 \times (1.0554 \times 10^{-34}\ \text{J s})^2}{2 \times 1.401 \times 10^{-46}\ \text{kg m}^2} = 7.943 \times 10^{-23}\ \text{J}$$

P19.38   In Problem P19.29 you obtained the result

$$J_{max} = 1/2\left[\sqrt{4I k_B T / \hbar^2} - 1\right]$$

Using this result, estimate T for the simulated $^1\text{H}^{35}\text{Cl}$ rotational spectra shown in the following figure. Give realistic estimates of the precision with which you can determine T from the spectra. In generating the simulation, we assumed that the intensity of the individual peaks is solely determined by the population in the originating state and that it does not depend on the initial and final J values.

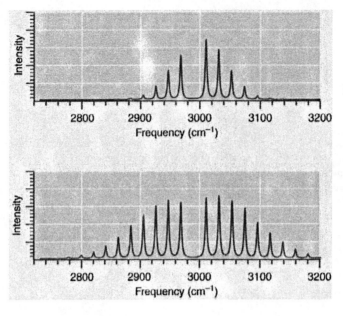

$$J_{max} = \frac{1}{2}\left[\sqrt{\frac{4I k_B T}{\hbar^2}} - 1\right]; \quad T = \frac{(2J_{max} + 1)^2 \hbar^2}{4I k_B}$$

$$T = \frac{(2J_{max} + 1)^2 \hbar^2}{4I k_B}$$

For the upper spectrum, $J_{max} = 2$ in both the P and R branches.

$$T = \frac{5^2 \times (1.054 \times 10^{-34} \text{ J s})^2}{4 \times \dfrac{1.0078 \text{ amu} \times 34.9689 \text{ amu}}{1.0078 \text{ amu} + 34.9689 \text{ amu}} \times 1.661 \times 10^{-27} \text{ kg amu}^{-1} \times (1.2746 \times 10^{-10} \text{ m})^2 \times 1.3807 \times 10^{-23} \text{ J K}^{-1}}$$

$$T = 190 \text{ K}$$

Setting $J_{max} = 1$ and 3 gives temperatures of 68 K and 372 K. Given an uncertainty of ± 0.5 in J_{max}, a reasonable estimate of the temperature in the upper spectrum is 190 K $\pm$ 70 K.

For the lower spectrum, $J_{max} = 3$ in both the P and R branches, giving $T = 372$ K. Setting $J_{max} = 2$ and 4 gives temperatures of 190 K and 616 K. Given an uncertainty of ± 0.5 in J_{max}, a reasonable estimate of the temperature in the lower spectrum is 372 K $\pm$ 120 K.

Evaluating the intensities for all values of J and not just J_{max} would reduce the uncertainty considerably. The spectra were calculated using values of 150 K and 320 K.

P19.41   Calculate the angular momentum of $^7Li_2$ in the $J = 5$ state.

$$|J| = \sqrt{J(J + 1)}\hbar = \sqrt{5 \times 6} \times 1.0554 \times 10^{-34} \text{ J s} = 5.775 \times 10^{-34} \text{ J s}$$

20 The Hydrogen Atom

Numerical Problems

P20.1 Calculate the wave number corresponding to the most and least energetic spectral lines in the Lyman, Balmer, and Paschen series for the hydrogen atom.

$$\tilde{v} = 109{,}678\left(\frac{1}{n_{initial}^2} - \frac{1}{n_{final}^2}\right)\text{cm}^{-1}$$ and the most energetic line corresponds to $n \rightarrow \infty$.

$n_{initial}$ is 1, 2, and 3 for the Lyman, Balmer, and Paschen series, respectively.

$\tilde{v} = 109{,}678\ \text{cm}^{-1}$ for the Lyman series,

$$\tilde{v} = 109{,}678\left(\frac{1}{4}\right)\text{cm}^{-1} = 27419.5\,\text{cm}^{-1}$$ for the Balmer series, and

$$\tilde{v} = 109{,}678\left(\frac{1}{9}\right)\text{cm}^{-1} = 12186.4\,\text{cm}^{-1}$$ for the Paschen series.

The least energetic transition corresponds to $n_{initial} \rightarrow n_{initial} + 1$.

$$\tilde{v} = 109{,}678\left(1 - \frac{1}{4}\right)\text{cm}^{-1} = 82258.5\,\text{cm}^{-1},$$

$$\tilde{v} = 109{,}678\left(\frac{1}{4} - \frac{1}{9}\right)\text{cm}^{-1} = 15233.1\,\text{cm}^{-1},$$

and

$$\tilde{v} = 109{,}678\left(\frac{1}{9} - \frac{1}{16}\right)\text{cm}^{-1} = 5331.57\,\text{cm}^{-1}$$ for the Lyman, Balmer, and Paschen series, respectively.

P20.5 Calculate the probability that the $1s$ electron for H will be found between $r = a_0$ and $r = 2a_0$.

Let $u = r^2$ and $dv = e^{-\frac{r}{\alpha}}\,dr$

$du = 2r\,dr$ and $v = -\alpha e^{-\frac{r}{\alpha}}$

$$\int r^2 e^{-\frac{r}{\alpha}}\,dr = -\alpha r^2 e^{-\frac{r}{\alpha}} + 2\alpha \int r e^{-\frac{r}{\alpha}}\,dr$$

Integrating by parts,

$$\int r^2 e^{-\frac{r}{\alpha}}\,dr = -\alpha r^2 e^{-\frac{r}{\alpha}} + 2\alpha\left(-\alpha r e^{-\frac{r}{\alpha}} + \alpha \int e^{-\frac{r}{\alpha}}\,dr\right) = -\alpha r^2 e^{-\frac{r}{\alpha}} - 2\alpha\left(\alpha r e^{-\frac{r}{\alpha}} + \alpha^2 e^{-\frac{r}{\alpha}}\right)$$

$$\int r^2 e^{-\frac{r}{\alpha}}\,dr = e^{-\frac{r}{\alpha}}\left(-\alpha r^2 - 2\alpha^2 r - 2\alpha^3\right)$$

$$P = \frac{1}{\pi a_0^3} \int_0^{2\pi} d\phi \int_0^{\pi} \sin\theta \, d\theta \int_{a_0}^{2a_0} r^2 e^{-\frac{2r}{a_0}} dr = \frac{4}{a_0^3} \int_{a_0}^{2a_0} r^2 e^{-\frac{2r}{a_0}} dr$$

$$= \frac{4}{a_0^3} \left[e^{-\frac{2r}{a_0}} \left(-\frac{a_0}{2} r^2 - 2\left\{ \frac{a_0}{2} \right\}^2 r - 2\left\{ \frac{a_0}{2} \right\}^3 \right) \right]_{a_0}^{2a_0}$$

$$= \frac{4}{a_0^3} \left[e^{-4} \left(-\frac{a_0}{2} 4a_0^2 - 2\left\{ \frac{a_0}{2} \right\}^2 2a_0 - 2\left\{ \frac{a_0}{2} \right\}^3 \right) \right] - \frac{4}{a_0^3} \left[e^{-2} \left(-\frac{a_0}{2} a_0^2 - 2\left\{ \frac{a_0}{2} \right\}^2 a_0 - 2\left\{ \frac{a_0}{2} \right\}^3 \right) \right]$$

$$= \frac{4}{a_0^3} \left[e^{-4} \left(-2a_0^3 - a_0^3 - \frac{a_0^3}{4} \right) \right] - \frac{4}{a_0^3} \left[e^{-2} \left(-\frac{1}{2} a_0^3 - \frac{1}{2} a_0^3 - \frac{a_0^3}{4} \right) \right]$$

$$= 4 \left(-\frac{13}{4} e^{-4} + \frac{5}{4} e^{-2} \right) = 0.439$$

P20.6   Calculate the distance from the nucleus for which the radial distribution function for the 2p orbital has its main and subsidiary maxima.

$$r^2 R_{21}^2(r) = \frac{1}{24} \left(\frac{1}{a_0} \right)^3 \frac{r^4}{a_0^2} e^{-(r/a_0)}$$

The maxima and minima are found by setting the derivative of this function equal to zero.

$$\frac{d}{dr} \left(\frac{r^4}{a_0^2} e^{-r/a_0} \right) = \frac{4r^3}{a_0^2} e^{-r/a_0} - \frac{r^4}{a_0^3} e^{-r/a_0} = \frac{r^3 e^{-r/a_0}}{a_0^2} \left(4 - \frac{r}{a_0} \right) = 0$$

The solutions are $r = 0$, $r = 4a_0$.

There are no subsidiary maxima. The solution $r = 0$ corresponds to a minimum.

P20.12   In this problem, you will calculate the probability of finding an electron within a sphere of radius r for the H atom in its ground state.

a.   Show, using integration by parts, $\int u \, dv = uv - \int v \, du$, that $\int r^2 e^{-r/\alpha} dr = e^{-r/\alpha}(-2\alpha^3 - 2\alpha^2 r - \alpha r^2)$.

b.   Using this result, show that the probability density of finding the electron within a sphere of radius r for the hydrogen atom in its ground state is

$$1 - e^{-2r/a_0} - \frac{2r}{a_0} \left(1 + \frac{r}{a_0} \right) e^{-2r/a_0}$$

c.   Evaluate this probability density for $r = 0.25\, a_0$, $r = 2.25\, a_0$, and $r = 5.5\, a_0$.

(a)   Let $u = r^2$ and $dv = e^{-\frac{r}{\alpha}} dr$

$du = 2r \, dr$ and $v = -\alpha e^{-\frac{r}{\alpha}}$

$$\int r^2 e^{-\frac{r}{\alpha}} dr = -\alpha r^2 e^{-\frac{r}{\alpha}} + 2\alpha \int r e^{-\frac{r}{\alpha}} dr$$

Integrating by parts,

$$\int r^2 e^{-\frac{r}{\alpha}} dr = -\alpha r^2 e^{-\frac{r}{\alpha}} + 2\alpha \left(-\alpha r e^{-\frac{r}{\alpha}} + \alpha \int e^{-\frac{r}{\alpha}} dr \right) = -\alpha r^2 e^{-\frac{r}{\alpha}} - 2\alpha \left(\alpha r e^{-\frac{r}{\alpha}} + \alpha^2 e^{-\frac{r}{\alpha}} \right)$$

$$\int r^2 e^{-\frac{r}{\alpha}} dr = e^{-\frac{r}{\alpha}} \left(-\alpha r^2 - 2\alpha^2 r - 2\alpha^3 \right)$$

(b) $\quad P = \dfrac{1}{\pi a_0^3} \int\limits_0^{2\pi} d\phi \int\limits_0^{\pi} \sin\theta \, d\theta \int\limits_0^{2a_0} r'^2 e^{-\frac{2r'}{a_0}} \, dr = \dfrac{4}{a_0^3} \int\limits_0^{r} r'^2 e^{-\frac{2r'}{a_0}} \, dr'$

$\qquad = \dfrac{4}{a_0^3}\left[e^{-\frac{2r}{a_0}}\left(-\dfrac{a_0}{2}r^2 - 2\left\{\dfrac{a_0}{2}\right\}^2 r - 2\left\{\dfrac{a_0}{2}\right\}^3 \right) \right]_0^r$

$\qquad\qquad = \dfrac{4}{a_0^3}\left[e^{-\frac{2r}{a_0}}\left(-\dfrac{a_0}{2}r^2 - 2\left\{\dfrac{a_0}{2}\right\}^2 r - 2\left\{\dfrac{a_0}{2}\right\}^3 \right) \right] - \dfrac{4}{a_0^3}\left[e^{-0}\left(-2\left\{\dfrac{a_0}{2}\right\}^3 \right) \right]$

$\qquad\qquad = \dfrac{4}{a_0^3}\left[e^{-\frac{2r}{a_0}}\left(-\dfrac{a_0}{2}r^2 - \dfrac{a_0^2}{2} r - \dfrac{a_0^3}{4} \right) + \dfrac{a_0^3}{4} \right]$

$\qquad\qquad = 1 - e^{-\frac{2r}{a_0}} - 2\dfrac{r}{a_0}\left(1 + \dfrac{r}{a_0} \right) e^{-\frac{2r}{a_0}}$

(c)　　Evaluate this probability density for $r = 0.25\, a_0$, $r = 2.25\, a_0$, and $r = 5.5\, a_0$.

The probability evaluated at $r = 0.25\, a_0 = 1.4 \times 10^{-2}$. The probability evaluated at $r = 2.25\, a_0 = 0.83$. The probability evaluated at $r = 5.5\, a_0 = 0.999$.

P20.14　　Use the result of P20.13 to

a.　　Calculate the mass density of the H atom.

b.　　Compare your answer with the nuclear density assuming a nuclear radius of 1.0×10^{-15} m.

c.　　Calculate the mass density of the H atom outside of the nucleus.

(a)　$\rho_{atom} = \dfrac{m_H}{(4/3)\pi r^3} = \dfrac{1.0078 \text{ amu} \times 1.6605 \times 10^{-27} \text{ kg amu}^{-1}}{(4/3)\pi \times (2.65 \times 0.529 \times 10^{-10} \text{ m})^3}$

$\qquad\qquad = 145 \text{ kg m}^{-3}$

(b)　$\rho_{nucleus} = \dfrac{m_{nucleus}}{(4/3)\pi r^3} = \dfrac{1.0078 \text{ amu} \times 1.6605 \times 10^{-27} \text{ kg amu}^{-1} - 9.109 \times 10^{-31} \text{ kg}}{(4/3)\pi \times (1.0 \times 10^{-15} \text{ m})^3}$

$\qquad\qquad = 4.0 \times 10^{17} \text{ kg m}^{-3}$

(c)　$\rho = \dfrac{m_e}{(4/3)\pi r^3} = \dfrac{9.109 \times 10^{-31} \text{ kg}}{(4/3)\pi \times (2.65 \times 0.529 \times 10^{-10} \text{ m})^3}$

$\qquad\qquad = 0.0789 \text{ kg m}^{-3}$

P20.16　　In spherical coordinates, $z = r\cos\theta$. Calculate $\langle z \rangle$ and $\langle z^2 \rangle$ for the H atom in its ground state. Without doing the calculation, what would you expect for $\langle x \rangle$ and $\langle y \rangle$, and $\langle x^2 \rangle$ and $\langle y^2 \rangle$? Why?

$$\langle z \rangle = \dfrac{1}{\pi a_0^3} \int\limits_0^{2\pi} d\phi \int\limits_0^{\pi} \cos\theta \sin\theta \, d\theta \int\limits_0^{\infty} r^3 e^{-\frac{2r}{a_0}} \, dr$$

$$\langle z \rangle = \dfrac{2\pi}{\pi a_0^3} \left[-\dfrac{\cos^2\theta}{2} \right]_0^{\pi} \int\limits_0^{\infty} r^3 e^{-\frac{r}{a_0}} \, dr = 0$$

because the integral over θ is zero.

$$\left\langle z^2 \right\rangle = \frac{1}{\pi a_0^3} \int\limits_0^{2\pi} d\phi \int\limits_0^{\pi} \cos^2\theta \sin\theta \, d\theta \int\limits_0^{\infty} r^4 e^{-\frac{2r}{a_0}} dr$$

$$\left\langle z^2 \right\rangle = \frac{2\pi}{\pi a_0^3} \left[-\frac{\cos^3\theta}{3} \right]_0^{\pi} \int\limits_0^{\infty} r^4 e^{-\frac{2r}{a_0}} dr = \frac{4}{3a_0^3} \int\limits_0^{\infty} r^4 e^{-\frac{2r}{a_0}} dr$$

Using the standard integral $\int\limits_0^{\infty} r^n e^{-\alpha r} = \dfrac{n!}{\alpha^{n+1}}$,

$$\left\langle z^2 \right\rangle = \frac{4}{3a_0^3} \frac{24 a_0^5}{32} = a_0^2$$

Because the H atom is spherically symmetrical, $\langle x \rangle$ and $\langle y \rangle$, $\langle x^2 \rangle$, and $\langle y^2 \rangle$ will have the same values as $\langle z \rangle$ and $\langle z^2 \rangle$.

P20.17 The force acting between the electron and the proton in the H atom is given by $F = -e^2/4\pi\varepsilon_0 r^2$. Calculate the expectation value $\langle F \rangle$ for the $1s$ and $2p_z$ states of the H atom in terms of e, ε_0, and a_0.

$$\left\langle F \right\rangle_{1s} = -\frac{e^2}{4\pi\varepsilon_0} \int \psi^*(\tau) \frac{1}{r^2} \psi(\tau) \, d\tau$$

$$\left\langle F \right\rangle_{1s} = -\frac{e^2}{4\pi\varepsilon_0} \frac{1}{\pi a_0^3} \int\limits_0^{2\pi} d\phi \int\limits_0^{\pi} \sin\theta \, d\theta \int\limits_0^{\infty} \left[e^{-r/a_0} \right] \left(\frac{1}{r^2} \right) \left[e^{-r/a_0} \right] r^2 dr$$

$$\left\langle F \right\rangle_{1s} = -\frac{e^2}{4\pi\varepsilon_0} \frac{4}{a_0^3} \int\limits_0^{\infty} e^{-2r/a_0} dr = -\frac{e^2}{4\pi\varepsilon_0} \frac{4}{a_0^3} \left[-\frac{a_0}{2} e^{-2r/a_0} \right]_0^{\infty} = -\frac{e^2}{2\pi\varepsilon_0 a_0^2}$$

$$\left\langle F \right\rangle_{2pz} = -\frac{e^2}{4\pi\varepsilon_0} \int \psi^*(\tau) \frac{1}{r^2} \psi(\tau) \, d\tau$$

$$\left\langle F \right\rangle_{2pz} = -\frac{e^2}{4\pi\varepsilon_0} \frac{1}{32\pi a_0^3} \int\limits_0^{2\pi} d\phi \int\limits_0^{\pi} \cos^2\theta \sin\theta \, d\theta \int\limits_0^{\infty} \left(\frac{r}{a_0} \right)^2 \left[e^{-r/a_0} \right] \left(\frac{1}{r^2} \right) r^2 dr$$

$$\left\langle F \right\rangle_{2pz} = -\frac{e^2}{4\pi\varepsilon_0} \frac{1}{16 a_0^5} \left[\frac{\cos^3\theta}{3} \right]_0^{\pi} \times \int\limits_0^{\infty} r^2 e^{-r/a_0} dr$$

$$\left\langle F \right\rangle_{2pz} = -\frac{e^2}{4\pi\varepsilon_0} \frac{1}{24 a_0^5} \int\limits_0^{\infty} r^2 e^{-r/a_0} dr$$

Using the standard integral $\int\limits_0^{\infty} r^n e^{-\alpha r} = \dfrac{n!}{\alpha^{n+1}}$,

$$\left\langle F \right\rangle_{2pz} = -\frac{e^2}{4\pi\varepsilon_0} \frac{1}{24 a_0^5} \times 2 a_0^3 = -\frac{e^2}{48\pi\varepsilon_0 a_0^2}$$

P20.18 The d orbitals have the nomenclature $d_{z^2}, d_{xy}, d_{xz}, d_{yz}$, and $d_{x^2-y^2}$. Show how the d orbital

$$\psi_{3d_{yz}}(r, \theta, \phi) = \frac{\sqrt{2}}{81\sqrt{\pi}} \left(\frac{1}{a_0} \right)^{3/2} \frac{r^2}{a_0^2} e^{-r/3a_0} \sin\theta \cos\theta \sin\phi$$

can be written in the form $yzF(r)$.

In spherical coordinates, $x = r\sin\theta\cos\phi$, $y = r\sin\theta\sin\phi$, and $z = r\cos\theta$. Therefore,

$$\psi_{3d_{yz}}(r,\theta,\phi) = \frac{\sqrt{2}}{81\sqrt{\pi}}\left(\frac{1}{a_0}\right)^{3/2}\frac{r^2}{a_0^2}e^{-r/3a_0}\sin\theta\cos\theta\sin\phi$$

$$= \frac{\sqrt{2}}{81\sqrt{\pi}}\left(\frac{1}{a_0}\right)^{3/2}\frac{1}{a_0^2}e^{-r/3a_0}(r\cos\theta)(r\sin\theta\sin\phi) = \frac{\sqrt{2}}{81\sqrt{\pi}}\left(\frac{1}{a_0}\right)^{3/2}\frac{1}{a_0^2}e^{-r/3a_0}(yz)$$

P20.19 Calculate the expectation value of the moment of inertia of the H atom in the $2s$ and $2p_z$ states in terms of μ and a_0. For the $2s$ state,

$$\langle I \rangle = \left\langle \mu r^2 \right\rangle = \mu\,\frac{1}{32\pi\,a_0^3}\int_0^{2\pi}d\phi\int_0^{\pi}\sin\theta\,d\theta\int_0^{\infty}r^4\left(2-\frac{r}{a_0}\right)^2 e^{-r/a_0}\,dr$$

$$= \mu\frac{1}{8\,a_0^3}\int_0^{\infty}\left(4r^4 - \frac{4r^5}{a_0} + \frac{r^6}{a_0^2}\right)e^{-r/a_0}\,dr = \mu\frac{1}{8\,a_0^3}\left(4\int_0^{\infty}r^4 e^{-r/a_0}\,dr - \frac{4}{a}\int_0^{\infty}r^5\,e^{-r/a_0}\,dr + \frac{1}{a^2}\int_0^{\infty}r^6 e^{-r/a_0}\,dr\right)$$

Using the standard integral $\int_0^{\infty}r^n e^{-\alpha r} = \frac{n!}{\alpha^{n+1}}$,

$$\langle I \rangle = \mu\frac{1}{8\,a_0^3}\left(4\times 4!a_0^5 - \frac{4}{a_0}\times 5!\times a_0^6 + \frac{1}{a_0^2}\times 6!\times a_0^7\right) = 42\mu a_0^2$$

For the $2p_z$ state,

$$\langle I \rangle = \left\langle \mu r^2 \right\rangle = \mu\,\frac{1}{32\pi a_0^3}\int_0^{2\pi}d\phi\int_0^{\pi}\cos^2\theta\sin\theta\,d\theta\int_0^{\infty}r^4\left(\frac{r}{a_0}\right)^2 e^{-r/a_0}\,dr$$

$$= \mu\frac{1}{16\,a_0^5}\left[\frac{\cos^3\theta}{3}\right]_0^{\pi}\times\int_0^{\infty}r^6 e^{-r/a_0}\,dr = \mu\frac{1}{24\,a_0^5}6!a_0^7 = 30\mu a_0^2$$

P20.22 The total energy eigenvalues for the hydrogen atom are given by $E_n = -e^2/(8\pi\varepsilon_0 a_0 n^2)$, $n = 1, 2, 3, 4, \ldots$, and the three quantum numbers associated with the total energy eigenfunctions are related by $n = 1,2,3,4,\ldots.; l = 0,1,2,3, \ldots, n-1;$ and $m_l = 0,\pm 1,\pm 2,\pm 3,\ldots \pm l$, $n = 1,2,3,4,\ldots.; l = 0,1,2,3, \ldots, n-1;$ and $m_l = 0,\pm 1,\pm 2,\pm 3,\ldots,\pm l$. Using the nomenclature ψ_{nlm_l}, list all eigenfunctions that have the following total energy eigenvalues:

a. $\quad E = -\dfrac{e^2}{32\,\pi\varepsilon_0 a_0}$

b. $\quad E = -\dfrac{e^2}{72\,\pi\varepsilon_0 a_0}$

c. $\quad E = -\dfrac{e^2}{128\,\pi\varepsilon_0 a_0}$

d. $\quad$ What is the degeneracy of each of these energy levels?

(a) $\quad E = -\dfrac{e^2}{32\,\pi\varepsilon_0 a_0}$ corresponds to $n = 2$. The total energy eigenfunctions are ψ_{200}, ψ_{210}, ψ_{211}, ψ_{21-1}. The level has a degeneracy of 4.

(b) $\quad E = -\dfrac{e^2}{72\,\pi\varepsilon_0 a_0}$ corresponds to $n = 3$. The total energy eigenfunctions are ψ_{300}, ψ_{310}, ψ_{311}, ψ_{31-1},

ψ_{320}, ψ_{321}, ψ_{32-1}, ψ_{322}, ψ_{32-2}. The level has a degeneracy of 9.

(c) $E = -\dfrac{e^2}{128\,\pi\varepsilon_0 a_0}$ corresponds to $n = 4$. The total energy eigenfunctions are ψ_{400}, ψ_{410}, ψ_{411}, ψ_{41-1},

ψ_{420}, ψ_{421}, ψ_{42-1}, ψ_{422}, ψ_{42-2}, ψ_{43-2}, ψ_{43-1}, ψ_{430}, ψ_{431}, ψ_{432}, ψ_{433}, ψ_{43-3}. The level has a degeneracy of 16.

P20.25   Show by substitution that $\psi_{100}(r, \theta, \phi) = 1/\sqrt{\pi}\,(1/a_0)^{3/2}e^{-r/a_0}$ is a solution of

$$
-\frac{\hbar^2}{2m_e}
\left[
\begin{array}{l}
\dfrac{1}{r^2}\dfrac{\partial}{\partial r}\left(r^2\dfrac{\partial \psi(r,\theta,\phi)}{\partial r}\right) + \dfrac{1}{r^2\sin\theta}\dfrac{\partial}{\partial \theta}\left(\sin\theta\dfrac{\partial \psi(r,\theta,\phi)}{\partial \theta}\right) \\[3mm]
+\dfrac{1}{r^2\sin\theta}\dfrac{\partial^2 \psi(r,\theta,\phi)}{\partial \phi^2}
\end{array}
\right]
$$
$$
-\frac{e^2}{4\pi\varepsilon_0 r}\psi(r,\theta,\phi) = E\psi(r,\theta,\phi)
$$

What is the eigenvalue for total energy? Use the relation $a_0 = \varepsilon_0 h^2/(\pi m_e e^2)$.

Because the wave function does not depend on the angles, we need not consider the portion of the Schrödinger equation that involves partial derivatives with respect to θ and ϕ.

$$
-\frac{\hbar^2}{2m_e}\left[\frac{1}{r^2}\frac{\partial}{\partial r}\left(r^2\frac{\partial \psi(r,\theta,\phi)}{\partial r}\right) + \frac{1}{r^2\sin\theta}\frac{\partial}{\partial \theta}\left(\sin\theta\frac{\partial \psi(r,\theta,\phi)}{\partial \theta}\right) + \frac{1}{r^2\sin\theta}\frac{\partial^2 \psi(r,\theta,\phi)}{\partial \phi^2}\right]
$$
$$
-\frac{e^2}{4\pi\varepsilon_0 |\vec{r}|}\psi(r,\theta,\phi)
$$

$$
= -\frac{1}{\sqrt{\pi}}\left(\frac{1}{a_0}\right)^{3/2}\frac{\hbar^2}{2m_e}\left[\frac{1}{r^2}\frac{\partial}{\partial r}\left(r^2\frac{\partial e^{-r/a_0}}{\partial r}\right)\right] - \frac{1}{\sqrt{\pi}}\left(\frac{1}{a_0}\right)^{3/2}\frac{e^2}{4\pi\varepsilon_0 |\vec{r}|}e^{-r/a_0}
$$

$$
-\frac{1}{\sqrt{\pi}}\left(\frac{1}{a_0}\right)^{3/2}\frac{\hbar^2}{2m_e}\left[\frac{1}{r^2}e^{-r/a_0}\left(\frac{r^2}{a_0^2} - \frac{2r}{a_0}\right)\right] - \frac{1}{\sqrt{\pi}}\left(\frac{1}{a_0}\right)^{3/2}\frac{e^2}{4\pi\varepsilon_0 |\vec{r}|}e^{-r/a_0}
$$

$$
= -\frac{1}{\sqrt{\pi}}\left(\frac{1}{a_0}\right)^{3/2}\left(\frac{\hbar^2}{2m_e a_0^2}\right)e^{-r/a_0} - \frac{1}{\sqrt{\pi}}\left(\frac{1}{a_0}\right)^{3/2}\left[-\frac{\hbar^2}{m_e a_0} + \frac{e^2}{4\pi\varepsilon_0 r}\right]\frac{1}{r}e^{-r/a_0}
$$

$$
= -\frac{1}{\sqrt{\pi}}\left(\frac{1}{a_0}\right)^{3/2}\left(\frac{\hbar^2}{2m_e a_0^2}\right)e^{-r/a_0} - \frac{1}{\sqrt{\pi}}\left(\frac{1}{a_0}\right)^{3/2}\left[-\frac{\hbar^2\pi m_e e^2}{m_e \varepsilon_0 h^2} + \frac{e^2}{4\pi\varepsilon_0 r}\right]\frac{1}{r}e^{-r/a_0}
$$

$$
= -\frac{1}{\sqrt{\pi}}\left(\frac{1}{a_0}\right)^{3/2}\left(\frac{\hbar^2}{2m_e a_0^2}\right)e^{-r/a_0} - \frac{1}{\sqrt{\pi}}\left(\frac{1}{a_0}\right)^{3/2}\left[-\frac{e^2}{4\pi\varepsilon_0} + \frac{e^2}{4\pi\varepsilon_0}\right]\frac{1}{r}e^{-r/a_0}
$$

$$
= -\frac{1}{\sqrt{\pi}}\left(\frac{1}{a_0}\right)^{3/2}\left(\frac{\hbar^2}{2m_e a_0}\frac{\pi m_e e^2}{4\pi^2\hbar^2\varepsilon_0}\right)e^{-r/a_0} = \frac{1}{\sqrt{\pi}}\left(\frac{1}{a_0}\right)^{3/2}\left(\frac{-e^2}{8\pi\varepsilon_0 a_0^2}\right)e^{-r/a_0}
$$

The function is an eigenfunction of the Schrödinger equation with the eigenvalue $E = \dfrac{-e^2}{8\pi\varepsilon_0 a_0}$.

21 Many-Electron Atoms

Numerical Problems

P21.1   Is $\psi(1,2) = 1s(1)\alpha(1)1s(2)\beta(2) + 1s(2)\alpha(2)1s(1)\beta(1)$ an eigenfunction of the operator $\hat{S}_z$? If so, what is its eigenvalue M_S?

This problem uses a notation that is not explained in this chapter, but is explained in Chapter 22.

$$
\begin{aligned}
&\hat{S}_z\left[1s(1)\alpha(1)1s(2)\beta(2) + 1s(2)\alpha(2)1s(1)\beta(1)\right] \\
&= (\hat{s}_{z1} + \hat{s}_{z2})\left[1s(1)\alpha(1)1s(2)\beta(2) + 1s(2)\alpha(2)1s(1)\beta(1)\right] \\
&= \frac{\hbar}{2}\left[1s(1)\alpha(1)1s(2)\beta(2) - 1s(2)\alpha(2)1s(1)\beta(1)\right] \\
&\quad + \frac{\hbar}{2}\left[-1s(1)\alpha(1)1s(2)\beta(2) + 1s(2)\alpha(2)1s(1)\beta(1)\right] \\
&= \frac{\hbar}{2}(0)
\end{aligned}
$$

The function is an eigenfunction of $\hat{S}_z$ with the eigenvalue $M_S = 0$.

P21.4   In this problem you will prove that the ground-state energy for a system obtained using the variational method is greater than the true energy.

a.   The approximate wave function Φ can be expanded in the true (but unknown) eigenfunctions ψ_n of the total energy operator in the form $\Phi = \sum_n c_n \psi_n$. Show that by substituting $\Phi = \sum_n c_n \psi_n$ in the equation

$$
E = \frac{\int \Phi^* \hat{H} \Phi \, d\tau}{\int \Phi^* \Phi \, d\tau}
$$

you obtain the result

$$
E = \frac{\sum_n \sum_m \int (c_n^* \psi_n^*) \hat{H}(c_m \psi_m) \, d\tau}{\sum_n \sum_m \int (c_n^* \psi_n^*)(c_m \psi_m) \, d\tau}
$$

b.   Because the ψ_n are eigenfunctions of $\hat{H}$, they are orthonormal and $\hat{H}\psi_n = E_n \psi_n$. Show that this information allows us to simplify the expression for D from part (a) to

$$
E = \frac{\sum_m E_m c_m^* c_m}{\sum_m c_m^* c_m}
$$

c. Arrange the terms in the summation such that the first energy is the true ground-state energy E_0 and the energy increases with the summation index m. Why can you conclude that $E - E_0 \geq 0$?

(a)
$$E = \frac{\int \Phi^* \hat{H} \Phi \, d\tau}{\int \Phi^* \Phi \, d\tau} = \frac{\int \left(\Phi = \sum_n c_n \psi_n\right)^* \hat{H} \left(\Phi = \sum_m c_m \psi_m\right) d\tau}{\int \left(\Phi = \sum_n c_n \psi_n\right)^* \left(\Phi = \sum_m c_m \psi_m\right) d\tau} = \frac{\sum_n \sum_m \int (c_n^* \psi_n^*) \hat{H} (c_m \psi_m) \, d\tau}{\sum_n \sum_m \int (c_n^* \psi_n^*)(c_m \psi_m) \, d\tau}$$

(b) Simplify the previous expression for E from part (a) to $E = \dfrac{\sum\limits_m E_m c_m^* c_m}{\sum\limits_m c_m^* c_m}$.

$$E = \frac{\sum_n \sum_m \int (c_n^* \psi_n^*) \hat{H} (c_m \psi_m) \, d\tau}{\sum_n \sum_m \int (c_n^* \psi_n^*)(c_m \psi_m) \, d\tau} = \frac{\sum_n \sum_m E_m \int (c_n^* \psi_n^*)(c_m \psi_m) \, d\tau}{\sum_n \sum_m \int (c_n^* \psi_n^*)(c_m \psi_m) \, d\tau}$$

$$= \frac{\sum_m E_m c_m^* c_m \int \psi_m^* \psi_m \, d\tau}{\sum_m c_m^* c_m \int \psi_m^* \psi_m \, d\tau} = \frac{\sum_m E_m c_m^* c_m}{\sum_m c_m^* c_m}$$

(c)
$$E - E_0 = \frac{\sum\limits_m E_m c_m^* c_m}{\sum\limits_m c_m^* c_m} - \frac{E_0 \sum\limits_m c_m^* c_m}{\sum\limits_m c_m^* c_m} = \frac{\sum\limits_m (E_m - E_0) c_m^* c_m}{\sum\limits_m c_m^* c_m} \geq 0. \text{ Both } (E_m - E_0) \text{ and } c_m^* c_m \text{ are greater}$$

than zero. Therefore, $E - E_0 \geq 0$.

P21.6 The operator for the square of the total spin of two electrons is $\hat{S}_{total}^2 = (\hat{S}_1 + \hat{S}_2)^2 = \hat{S}_1^2 + \hat{S}_2^2 + 2(\hat{S}_{1x}\hat{S}_{2x} + \hat{S}_{1y}\hat{S}_{2y} + \hat{S}_{1z}\hat{S}_{2z})$. Given that

$$\hat{S}_x \alpha = \frac{\hbar}{2}\beta, \quad \hat{S}_y \alpha = \frac{i\hbar}{2}\beta, \quad \hat{S}_z \alpha = \frac{\hbar}{2}\alpha,$$

$$\hat{S}_x \beta = \frac{\hbar}{2}\alpha, \quad \hat{S}_y \beta = -\frac{i\hbar}{2}\alpha, \quad \hat{S}_z \beta = -\frac{\hbar}{2}\beta$$

show that $\alpha(1)\alpha(2)$ and $\beta(1)\beta(2)$ are eigenfunctions of the operator $\hat{S}_{total}^2$. What is the eigenvalue in each case?

This problem uses a notation that is not explained in this chapter, but is explained in Chapter 22.

$\hat{S}_{total}^2 \alpha(1)\alpha(2)$

$= \hat{S}_1^2 \alpha(1)\alpha(2) + \hat{S}_2^2 \alpha(1)\alpha(2) + 2(\hat{S}_{1x}\hat{S}_{2x}\alpha(1)\alpha(2) + \hat{S}_{1y}\hat{S}_{2y}\alpha(1)\alpha(2) + \hat{S}_{1z}\hat{S}_{2z}\alpha(1)\alpha(2))$

$= \alpha(2)\hat{S}_1^2 \alpha(1) + \alpha(1)\hat{S}_2^2 \alpha(2) + 2(\hat{S}_{1x}\alpha(1)\hat{S}_{2x}\alpha(2) + \hat{S}_{1y}\alpha(1)\hat{S}_{2y}\alpha(2) + \hat{S}_{1z}\alpha(1)\hat{S}_{2z}\alpha(2))$

$= \frac{3\hbar^2}{4}\alpha(1)\alpha(2) + \frac{3\hbar^2}{4}\alpha(1)\alpha(2) + 2(\hat{S}_{1x}\alpha(1)\hat{S}_{2x}\alpha(2) + \hat{S}_{1y}\alpha(1)\hat{S}_{2y}\alpha(2) + \hat{S}_{1z}\alpha(1)\hat{S}_{2z}\alpha(2))$

$= \frac{3\hbar^2}{4}\alpha(1)\alpha(2) + \frac{3\hbar^2}{4}\alpha(1)\alpha(2) + 2 \times \frac{\hbar}{2}(\hat{S}_{1x}\alpha(1)\beta(2) + i\hat{S}_{1y}\alpha(1)\beta(2) + \hat{S}_{1z}\alpha(1)\alpha(2))$

$= \frac{3\hbar^2}{4}\alpha(1)\alpha(2) + \frac{3\hbar^2}{4}\alpha(1)\alpha(2) + 2 \times \left(\frac{\hbar}{2}\right)^2 (\beta(1)\beta(2) + i^2\beta(1)\beta(2) + \alpha(1)\alpha(2))$

$= \frac{3\hbar^2}{4}\alpha(1)\alpha(2) + \frac{3\hbar^2}{4}\alpha(1)\alpha(2) + \frac{2\hbar^2}{4}\alpha(1)\alpha(2) = 2\hbar^2 \alpha(1)\alpha(2)$

The eigenvalue is $2\hbar^2$.

$$\hat{S}^2_{total}\beta(1)\beta(2)$$

$$= \hat{S}^2_1\beta(1)\beta(2) + \hat{S}^2_2\beta(1)\beta(2) + 2(\hat{S}_{1x}\hat{S}_{2x}\beta(1)\beta(2) + \hat{S}_{1y}\hat{S}_{2y}\beta(1)\beta(2) + \hat{S}_{1z}\hat{S}_{2z}\beta(1)\beta(2))$$

$$= \alpha(2)\hat{S}^2_1\beta(1) + \beta(1)\hat{S}^2_2\beta(2) + 2(\hat{S}_{1x}\beta(1)\hat{S}_{2x}\beta(2) + \hat{S}_{1y}\beta(1)\hat{S}_{2y}\beta(2) + \hat{S}_{1z}\beta(1)\hat{S}_{2z}\beta(2))$$

$$= \frac{3\hbar^2}{4}\beta(1)\beta(2) + \frac{3\hbar^2}{4}\beta(1)\beta(2) + 2(\hat{S}_{1x}\beta(1)\hat{S}_{2x}\beta(2) + \hat{S}_{1y}\beta(1)\hat{S}_{2y}\beta(2) + \hat{S}_{1z}\beta(1)\hat{S}_{2z}\beta(2))$$

$$= \frac{3\hbar^2}{4}\alpha(1)\alpha(2) + \frac{3\hbar^2}{4}\beta(1)\beta(2) + 2\times\frac{\hbar}{2}(\hat{S}_{1x}\beta(1)\alpha(2) - i\hat{S}_{1y}\beta(1)\alpha(2) - \hat{S}_{1z}\beta(1)\beta(2))$$

$$= \frac{3\hbar^2}{4}\alpha(1)\alpha(2) + \frac{3\hbar^2}{4}\beta(1)\beta(2) + 2\times\left(\frac{\hbar}{2}\right)^2(\alpha(1)\alpha(2) + i^2\alpha(1)\alpha(2) + \beta(1)\beta(2))$$

$$= \frac{3\hbar^2}{4}\alpha(1)\alpha(2) + \frac{3\hbar^2}{4}\beta(1)\beta(2) + \frac{2\hbar^2}{4}\beta(1)\beta(2) = 2\hbar^2\beta(1)\beta(2)$$

The eigenvalue is $2\hbar^2$.

P21.11   Write the Slater determinant for the ground-state configuration of Be.

$$\psi_{Be} = \frac{1}{\sqrt{4!}}\begin{vmatrix} 1s(1)\alpha(1) & 1s(1)\beta(1) & 2s(1)\alpha(1) & 2s(1)\beta(1) \\ 1s(2)\alpha(2) & 1s(2)\beta(2) & 2s(2)\alpha(2) & 2s(2)\beta(2) \\ 1s(3)\alpha(3) & 1s(3)\beta(3) & 2s(3)\alpha(3) & 2s(3)\beta(3) \\ 1s(4)\alpha(4) & 1s(4)\beta(4) & 2s(4)\alpha(4) & 2s(4)\beta(4) \end{vmatrix}$$

22 Quantum States for Many-Electron Atoms and Atomic Spectroscopy

Numerical Problems

P22.1   The principal line in the emission spectrum of sodium is yellow. On close examination, the line is seen to be a doublet with wavelengths of 589.0 and 589.6 nm. Explain the source of this doublet.

The lower-lying state is the level and state $^2S_{1/2}$, and the upper level is ^2P, which contains the states $^2P_{1/2}$ and $^2P_{3/2}$. Both transitions are allowed, and the 589.0 nm wavelength corresponds to the transition between the ground state and $^2P_{3/2}$ because states with higher J lie lower in energy. The 589.6 nm wavelength corresponds to the transition between the ground state and $^2P_{1/2}$.

P22.5   What J values are possible for a ^6H term? Calculate the number of states associated with each level and show that the total number of states is the same as that calculated from the term symbol.

$$S = 5/2, \quad L = 5$$

J lies between $|L + S|$ and $|L - S|$ and can have the values $15/2, 13/2, 11/2, 9/2, 7/2$, and $5/2$. The number of states is $2J + 1$ or 16, 14, 12, 10, 8, and 6, respectively.

This gives a total number of states of $16 + 14 + 12 + 10 + 8 + 6 = 66$.

$(2s + 1)(2L + 1) = 6 \times 11 = 66$ also.

P22.12   Calculate the wavelengths of the first three lines of the Lyman, Balmer, and Paschen series, and the series limit (the shortest wavelength) for each series.

Lyman Series: $E_n = R_H \left(\dfrac{1}{1^2} - \dfrac{1}{n^2} \right)$

$$n = 2 \quad E_2 = R_H \left(1 - \frac{1}{4} \right) = \frac{3}{4} R_H = 82258 \, \text{cm}^{-1} \qquad \lambda = 121.6 \, \text{nm}$$

$$n = 3 \quad E_2 = R_H \left(1 - \frac{1}{9} \right) = \frac{8}{9} R_H = 97491 \, \text{cm}^{-1} \qquad \lambda = 102.6 \, \text{nm}$$

$$n = 4 \quad E_2 = R_H \left(1 - \frac{1}{16} \right) = \frac{15}{16} R_H = 102823 \, \text{cm}^{-1} \qquad \lambda = 97.255 \, \text{nm}$$

$$n = \infty \quad E_2 = R_H \left(1 - \frac{1}{\infty} \right) = R_H = 109678 \, \text{cm}^{-1} \qquad \lambda = 91.18 \, \text{nm}$$

Balmer Series: $E_n = R_H \left(\dfrac{1}{2^2} - \dfrac{1}{n^2} \right)$

$n = 3 \quad E_2 = R_H \left(\dfrac{1}{4} - \dfrac{1}{9} \right) = \dfrac{5}{36} R_H = 15233 \, \text{cm}^{-1} \qquad \lambda = 656.5 \, \text{nm}$

$n = 4 \quad E_2 = R_H \left(\dfrac{1}{4} - \dfrac{1}{16} \right) = \dfrac{3}{16} R_H = 20565 \, \text{cm}^{-1} \qquad \lambda = 486.3 \, \text{nm}$

$n = 5 \quad E_2 = R_H \left(\dfrac{1}{4} - \dfrac{1}{25} \right) = \dfrac{21}{100} R_H = 23032 \, \text{cm}^{-1} \qquad \lambda = 434.2 \, \text{nm}$

$n = \infty \quad E_2 = R_H \left(\dfrac{1}{4} - \dfrac{1}{\infty} \right) = \dfrac{1}{4} R_H = 27419 \, \text{cm}^{-1} \qquad \lambda = 364.7 \, \text{nm}$

Paschen Series: $E_n = R_H \left(\dfrac{1}{3^2} - \dfrac{1}{n^2} \right)$

$n = 4 \quad E_2 = R_H \left(\dfrac{1}{9} - \dfrac{1}{16} \right) = \dfrac{7}{144} R_H = 5331.5 \, \text{cm}^{-1} \quad \lambda = 1876 \, \text{nm}$

$n = 5 \quad E_2 = R_H \left(\dfrac{1}{9} - \dfrac{1}{25} \right) = \dfrac{16}{225} R_H = 7799.3 \, \text{cm}^{-1} \quad \lambda = 1282 \, \text{nm}$

$n = 6 \quad E_2 = R_H \left(\dfrac{1}{9} - \dfrac{1}{36} \right) = \dfrac{1}{12} R_H = 9139.8 \, \text{cm}^{-1} \quad \lambda = 1094 \, \text{nm}$

$n = \infty \quad E_2 = R_H \left(\dfrac{1}{9} - \dfrac{1}{\infty} \right) = \dfrac{1}{9} R_H = 12186.4 \, \text{cm}^{-1} \quad \lambda = 820.6 \, \text{nm}$

P22.14 The inelastic mean free path of electrons in a solid, λ, governs the surface sensitivity of techniques such as AES and XPS. The electrons generated below the surface must make their way to the surface without losing energy in order to give elemental and chemical shift information. An empirical expression for elements that give λ as a function of the kinetic energy of the electron generated in AES or XPS is $\lambda = 538 E^{-2} + 0.41(lE)^{0.5}$. The units of λ are monolayers, E is the kinetic energy of the electron in eV, and l is the monolayer thickness in nanometers. On the basis of this equation, what kinetic energy maximizes the surface sensitivity for a monolayer thickness of 0.3 nm? An equation solver would be helpful in obtaining the answer.

$$\frac{d}{dE}(538 E^{-2} + 0.41(lE)^{0.5}) = -\frac{1076}{E^3} + \frac{0.1123}{\sqrt{E}}$$

Setting the derivative equal to zero and solving for E gives two complex roots and $E = 39 \, \text{eV}$.

P22.17 What are the levels that arise from a ^4F term? How many states are there in each level?

$S = 3/2$ and $L = 3$. M_L can range from $-L$ to $+L$ and can have the values $-3, -2, -1, 0, 1, 2,$ and 3 in this case. M_S can range from $-S$ to $+S$ and can have the values $-3/2, -1/2, 1/2,$ and $3/2$ in this case. J lies between $|L + S|$ and $|L - S|$ and can have the values $9/2, 7/2, 5/2,$ and $3/2$ for this case. Because the number of states is $2J + 1$, these levels have 10, 8, 6, and 4 states, respectively.

P22.23 Use the transition frequencies shown in Example Problem 22.7 to calculate the energy (in joules and electron-volts) of the six levels relative to the $3s \, ^2S_{1/2}$ level. State your answers with the correct number of significant figures.

$$E(3p \, ^2P_{1/2}) = \frac{hc}{\lambda} = \frac{6.626 \times 10^{-34} \, \text{J s} \times 2.998 \times 10^8 \, \text{m s}^{-1}}{589.6 \times 10^{-9} \, \text{m}} = 3.369 \times 10^{-19} \, \text{J} = 2.102 \, \text{eV}$$

$$E(3p \, ^2P_{3/2}) = \frac{hc}{\lambda} = \frac{6.626 \times 10^{-34} \, \text{J s} \times 2.998 \times 10^8 \, \text{m s}^{-1}}{589.0 \times 10^{-9} \, \text{m}} = 3.373 \times 10^{-19} \, \text{J} = 2.105 \, \text{eV}$$

$$E(4s\,^2S_{1/2}) = \frac{hc}{\lambda} = \frac{6.626 \times 10^{-34}\ \text{J s} \times 2.998 \times 10^8\ \text{m s}^{-1}}{589.6 \times 10^{-9}\ \text{m}} + \frac{6.626 \times 10^{-34}\ \text{J s} \times 2.998 \times 10^8\ \text{m s}^{-1}}{1183.3 \times 10^{-9}\ \text{m}}$$

$$= 5.048 \times 10^{-19}\ \text{J} = 3.150\ \text{eV}$$

$$E(5s\,^2S_{1/2}) = \frac{hc}{\lambda} = \frac{6.626 \times 10^{-34}\ \text{J s} \times 2.998 \times 10^8\ \text{m s}^{-1}}{589.0 \times 10^{-9}\ \text{m}} + \frac{6.626 \times 10^{-34}\ \text{J s} \times 2.998 \times 10^8\ \text{m s}^{-1}}{616.0 \times 10^{-9}\ \text{m}}$$

$$= 6.597 \times 10^{-19}\ \text{J} = 4.118\ \text{eV}$$

$$E(3d\,^2D_{3/2}) = \frac{hc}{\lambda} = \frac{6.626 \times 10^{-34}\ \text{J s} \times 2.998 \times 10^8\ \text{m s}^{-1}}{589.6 \times 10^{-9}\text{m}} + \frac{6.626 \times 10^{-34}\ \text{J s} \times 2.998 \times 10^8\ \text{m s}^{-1}}{818.3 \times 10^{-9}\ \text{m}}$$

$$= 5.797 \times 10^{-19}\ \text{J} = 3.618\ \text{eV}$$

$$E(4d\,^2D_{3/2}) = \frac{hc}{\lambda} = \frac{6.626 \times 10^{-34}\ \text{J s} \times 2.998 \times 10^8\ \text{m s}^{-1}}{589.0 \times 10^{-9}\ \text{m}} + \frac{6.626 \times 10^{-34}\ \text{J s} \times 2.998 \times 10^8\ \text{m s}^{-1}}{568.2 \times 10^{-9}\ \text{m}}$$

$$= 6.869 \times 10^{-19}\ \text{J} = 4.287\ \text{eV}$$

P22.25 The spectrum of the hydrogen atom reflects the splitting of the $1s\,^2S$ and $2p\,^2P$ terms into levels. The energy difference between the levels in each term is much smaller than the difference in energy between the terms. Given this information, how many spectral lines are observed in the $1s\,^2S \rightarrow 2p\,^2P$ transition? Are the frequencies of these transitions very similar or quite different?

The ^2S term has a single level, $^2S_{1/2}$. The ^2P term splits into two levels, $^2P_{1/2}$ and $^2P_{3/2}$. Therefore, there will be two closely spaced lines in the spectrum corresponding to the transitions $^2S_{1/2} \rightarrow {}^2P_{1/2}$ and $^2S_{1/2} \rightarrow {}^2P_{3/2}$. The energy spacing between the lines will be much smaller than the energy of the transition.

P22.27 What atomic terms are possible for the following electron configurations? Which of the possible terms has the lowest energy?

    a. ns^1np^1

    b. ns^1nd^1

    c. ns^2np^1

    d. ns^1np^2

    (a) ns^1np^1 L can only have the value 1, and S can have the values 0 and 1. The possible terms are ^1P and ^3P. Hund's rules predict that the ^3P term will have the lower energy.

    (b) ns^1nd^1 L can only have the value 2, and S can have the values 0 and 1. The possible terms are ^1D and ^3D. Hund's rules predict that the ^3D term will have the lower energy.

    (c) ns^2np^1 L can only have the value 1, and S can only have the value 1/2. The only possible term is ^2P.

    (d) ns^1np^2 A table such as the table in the text for the p^2 configuration will have three columns, one for each of the electrons, for M_L and M_S. Each of the 15 states for the p^2 configuration can be combined with $m_s = \pm\frac{1}{2}$ for the ns electron. This gives a total of 30 states. Working through the table gives ^2D, ^4P, ^2P, and ^2S terms. Hund's rules predict that the ^4P term will have the lowest energy.

P22.28 Two angular momenta with quantum numbers $j_1 = 3/2$ and $j_2 = 5/2$ are added. What are the possible values of J for the resultant angular momentum states?

$J = |J_1 + J_2|, |J_1 + J_2 - 1|, |J_1 + J_2 - 2|, \ldots, |J_1 - J_2|$, giving possible J values of 4, 3, 2, and 1.

P22.29   Derive the ground-state term symbols for the following configurations:

a. d^2

b. f^9

c. f^{12}

The method illustrated in Example Problem 22.4 is used for all parts.

(a)

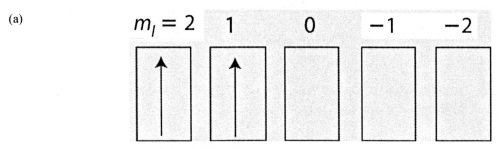

$M_{L\,max} = 3$ and $M_{S\,max} = 1$. Therefore, the ground-state term is ^{3}F.

(b)

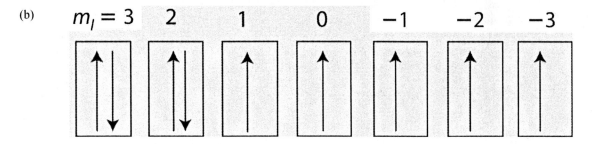

$M_{L\,max} = 5$ and $M_{S\,max} = 2.5$. Therefore, the ground-state term is ^{6}H.

(c)

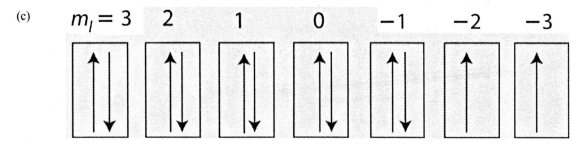

$M_{L\,max} = 5$ and $M_{S\,max} = 1$. Therefore, the ground-state term is ^{3}H.

P22.30   The first ionization potential of ground-state He is 24.6 eV. The wavelength of light associated with the $1s2p$ ^{1}P term is 58.44 nm. What is the ionization energy of the He atom in this excited state?

The photon energy is

$$E = \frac{hc}{\lambda} = \frac{6.626 \times 10^{-34} \text{ J s} \times 2.998 \times 10^8 \text{ m s}^{-1}}{58.44 \times 10^{-9} \text{ m}} \times \frac{1 \text{ eV}}{1.602 \times 10^{-19} \text{ J}} = 21.2 \text{ eV}$$

Therefore, the ionization energy of the He atom in this state is 24.6 eV − 21.2 eV = 3.4 eV.

P22.31 In the Na absorption spectrum, the following transitions are observed:

$$4p \, ^2\text{P} \rightarrow 3s \, ^2\text{S} \quad \lambda = 330.26 \text{ nm}$$

$$3p \, ^2\text{P} \rightarrow 3s \, ^2\text{S} \quad \lambda = 589.593 \text{ nm}, 588.996 \text{ nm}$$

$$5s \, ^2\text{S} \rightarrow 3p \, ^2\text{P} \quad \lambda = 616.073 \text{ nm}, 615.421 \text{ nm}$$

Calculate the energies of the $4p \, ^2\text{P}$ and $5s \, ^2\text{S}$ states with respect to the $3s \, ^2\text{S}$ ground state.

$$E(4p \, ^2\text{P}) = \frac{hc}{\lambda} = \frac{6.626 \times 10^{-34} \text{ J s} \times 2.998 \times 10^8 \text{ m s}^{-1}}{330.26 \times 10^{-9} \text{ m}} = 6.015 \times 10^{-19} \text{ J} = 3.754 \text{ eV}$$

By looking at the Grotrian diagram of Example Problem 22.7, it is seen that the $5s \, ^2\text{S}$ state is accessed by absorption of the photons of wavelength 588.996 nm and 616.073 nm.

$$E(5s \, ^2\text{S}) = \frac{hc}{\lambda} = \frac{6.626 \times 10^{-34} \text{ J s} \times 2.998 \times 10^8 \text{ m s}^{-1}}{588.996 \times 10^{-9} \text{ m}} + \frac{6.626 \times 10^{-34} \text{ J s} \times 2.998 \times 10^8 \text{ m s}^{-1}}{616.073 \times 10^{-9} \text{ m}}$$
$$= 6.597 \times 10^{-19} \text{ J} = 4.117 \text{ eV}$$

P22.33 List the quantum numbers L and S that are consistent with the following terms:

a. ^4S

b. ^4G

c. ^3P

d. ^2D

(a) ^4S: $L = 0$, $2S + 1 = 4$, $S = 3/2$

(b) ^4G: $L = 4$, $2S + 1 = 4$, $S = 3/2$

(c) ^3P: $L = 1$, $2S + 1 = 3$, $S = 1$

(d) ^2D: $L = 2$, $2S + 1 = 2$, $S = 1/2$

23 The Chemical Bond in Diatomic Molecules

Numerical Problems

P23.2 The overlap integral for ψ_g and ψ_u as defined in Section 23.3 is given by

$$S_{ab} = e^{-\zeta R/a_0}\left(1 + \zeta\frac{R}{a_0} + \frac{1}{3}\zeta^2\frac{R^2}{a_0^2}\right)$$

Plot S_{ab} as a function of R/a_0 for $\zeta = 0.8$, 1.0, and 1.2. Estimate the value of R/a_0 for which $S_{ab} = 0.4$ for each of these values of ζ.

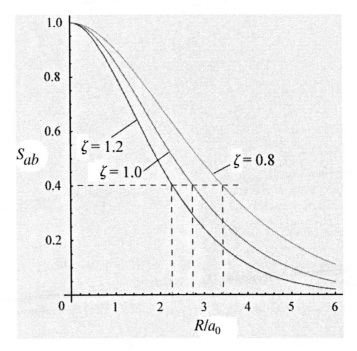

S_{ab} has the value of 0.4 at $R/a_0 = 3.42$, 2.68, and 2.25 for $\zeta = 0.8$, 1.0, and 1.2, respectively.

P23.3 Sketch out a molecular orbital energy diagram for CO and place the electrons in the levels appropriate for the ground state. The AO ionization energies are O2s: 32.3 eV; O2p: 15.8 eV; C2s: 19.4 eV; and C2p: 10.9 eV. The MO energies follow the sequence (from lowest to highest) $1\sigma, 2\sigma, 3\sigma, 4\sigma, 1\pi, 5\sigma, 2\pi, 6\sigma$. Connect each MO level with the level of the major contributing AO on each atom.

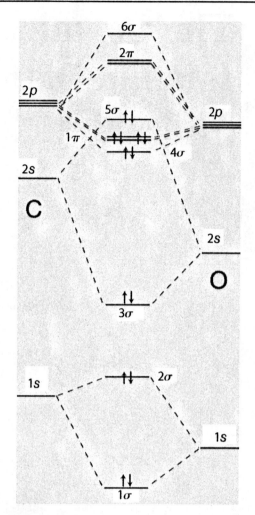

P23.9 Calculate the value for the coefficients of the AOs in Example Problem 23.4 for $S_{12} = 0.45$. How are they different from the values calculated in that problem for $S_{12} = 0.3$? Can you offer an explanation for the changes?

We first obtain the result $H_{HF} = -1.75\, S_{HF}\sqrt{H_{HH}H_{FF}} = -12.5\,\text{eV}$.

Solving for the energies of ε_1 and ε_2 (bonding and antibonding MOs, respectively) gives the values $\varepsilon_1 = -20.3\,\text{eV}$ and $\varepsilon_2 = -5.93\,\text{eV}$.

We calculate $\dfrac{c_{2H}}{c_{2F}}$ by substituting the values for ε_1 and ε_2 in the first of the Equations (24.10). Both equations give the same result.

$$c_{2H}(H_{\text{H}} - \varepsilon_2) + c_{2F}(H_{HF} - \varepsilon_2 S_{HF}) = 0$$

For $\varepsilon_2 = -5.93\,\text{eV}$, $c_{2H}(-13.6 + 5.93) + c_{2F}(-12.5 + 0.45 \times 5.93) = 0$.

$$\frac{c_{2H}}{c_{2F}} = -1.29$$

Using this result in the normalization equation $c_{2H}^2 + c_{2F}^2 + 2c_{2H}c_{2F}S_{HF} = 1$,

$$(-1.29c_{2F})^2 + c_{2F}^2 + 2(-1.29c_{2F}) \times 0.45 = 1$$

$$c_{2H} = 1.1, \text{and } c_{2F} = 0.82 \text{ and } \psi_2 = 1.1\phi_{H1s} - 0.82\phi_{F2p_z}$$

$$c_{1H}(H_{HH} - \varepsilon_1) + c_{1F}(H_{HF} - \varepsilon_1 S_{HF}) = 0$$

For $\varepsilon_1 = -20.3\,\text{eV}$, $c_{1H}(-13.6 + 20.3) + c_{1F}(-12.5 + 0.45 \times 20.3) = 0$.

$$\frac{c_{1H}}{c_{1F}} = 0.505$$

Using this result in the normalization equation $c_{1H}^2 + c_{1F}^2 + 2c_{1H}c_{1F}S_{HF} = 1$,

$$(0.505c_{1H})^2 + c_{1H}^2 + 2(0.505c_{1H}^2) \times 0.45 = 1$$

$$c_{1H} = 0.39 \text{ and } c_{1F} = 0.76, \text{ and } \psi_1 = 0.39\phi_{H1s} + 0.76\phi_{F\,2p_z}$$

The increase in the overlap results in $c_{1H} = 0.39$ and $c_{1F} = 0.76$ for $\varepsilon_1 = -20.3\,\text{eV}$ and in $c_{2H} = 1.1$ and $c_{2F} = -0.82$ for $\varepsilon_2 = -5.93\,\text{eV}$. As for $S_{12} = 0.3$, the coefficient on the lower-lying AO is greater for the bonding orbital and less for the antibonding orbital. Also as before, the signs of the coefficients are the same for the bonding orbital and opposite for the antibonding orbital. However, the magnitude of the coefficients is more nearly equal, due to the greater interaction that arises from a greater overlap, because H_{HF} increases linearly with S_{HF}. In the bonding MO, the electron is shared more equally by the two atoms for the greater overlap.

P23.10   Using the method of Mulliken, calculate the probabilities of finding an electron involved in the chemical bond on the H and F atoms for the bonding and antibonding MOs for Problem P23.9.

The individual terms in $\int \psi_1^* \psi_1 \, d\tau = (c_{1H})^2 + (c_{1F})^2 + 2c_{1H}c_{1F}S_{HF} = 1$ can be interpreted in the following way. For the bonding MO, we associate $(c_{1H})^2 = 0.15$ with the probability of finding the electron around the H atom, $(c_{1F})^2 = 0.59$ with the probability of finding the electron around the F atom, and $2c_{1H}c_{1F}S_{HF} = 0.27$ with the probability of finding the electron shared by the F and H atoms. We divide the shared probability equally between the atoms. This gives the probabilities of $(c_{1H})^2 + c_{1H}c_{1F}S_{HF} = 0.28$ and $(c_{1F})^2 + c_{1H}c_{1F}S_{HF} = 0.72$ for finding the electron on the H and F atoms, respectively.

For the antibonding MO, the probabilities are $(c_{2H})^2 + c_{2H}c_{2F}S_{HF} = 0.72$ and $(c_{2F})^2 + c_{2H}c_{2F}S_{HF} = 0.28$ for finding the electron on the H and F atoms, respectively.

P23.13   Images of molecular orbitals for LiH calculated using the minimal basis set are shown here. In these images, the smaller atom is H. The H1s AO has a lower energy than the Li2s AO. The energy of the MOs is (left to right) $-63.9\,\text{eV}, -7.92\,\text{eV}$, and $+2.14\,\text{eV}$. Make a molecular orbital diagram for this molecule, associate the MOs with the images, and designate the MOs in the images as filled or empty. Which MO is the HOMO? Which MO is the LUMO? Do you expect the dipole moment in this molecule to have the negative end on H or Li?

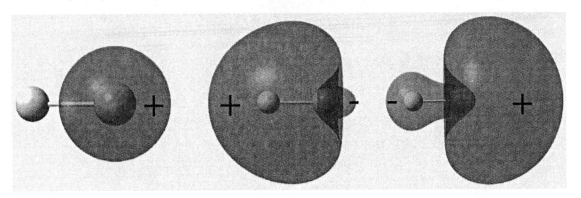

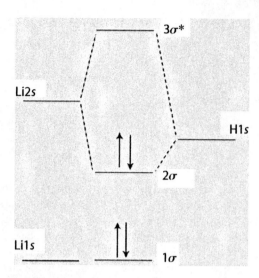

The left image corresponds to the 1σ MO, because the MO is localized on the Li atom. From the MO diagram, the 2σ MO should have a larger coefficient for the H1s AO than for the Li2s AO, following the rules outlined in Section 24.2. Therefore, the middle image is the 2σ MO. Following the same reasoning, the $3\sigma^*$ MO should have a larger coefficient for the Li2s AO. Therefore, the right image is the $3\sigma^*$ MO. The 1σ and 2σ MOs are filled, and the $3\sigma^*$ is empty.

Because the coefficient for the H1s AO is larger than that for the Li2s AO in the 2s MO, the bonding electrons have a higher probability of being on the H than on the Li. Therefore, the negative end of the dipole is on the H atom.

P23.15   Calculate the dipole moment of HF for the bonding MO in Equation (23.33). Use the method outlined in Section 23.8 to calculate the charge on each atom. The bond length in HF is 91.7 pm. The experimentally determined dipole moment of ground-state HF is 1.91 debye, where 1 debye $= 3.33 \times 10^{-30}$ C m. Compare your result with this value. Does the simple theory give a reliable prediction of the dipole moment?

As discussed in Section 23.8, the probabilities for finding the electron on the H and F atoms are $(c_{11})^2 + c_{11}c_{21}S_{12} = 0.21$ and $(c_{21})^2 + c_{11}c_{21}S_{12} = 0.79$ respectively. The magnitude of the dipole moment for this diatomic molecule is given by $\mu = e(z_2 - z_1)r$, where z_i is the charge on each atom, and r is the bond length.

$$\mu = 1.609 \times 10^{-19} \text{ C} \times 91.7 \times 10^{-12} \text{ m} \times \left|(0.79 - 0.21)\right|$$
$$= 8.67 \times 10^{-30} \text{ C m} = 2.6 \text{ D}$$

The error, although considerable, indicates that this very simple model is reasonable.

P23.16   Evaluate the energy for the two MOs generated by combining two H1s AOs. Carry out the calculation for $S_{12} = 0.15, 0.30,$ and 0.45 to mimic the effect of decreasing the atomic separation in the molecule. Use the parameters $H_{11} = H_{22} = -13.6$ eV and $H_{12} = -1.75 S_{12}\sqrt{H_{11}H_{22}}$. Explain the trend that you observe in the results.

$$E_1 = \frac{H_{11} + H_{12}}{1 + S_{12}} \quad \text{and} \quad E_2 = \frac{H_{11} - H_{12}}{1 - S_{12}}$$

S_{12}	H_{12}(eV)	ε_1(eV)	ε_2(eV)
0.15	−3.57	−14.9	−11.8
0.30	−7.14	−16.0	−9.23
0.45	−10.7	−16.8	−5.25

The increase in the overlap mimics a decrease in the bond length. As S_{12} increases, the orbitals overlap more and their interaction becomes greater. The result is that ε_1 becomes more strongly bonding and ε_2 becomes more strongly antibonding.

P23.17 Show that calculating E_u in the manner described by Equation (23.21) gives the result $E_u = (H_{aa} - H_{ab})/(1 - S_{ab})$.

$$E_u = \frac{\int \psi_u^* \hat{H} \psi_u \, d\tau}{\int \psi_u^* \psi_u \, d\tau}$$

$$= \frac{1}{2(1 - S_{ab})} \left(\int \psi_{H1s_a}^* \hat{H} \psi_{H1s_a} \, d\tau + \int \psi_{H1s_b}^* \hat{H} \psi_{H1s_b} \, d\tau - \int \psi_{H1s_b}^* \hat{H} \psi_{H1s_a} \, d\tau - \int \psi_{H1s_a}^* \hat{H} \psi_{H1s_b} \, d\tau \right)$$

$$= \frac{1}{2(1 - S_{ab})} (H_{aa} + H_{bb} - H_{ba} - H_{ab})$$

$$= \frac{H_{aa} - H_{ab}}{1 + S_{ab}}$$

P23.23 Follow the procedure outlined in Section 23.2 to determine c_u in Equation (23.17).

$$1 = \int c_u^* (\psi_{H1s_a}^* - \psi_{H1s_b}^*) c_u (\psi_{H1s_a} - \psi_{H1s_b}) \, d\tau$$

$$= c_u^2 \left(\int \psi_{H1s_a}^* \psi_{H1s_a} \, d\tau + \int \psi_{H1s_b}^* \psi_{H1s_b} \, d\tau - 2 \int \psi_{H1s_b}^* \psi_{H1s_a} \, d\tau \right)$$

$$= c_u^2 (2 - 2 S_{ab}) \quad \text{and}$$

$$c_u = \frac{1}{\sqrt{2 - 2 S_{ab}}}$$

24 Molecular Structure and Energy Levels for Polyatomic Molecules

Numerical Problems

P24.2 Predict whether LiH_2^+ and NH_2^- should be linear or bent based on the Walsh correlation diagram in Figure 24.11. Explain your answers.

The LiH_2^+ molecular ion has two valence electrons. These fill only the $1a_1$ MO. The correlation diagram shows that the energy of this MO is lowered if the molecule is bent. The molecule NH_2^- has eight valence electrons, as does H_2O, and is bent for the same reason.

P24.5 Use the method described in Example Problem 24.3 to show that the *sp*-hybrid orbitals $\psi_a = 1/\sqrt{2}(-\phi_{2s} + \phi_{2p_z})$ and $\psi_b = 1/\sqrt{2}(-\phi_{2s} - \phi_{2p_z})$ are oriented 180° apart.

We differentiate ψ_a with respect to θ and set the derivative equal to zero.

$$\frac{d\psi_a}{d\theta} = \frac{1}{\sqrt{2}} \frac{d}{d\theta}\left[-\frac{1}{\sqrt{32\pi}}\left(\frac{\zeta}{a_0}\right)^{3/2}\left(2 - \frac{r}{a_0}\right)e^{-r/a_0} + \frac{1}{\sqrt{32\pi}}\left(\frac{\zeta}{a_0}\right)^{3/2}\frac{r}{a_0}e^{-\zeta r/a_0}\cos\theta \right]$$

$$= -\frac{1}{\sqrt{32\pi}}\left(\frac{\zeta}{a_0}\right)^{3/2}\frac{r}{a_0}e^{-\zeta r/a_0}\sin\theta = 0$$

This equation is satisfied for $\theta = 0$ and $\theta = 180°$. The second derivative is used to establish which of these corresponds to a maximum.

$$\frac{d^2\psi_a}{d\theta^2} = -\frac{1}{\sqrt{32\pi}}\left(\frac{\zeta}{a_0}\right)^{3/2}\left(\frac{r}{a_0}\right)e^{-\zeta r/a_0}\cos\theta$$

$$= -\frac{1}{\sqrt{32\pi}}\left(\frac{\zeta}{a_0}\right)^{3/2}\left(\frac{r}{a_0}\right)e^{-\zeta r/a_0} \text{ at } \theta = 0$$

$$= \frac{1}{\sqrt{32\pi}}\left(\frac{\zeta}{a_0}\right)^{3/2}\left(\frac{r}{a_0}\right)e^{-\zeta r/a_0} \text{ at } \theta = 180°$$

At a maximum, $\frac{d^2\psi_a}{d\theta^2} < 0$ and at a minimum, $\frac{d^2\psi_a}{d\theta^2} > 0$. Therefore, the maximum is at 0°, and a minimum at 180°. Applying the same procedure to ψ_b shows that the maximum is at 0°. Therefore, ψ_a and ψ_b point in opposite directions separated by 180°.

P24.11 Use the Boltzmann distribution to answer parts (a) and (b):

a. Calculate the ratio of the number of electrons at the bottom of the conduction band to those at the top of the valence band for pure Si at 300. K. The Si band gap is 1.1 eV.

b. Calculate the ratio of the number of electrons at the bottom of the conduction band to those at the top of the dopant band for P-doped Si at 300. K. The top of the dopant band lies 0.040 eV below the bottom of the Si conduction band.

Assume for these calculations that the ratio of the degeneracies is unity. What can you conclude about the room temperature conductivity of these two materials on the basis of your calculations?

Assuming that $\frac{g_2}{g_1} = 1$,

$$\frac{n_2}{n_1} = e^{-\Delta E / k_B T}$$

(a) $\quad \frac{n_2}{n_1} = \exp\left[-\frac{1.1\,eV}{1.381 \times 10^{-23}\,J\,K^{-1} \times 300.\,K} \times \frac{1\,J}{6.242 \times 10^{18}\,eV} \right]$

$$= \exp[-42.54]$$

$$\frac{n_2}{n_1} = 3.34 \times 10^{-19}$$

Pure Si at 300. K is an insulator.

(b) $\quad \frac{n_2}{n_1} = \exp\left[-\frac{0.040\,eV}{1.381 \times 10^{-23}\,J\,K^{-1} \times 300.\,K} \times \frac{1\,J}{6.242 \times 10^{18}\,eV} \right]$

$$= \exp[-1.547]$$

$$= 0.21$$

Doped Si at 300. K is a semiconductor.

P24.13 In P24.3, the hybrid bonding orbitals for ozone were derived. Use the framework described in Section 24.3 to derive the normalized hybrid lone pair orbital on the central oxygen in O_3 that is derived from $2s$ and $2p$ atomic orbitals. The bond angle in ozone is 116.8°.

Ozone has only one lone pair (instead of two like H_2O). The lone pair lies along the z axis and therefore is a combination of the $2s$ and $2p_z$ orbitals. Therefore

$$\psi_{lp} = d_1 \phi_{2p_z} + d_2 \phi_{2s}$$

Normalization requires that $d_1^2 + d_2^2 = 1$. Secondly, conservation of probability requires that the sum of the squares of the coefficients of ψ_{2p_z} in the three hybrids must equal one.

Thus

$$1 = d_1^2 + 0.4350^2 + 0.4350^2$$

or

$$d_1 = \sqrt{1 - 2 \times 0.4350^2}$$
$$d_1 = \pm 0.7884$$

and

$$1 = d_2^2 + 0.5575^2 + 0.5575^2$$

or

$$d_2 = \sqrt{1 - 2 \times 0.5575^2}$$
$$d_2 = \pm 0.6151$$

Because of the node in the $2s$ orbital, we choose $d_2 = -0.6151$, which gives the amplitude of the orbital a positive sign in the region of interest. We choose $d_1 = -0.7884$ because this makes ψ_{ep} orthogonal to ψ_a and ψ_b as shown here.

$$\psi_{ep} = -0.7884\phi_{2p_z} - 0.615\phi_{2s}$$

$$\int \psi_{ep}^* \psi_a \, d\tau = \int (-0.7884\phi_{2p_z} - 0.615\phi_{2s})(0.4350\phi_{2p_z} - 0.7071\phi_{2p_y} - 0.5575\phi_{2s}) \, d\tau$$

$$= -(0.7884)(0.4350) \int \left(\phi_{2p_z}\right)^2 d\tau + (0.6151)(0.5575) \int \phi_{2s}^2 \, d\tau$$

$$= -(0.7884)(0.4350) + (0.6151)(0.5575)$$

$$= 0.00004 \approx 0$$

With this choice of coefficients, ψ_{ep} is normalized as shown below.

$$\int \psi_{ep}^* \psi_{ep} \, d\tau = \int (-0.7884\phi_{2p_z} - 0.6151\phi_{2s})^2 \, d\tau$$

$$= (0.7884)^2 \int (\phi_{2p_z})^2 d\tau + 2 \times (0.7884)(0.6151) \int \phi_{2p_z}\phi_{2s} \, d\tau$$

$$+ (0.6151)^2 \int (\phi_{2s})^2 \, d\tau$$

$$= (0.7884)^2 + (0.6151)^2$$

$$= 0.9999 \approx 1$$

P24.17 Predict whether the ground state or the first excited state of CH_2 should have the larger bond angle on the basis of the Walsh correlation diagram shown in Figure 24.11. Explain your answer.

CH_2 has six valence electrons. In the ground state, the HOMO is the $2a_1$ MO and in the excited state, the HOMO is the $1b_1$ MO. The promotion of the electron from the doubly occupied $2a_1$ MO to the nonbonding $1b_1$ will increase the energy and shift the shape to a more linear orientation. Thus the first excited state of CH_2 will have a larger bond angle.

P24.20 Use the geometrical construction shown in Example Problem 24.8 to derive the π electron MO levels for cyclobutadiene. What is the total π energy of the molecule? How many unpaired electrons will the molecule have?

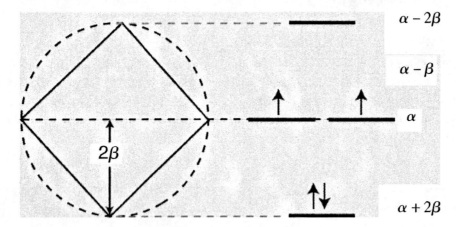

There are two unpaired electrons.

The total π energy is $E_\pi = 2(\alpha + 2\beta) + 2(\alpha) = 4\alpha + 4\beta$.

P24.22 Use the geometrical construction shown in Example Problem 24.8 to derive the π electron MO levels for the cyclopentadienyl radical. What is the total π energy of the molecule? How many unpaired electrons will the molecule have?

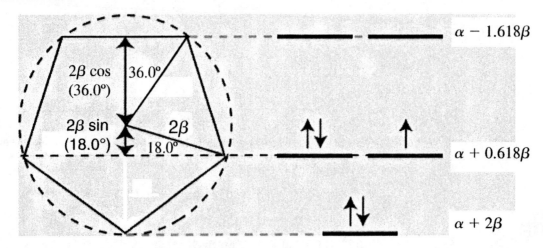

$$2\beta \sin(18.0°) = 0.618$$
$$2\beta \cos(36.0°) = 1.618$$

There is one unpaired electron. The π energy is

$$E_\pi = 2[\alpha + 2\beta] + 3(\alpha + 0.618\beta)$$
$$E_\pi = 5\alpha + 5.85\beta$$

P24.28 $S-p$ hybridization on each Ge atom in planar *trans*-digermane has been described as $sp^{1.5}$ for the Ge—Ge sigma bond and $sp^{1.8}$ for the Ge—H bond. Calculate the H—Ge—Ge bond angle based on this information. Note that the $4p_x$ and $4p_y$ orbitals are proportional to $\cos(\theta)$ and $\sin(\theta)$, respectively, and use the coefficients determined in Problem P24.27 to solve this problem.

We need to solve for the angle between the $\psi_a(sp^{1.5})$ and $\psi_b(sp^{1.8})$ hybrids

$$\psi_b(sp^{1.8}) = N((\cos\theta)\phi_{4pz} + (\sin\theta)\phi_{4px}) + c_{bs}\phi_{4s}$$
$$N\cos\theta = -0.488; \quad N\sin\theta = 0.636$$

Because $\sin^2\theta + \cos^2\theta = 1$,

$$\sin\theta = \frac{0.636}{\sqrt{0.636^2 + 0.488^2}}$$
$$\theta = 127°$$

P24.31 The density of states (DOS) of pyrite, crystalline FeS_2 (as calculated by Eyert et al., *Physical Review B* **55** (1998): 6350, is shown next. The highest occupied energy level corresponds to zero energy. Based on the DOS, is pyrite an insulator, a conductor, or a semiconductor? Also, how does the DOS support the localized-bonding view that some iron valence orbitals are nonbonding?

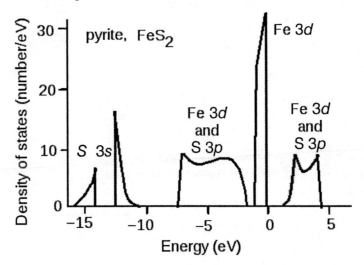

The band gap is approximately 1 eV, so pyrite is not a conductor.

$k_B T = (1.38 \times 10^{-23} \text{ J/K})(1444 \text{ K}) = 1.99 \times 10^{-20}$ J, which is only eight times smaller than the band gap (1 eV $= 1.6 \times 10^{-19}$ J), so pyrite is a semiconductor rather than an insulator. The narrow, localized Fe $3d$ band near zero energy apparently comes from nonbonding iron orbitals. Other iron orbitals participate in the bonding and conduction bands at lower and higher energy.

25 Electronic Spectroscopy

Numerical Problems

P25.3 Ozone (O_3) has an absorptivity at 300. nm of 0.000500 $torr^{-1}$ cm^{-1}. In atmospheric chemistry the amount of ozone in the atmosphere is quantified using the Dobson unit (DU), where 1 DU is equivalent to a 10^{-2} mm thick layer of ozone at 1 atm and 273.15 K.

a. Calculate the absorbance of the ozone layer at 300. nm for a typical coverage of 300. DU.

b. Seasonal stratospheric ozone depletion results a decrease in ozone coverage to values as low as 120. DU. Calculate the absorbance of the ozone layer at this reduced coverage.

In each part, also calculate the transmission from the absorbance using Beer's law.

(a) At 300. DU the effective thickness of the ozone layer is 3.00 mm. With this thickness, the absorbance is:

$$A = \varepsilon bc = (0.000500\ torr^{-1}\ cm^{-1})\left(3.00\ mm \times \frac{1\ cm}{mm}\right)\left(1.00\ atm \times \frac{760\ torr}{atm}\right)$$

$$= 1.14$$

This corresponds to 7.24% transmission of the incident solar radiation at 300 nm through the ozone layer to the Earth's surface.

(b) Performing a calculation identical to that performed in part (a), but at 120. DU or a thickness of 1.2 mm:

$$A = \varepsilon bc = (0.000500\ torr^{-1}\ cm^{-1})\left(1.20\ mm \times \frac{1\ cm}{mm}\right)\left(1.00\ atm \times \frac{760\ torr}{atm}\right)$$

$$= 0.456$$

This corresponds to 35% transmission of the incident solar radiation at 300 nm, or a five-fold increase relative to transmission with typical ozone coverage.

P25.8 Electronic spectroscopy of the Hg—Ar van der Waals complex was performed to determine the dissociation energy of the complex in the first excited state (Quayle, C. J. K. et al. *Journal of Chemical Physics* 99 (1993): 9608). As described in P25.6, the following data regarding ΔG versus n were obtained:

n	ΔG (cm^{-1})
1	37.2
2	34
3	31.6
4	29.2
5	26.8

a. Construct a Birge–Sponer plot (ΔG versus $n + 1/2$) using the given data, and using a best fit to a straight line, determine the value of n where $\Delta G = 0$. This is the vibrational quantum number at dissociation in the excited state. $\Delta G_{1/2}$ can also be determined from the plot.

b.   The area under the Birge–Sponer plot is equal to the dissociation energy, D_0 for this van der Waals complex in the excited state. The area can be determined by summing the ΔG values from $n = 0$ to n at dissociation [determined in part (a)]. You can also integrate the best-fit equation to determine D_0.

(a)   Best fit of the data to a straight line yields:

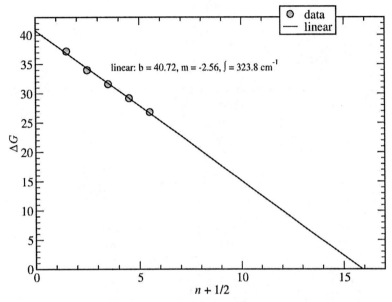

linear: b = 40.72, m = -2.56, $\int$ = 323.8 cm^{-1}

Setting y equal to 0 in the best-fit equation, the value of n at dissociation is:

$$0 = -2.56\left(n + \frac{1}{2}\right) + 40.72$$

$$15.9 = n + \frac{1}{2}$$

$$15 \approx n$$

(b)   This calculation is best performed in a spreadsheet program such as Excel. Remembering to include $n = 0$ (with ΔG value determined using the y intercept from the best-fit equation), the following is obtained:

n	$n + 1/2$	ΔG (cm^{-1})
0	0.5	40.7
1	1.5	37.2
2	2.5	34
3	3.5	31.6
4	4.5	29.2
5	5.5	26.8
6	6.5	24.08
7	7.5	21.52
8	8.5	18.96
9	9.5	16.4
10	10.5	13.84
11	11.5	11.28
12	12.5	8.72
13	13.5	6.16
14	14.5	3.6
15	15.5	1.04
	SUM	325.1

Therefore, the dissociation energy is 325 cm^{-1}.

27 Molecular Symmetry

Numerical Problems

P27.2   Use the 3×3 matrices for the C_{2v} group in Equation (27.3) to verify the group multiplication table for the following successive operations:

    a. $\hat{\sigma}_v \hat{\sigma}'_v$ b. $\hat{\sigma}_v \hat{C}_2$ c. $\hat{C}_2 \hat{C}_2$

$$\hat{E} = \begin{pmatrix} 1 & 0 & 0 \\ 0 & 1 & 0 \\ 0 & 0 & 1 \end{pmatrix} \quad \hat{C}_2 = \begin{pmatrix} -1 & 0 & 0 \\ 0 & -1 & 0 \\ 0 & 0 & 1 \end{pmatrix}$$

$$\hat{\sigma}_v = \begin{pmatrix} 1 & 0 & 0 \\ 0 & -1 & 0 \\ 0 & 0 & 1 \end{pmatrix} \quad \hat{\sigma}'_v = \begin{pmatrix} -1 & 0 & 0 \\ 0 & 1 & 0 \\ 0 & 0 & 1 \end{pmatrix}$$

(a) $\hat{\sigma}_v \hat{\sigma}'_v = \begin{pmatrix} 1 & 0 & 0 \\ 0 & -1 & 0 \\ 0 & 0 & 1 \end{pmatrix} \begin{pmatrix} -1 & 0 & 0 \\ 0 & 1 & 0 \\ 0 & 0 & 1 \end{pmatrix} = \begin{pmatrix} -1 & 0 & 0 \\ 0 & -1 & 0 \\ 0 & 0 & 1 \end{pmatrix}$

        $\hat{\sigma}_v \hat{\sigma}'_v = \hat{C}_2$

(b) $\hat{\sigma}_v \hat{C}_2 = \begin{pmatrix} 1 & 0 & 0 \\ 0 & -1 & 0 \\ 0 & 0 & 1 \end{pmatrix} \begin{pmatrix} -1 & 0 & 0 \\ 0 & -1 & 0 \\ 0 & 0 & 1 \end{pmatrix} = \begin{pmatrix} -1 & 0 & 0 \\ 0 & 1 & 0 \\ 0 & 0 & 1 \end{pmatrix}$

        $\hat{\sigma}_v \hat{C}_2 = \hat{\sigma}'_v$

(c) $\hat{C}_2 \hat{C}_2 = \begin{pmatrix} -1 & 0 & 0 \\ 0 & -1 & 0 \\ 0 & 0 & 1 \end{pmatrix} \begin{pmatrix} -1 & 0 & 0 \\ 0 & -1 & 0 \\ 0 & 0 & 1 \end{pmatrix} = \begin{pmatrix} 1 & 0 & 0 \\ 0 & 1 & 0 \\ 0 & 0 & 1 \end{pmatrix}$

        $\hat{C}_2 \hat{C}_2 = \hat{E}$

P27.4   The D_3 group has the following classes: E, $2C_3$, and $3C_2$. How many irreducible representations does this group have and what is the dimensionality of each?

The D_3 group has three classes and six elements ($1E$, $2C_3$, $3C_2$).

There are three irreducible representations because there are three classes. Their dimensions are determined by the equation

$$d_1^2 + d_2^2 + d_3^2 = 6$$

This can only be satisfied by $d_1 = 2$, $d_2 = 1$, $d_3 = 1$. Thus there are three irreducible representations; two one-dimensional and one two-dimensional.

P27.7 XeF_4 belongs to the D_{4h} point group with the following symmetry elements: E, C_4, C_4^2, C_2, C_2', C_2'', i, S_4, S_4^2, σ, $2\sigma'$, and $2\sigma''$. Make a drawing similar to Figure 27.1 showing these elements.

The four-fold axis is perpendicular to the plane of the molecule.

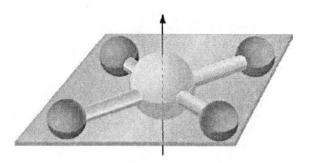

There are three two-fold axes, as shown below. Two pass through the opposed F atoms, and the third bisects the F—Xe—F angle.

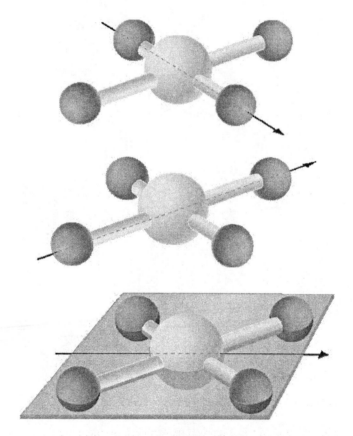

The mirror plane denoted σ lies in the plane of the molecule.

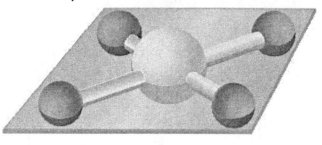

The two mirror planes denoted σ' are mutually perpendicular, and are perpendicular to σ. The Xe atom and two opposed F atoms lie at the intersection of σ' with σ.

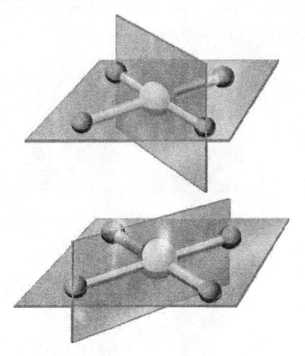

The two mutually perpendicular mirror planes σ'' contain the Xe, but no F atoms.

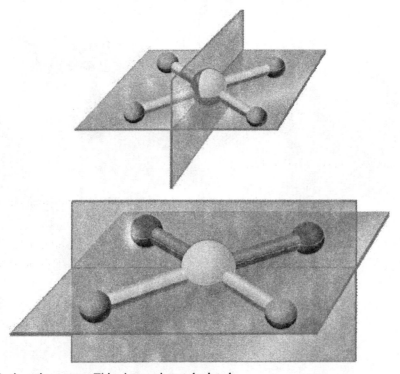

The Xe atom is the invasion center. This element is not depicted.

P27.13 Decompose the following reducible representation into irreducible representations of the C_{2v} group:

E	C_2	σ_v	σ'_v
4	0	0	0

We set up the following table and apply Equation (27.22). The reducible representation is in the first row.

	E	C_2	σ_v	σ_v'
Red	4	0	0	0
A_1	1	1	1	1
A_2	1	1	−1	−1
B_1	1	−1	1	−1
B_2	1	−1	−1	1

$$n_{A_1} = \frac{1}{4}[(1 \times 4 \times 1) + (1 \times 0 \times 1) + (1 \times 0 \times 1) + (1 \times 0 \times 1)] = 1$$

$$n_{A_2} = \frac{1}{4}[(1 \times 4 \times 1) + (1 \times 0 \times 1) + (1 \times 0 \times (-1)) + (1 \times 0 \times (-1))] = 1$$

$$n_{B_1} = \frac{1}{4}[(1 \times 4 \times 1) + (1 \times 0 \times (-1)) + (1 \times 0 \times 1) + (1 \times 0 \times (-1))] = 1$$

$$n_{B_2} = \frac{1}{4}[(1 \times 4 \times 1) + (1 \times 0 \times (-1)) + (1 \times 0 \times (-1)) + (1 \times 0 \times 1)] = 1$$

Thus $\Gamma_{reducible} = A_1 + A_2 + B_1 + B_2$.

P27.15 Use the logic diagram of Figure 27.2 to determine the point group for PCl_5. Indicate your decision-making process as was done in the text for NH_3.

(1) linear?   No

(2) C_n axis?   Yes C_3 axis $\Rightarrow z$ axis

(3) more than $1C_n$ axis   Yes

(4) more than $1C_n$ axis, $n > 2$?   No

(5) σ plane?   Yes

(6) $\sigma \perp$ to C_3?   Yes

We conclude that the point group is D_{3h}.

P27.20 Show that the presence of a C_2 axis and a mirror plane perpendicular to the rotation axis imply the presence of a center of inversion.

Under a 180° rotation around the z axis, $x, y, z \rightarrow -x, -y, z$. — The effect of the mirror plane perpendicular to the rotation axis is $-x, -y, z \rightarrow -x, -y, -z$, which is the same as the inversion operation.

29 Probability

Numerical Problems

P29.1   Suppose that you draw a card from a standard deck of 52 cards. What is the probability of drawing:

a.   an ace of any suit?

b.   the ace of spades?

c.   How would your answers to parts (a) and (b) change if you were allowed to draw three times, replacing the card drawn back into the deck after each draw?

(a)   In a deck of 52 cards there are four aces, therefore:

$$P_E = \frac{E}{N} = \frac{4}{52}$$

(b)   There is only one card that corresponds to the event of interest, therefore:

$$P_E = \frac{E}{N} = \frac{1}{52}$$

(c)   By replacing the card, the probability for each drawing is independent, and the total probability is the sum of probabilities for each event. Therefore, the probability will be $3P_E$ with P_E as given above.

P29.3   A pair of standard dice are rolled. What is the probability of observing the following:

a.   The sum of the dice is equal to 7.

b.   The sum of the dice is equal to 9.

c.   The sum of the dice is less than or equal to 7.

(a)   We are interested in the outcome where the sum of two dice is equal to 7. If any side of a die has an equal probability of being observed, then the probability of any number appearing is 1/6.

$$P_{sum=7} = [2 \times (P_1 \times P_6)] + [2 \times (P_2 \times P_5)] + [2 \times (P_3 \times P_4)]$$

$$= 3 \left[\frac{2}{36} \right] = \frac{1}{6}$$

(b)   Using the nomenclature developed above:

$$P_{sum=9} = [2 \times (P_3 \times P_6)] + [2 \times (P_4 \times P_5)]$$

$$= 2 \left[\frac{2}{36} \right] = \frac{1}{9}$$

(c)   Now, one has to sum all the probabilities that correspond to the event of interest:

$$P_{sum \leq 7} = [(P_1 \times P_1)] + [2 \times (P_1 \times P_2)] + [2 \times (P_1 \times P_3)] + [2 \times (P_1 \times P_4)] + [2 \times (P_1 \times P_5)] + [2 \times (P_1 \times P_6)]$$

$$+ [(P_2 \times P_2)] + [2 \times (P_2 \times P_3)][2 \times (P_2 \times P_4)] + [2 \times (P_2 \times P_5)] + [(P_3 \times P_3)] + [2 \times (P_3 \times P_4)]$$

$$= 21 \left[\frac{1}{36} \right] = \frac{21}{36}$$

P29.7   Atomic chlorine has two naturally occurring isotopes, ^{35}Cl and ^{37}Cl. If the molar abundance of these isotopes is 75.4% and 24.6%, respectively, what fraction of a mole of molecular chlorine (Cl_2) will have one of each isotope? What fraction will contain just the ^{35}Cl isotope?

The probabilities for observing a given isotopic composition of Cl_2 are equal to the product of probabilities for each isotope:

$$^{35}Cl^{35}Cl \quad P = 0.754 \times 0.754 = 0.569$$
$$^{37}Cl^{37}Cl \quad P = 0.246 \times 0.246 = 0.061$$
$$^{35}Cl^{37}Cl \quad P = 0.754 \times 0.246 = 0.186$$
$$^{37}Cl^{35}Cl \quad P = 0.186$$

Using the above probabilities, the fraction of a mole that will contain one of each isotope ($^{35}Cl^{37}Cl$ and $^{37}Cl^{35}Cl$) is $0.186 + 0.186 = 0.372$. The fraction of Cl_2 that will contain just the ^{35}Cl isotope is 0.569.

P29.11   Determine the numerical values for the following:

a.   The number of configurations employing all objects in a six-object set.

b.   The number of configurations employing four objects from a six-object set.

c.   The number of configurations employing no objects from a six-object set.

d. $C(50,10)$.

(a) $C(n,j) = C(6,6) = \left(\dfrac{n!}{j!(n-j)!} \right) = \dfrac{6!}{6!0!} = 1$

(b) $C(n,j) = C(6,4) = \left(\dfrac{n!}{j!(n-j)!} \right) = \dfrac{6!}{4!2!} = 15$

(c) $C(n,j) = C(6,0) = \left(\dfrac{n!}{j!(n-j)!} \right) = \dfrac{6!}{0!6!} = 1$

(d) $P(n,j) = P(50,10) = \left(\dfrac{n!}{j!(n-j)!} \right) = \dfrac{50!}{10!40!} \cong 1.03 \times 10^{10}$

P29.13   Four bases (A, C, T, and G) appear in DNA. Assume that the appearance of each base in a DNA sequence is random.

a.   What is the probability of observing the sequence AAGACATGCA?

b.   What is the probability of finding the sequence GGGGGAAAAA?

c.   How do your answers to parts (a) and (b) change if the probability of observing A is twice that of the probabilities used in parts (a) and (b) of this question when the preceding base is G?

(a)   There are four choices for each base, and the probability of observing any base is equal. Therefore, for a decamer the number of possible sequences is:

$$N_{total} = (4)^{10} \cong 1.05 \times 10^6$$

Since there is only one sequence that corresponds to the event of interest:

$$P_E = \frac{E}{N} \cong \frac{1}{1.05 \times 10^6} \cong 9.52 \times 10^{-7}$$

(b)   Identical to part (a).

(c)   In this case, the probability of observing a base at a given location is dependent on which base is present. If G appears in the sequence, then the probability of observing A is 1/2 while the probability of observing any other base is 1/6 (watch the normalization!). Therefore, the probability of observing the sequence in part (a) is:

$$P = \left(\frac{1}{4}\right)^3 \left(\frac{1}{2}\right)\left(\frac{1}{4}\right)^4 \left(\frac{1}{6}\right)\left(\frac{1}{4}\right) \cong 1.27 \times 10^{-6}$$

and the probability for the sequence in part (b) is:

$$P = \left(\frac{1}{4}\right)\left(\frac{1}{6}\right)^4 \left(\frac{1}{2}\right)\left(\frac{1}{4}\right)^4 \cong 3.77 \times 10^{-7}$$

P29.14   The natural abundance of ^{13}C is roughly 1%, and the abundance of deuterium (^2H or D) is 0.015%. Determine the probability of finding the following in a mole of acetylene:

a. $H\text{-}^{13}C\text{-}^{13}C\text{-}H$

b. $D\text{-}^{12}C\text{-}^{12}C\text{-}D$

c. $H\text{-}^{13}C\text{-}^{12}C\text{-}D$

The probability of observing each of these species is equal to the product of probabilities for observing the individual isotopes:

(a) $P = P(H) \times P(^{13}C) \times P(^{13}C) \times P(H) = P(H)^2 \times P(^{13}C)^2$

$= (0.99985)^2 \times (0.01)^2 = 9.997 \times 10^{-5} \approx 1 \times 10^{-4}$

(b) $P = P(D) \times P(^{12}C) \times P(^{12}C) \times P(D) = P(D)^2 \times P(^{12}C)^2$

$= (0.00015)^2 \times (0.99)^2 = 2.2 \times 10^{-8}$

(c) $P = 2(P(H) \times P(^{13}C) \times P(^{12}C) \times P(D)) =$

$= 2((0.99985) \times (0.01) \times (0.099) \times (0.00015)) = 2.970 \times 10^{-6} \approx 3 \times 10^{-6}$

The factor of 2 in part (c) arises from the fact that there are two configurations corresponding to the same isotopic composition.

P29.16   The Washington State Lottery consists of drawing five balls numbered 1 to 43, and a single ball numbered 1 to 23 from a separate machine.

a.   What is the probability of hitting the jackpot in which the values for all six balls are correctly predicted?

b.   What is the probability of predicting just the first five balls correctly?

c.   What is the probability of predicting the first five balls in the exact order they are picked?

(a)   The total probability is the product of probabilities for the five-ball outcome and the one-ball outcome. The five-ball outcome is derived by considering the configurations possible using 5 objects from a set of 43 total objects:

$$P_{fiveball} = [C(43,5)]^{-1} = \left(\frac{43!}{5!38!}\right)^{-1} \cong 1.04 \times 10^{-6}$$

The one-ball outcome is associated with the configurations possible using a single object from a set of 23 objects:

$$P_{oneball} = [C(23,1)]^{-1} = \left(\frac{23!}{1!22!}\right)^{-1} \cong 4.35 \times 10^{-2}$$

The total probability is the product of the above probabilities:

$$P_{total} = P_{fiveball} \times P_{oneball} \cong 4.52 \times 10^{-8}$$

(b)   The probability is that for the five-ball case determined above.

(c)   This case corresponds to a specific permutation of all permutations possible using 5 objects from a set of 43 objects:

$$P = [P(43,5)]^{-1} = \left(\frac{43!}{38!}\right)^{-1} \cong 8.66 \times 10^{-9}$$

P29.19   Imagine an experiment in which you flip a coin four times. Furthermore, the coin is balanced fairly such that the probability of landing heads or tails is equivalent. After tossing the coin 10 times, what is the probability of observing the following specific outcomes:

a.   no heads?

b.   two heads?

c.   five heads?

d.   eight heads?

(a)   The quantity of interest is the probability of observing a given number of successful trials (j) in a series of n trials in which the probability of observing a successful trial, P_E, is equal to 1/2:

$$P(j) = C(n,j)(P_E)^j(1 - P_E)^{n-j} = C(n,j)\left(\frac{1}{2}\right)^n$$

Substituting in for the specific case of $j = 0$ and $n = 10$ yields:

$$P(0) = C(10,0)\left(\frac{1}{2}\right)^{10} \cong 9.77 \times 10^{-4}$$

(b)   In this case, $j = 2$ and $n = 10$:

$$P(2) = C(10,2)\left(\frac{1}{2}\right)^{10} = \left(\frac{10!}{2!8!}\right)\left(\frac{1}{2}\right)^{10} \cong 0.044$$

(c)   In this case, $j = 5$ and $n = 10$:

$$P(5) = C(10,5)\left(\frac{1}{2}\right)^{10} = \left(\frac{10!}{5!5!}\right)\left(\frac{1}{2}\right)^{10} \cong 0.246$$

(d)   In this case, $j = 8$ and $n = 10$:

$$P(8) = C(10,8)\left(\frac{1}{2}\right)^{10} = \left(\frac{10!}{8!2!}\right)\left(\frac{1}{2}\right)^{10} \cong 0.044$$

P29.23   Another form of Stirling's approximation is:

$$N! = \sqrt{2\pi N}\left(\frac{N}{e}\right)^N$$

Use this approximation for $N = 10, 50, 100$ and compare your results to those given in Example Problem 12.9.

N	$N!$	Ln $N!$	Ln $N!$ (ex. problem)
10	1.5×10^8	18.82368	15.1
50	2.4×10^{69}	159.6399	148.5
100	3.7×10^{166}	383.5383	363.7

Notice that this form provides a slight overestimation of the actual value. In contrast, the version of Stirling's approximation given in the text provides a slight underestimation of the actual value.

P29.25   Radioactive decay can be thought of as an exercise in probability theory. Imagine that you have a collection of radioactive nuclei at some initial time (N_0) and are interested in how many nuclei will still remain at a later time (N). For first-order radioactive decay, $N/N_0 = e^{-kt}$. In this expression, k is known as the decay constant and t is time.

     a.   What is the variable of interest in describing the probability distribution?

     b.   At what time will the probability of nuclei undergoing radioactive decay be 0.50?

     (a)   The variable (k) defines the width of the distribution of population versus time.

     (b)

$$\frac{N}{N_0} = 0.5 = e^{-kt}$$

$$\ln(0.5) = -kt$$

$$\frac{-\ln(0.5)}{k} = \frac{\ln(2)}{k} = t$$

P29.27   In a subsequent chapter we will encounter the energy distribution $P(\varepsilon) = Ae^{-\varepsilon/kT}$, where $P(\varepsilon)$ is the probability of a molecule occupying a given energy state, ε is the energy of the state, k is a constant equal to 1.38×10^{-23} J K^{-1}, and T is temperature. Imagine that there are three energy states at 0, 100., and 500. J mol^{-1}.

     a.   Determine the normalization constant for this distribution.

     b.   What is the probability of occupying the highest energy state at 298 K?

     c.   What is the average energy at 298 K?

     d.   Which state makes the largest contribution to the average energy?

     (a)   Since the energies are given in units of J mol^{-1}, dividing by Avogadro's number will convert this energy to a per particle unit. Alternatively, Avogadro's number can be included with Boltzmann's constant, resulting in $k \times N_a = R$, where $R = 8.314$ J mol^{-1} K^{-1}.

     Using this relationship:

$$P_{500} = e^{-500 \text{ J mol}^{-1}/(8.314 \text{ J mol}^{-1} \text{ K}^{-1})(298 \text{ K})} = 0.817$$

$$P_{100} = e^{-100 \text{ J mol}^{-1}/(8.314 \text{ J mol}^{-1} \text{ K}^{-1})(298 \text{ K})} = 0.960$$

$$P_0 = e^{-0 \text{ J mol}^{-1}/(8.314 \text{ J mol}^{-1} \text{ K}^{-1})(298 \text{ K})} = 1$$

     Using these probabilities, the normalization constant becomes:

$$A = \frac{1}{\sum_{i=1}^{3} P_i} = \frac{1}{P_0 + P_{100} + P_{500}} \cong 0.360$$

     (b)   With normalization, the normalized probabilities are given by the product of the probabilities determined in part (a) of this question and the normalization constant:

$$P_{500,norm} = A \times P_{500} = 0.360 \times 0.817 = 0.294$$

$$P_{100,norm} = A \times P_{100} = 0.360 \times 0.960 = 0.345$$

$$P_{0,norm} = A \times P_0 = 0.360$$

     (c)   The average energy is given by:

$$\langle E \rangle = \sum_{i=1}^{3} E_i P_{i,norm} = (0 \text{ J mol}^{-1})(0.360) + (100 \text{ J mol}^{-1})(0.345) + (500 \text{ J mol}^{-1})(0.294) = 182 \text{ J mol}^{-1}$$

     (d)   Inspection of part (c) of this question illustrates that the highest-energy state makes the largest contribution to the average energy.

P29.29   Consider the following probability distribution corresponding to a particle located between point $x = 0$ and $x = a$:

$$P(x)\, dx = C \sin^2\left[\frac{\pi x}{a}\right] dx$$

    a.   Determine the normalization constant, C.

    b.   Determine $\langle x \rangle$.

    c.   Determine $\langle x^2 \rangle$.

    d.   Determine the variance.

(a) $$1 = C \int_0^a \sin^2\left(\frac{\pi x}{a}\right) dx = C\left(\frac{a}{2}\right)$$

    $$C = \frac{2}{a}$$

(b) $$\langle x \rangle = \int_0^a (x)\left(\frac{2}{a}\sin^2\left(\frac{\pi x}{a}\right)\right) dx = \frac{2}{a}\int_0^a x\sin^2\left(\frac{\pi x}{a}\right) dx$$

    $$= \frac{2}{a}\left(\frac{a^2}{4}\right) = \frac{a}{2}$$

(c) $$\langle x^2 \rangle = \int_0^a (x^2)\left(\frac{2}{a}\sin^2\left(\frac{\pi x}{a}\right)\right) dx = \frac{2}{a}\int_0^a x^2 \sin^2\left(\frac{\pi x}{a}\right) dx$$

    $$= \left(\frac{2}{a}\right)\left[\left(\frac{a^3}{6}\right)-\left(\frac{a^3}{4\pi^2}\right)\right] = a^2\left(\frac{1}{3}-\frac{1}{2\pi^2}\right)$$

(d) $$\sigma^2 = \langle x^2 \rangle - \langle x \rangle^2 = a^2\left(\frac{1}{12}-\frac{1}{2\pi^2}\right)$$

P29.31   One classic problem in quantum mechanics is the "harmonic oscillator." In this problem a particle is subjected to a one-dimensional potential (taken to be along x) of the form $V(x) \propto x^2$, where $-\infty \le x \le \infty$. The probability distribution function for the particle in the lowest-energy state is:

$$P(x) = C e^{-ax^2/2}$$

Determine the expectation value for the particle along x (that is, $\langle x \rangle$). Can you rationalize your answer by considering the functional form of the potential energy?

$$\langle x \rangle = \int_{-\infty}^{\infty} xP(x)\, dx = \int_{-\infty}^{\infty} x(Ce^{-ax^2/2})\, dx$$

$$= C\int_{-\infty}^{\infty} xe^{-ax^2/2} dx = 0$$

The last equality arises from x being an odd function while the exponential term is even. The integral of the product of an even and odd function is zero. This result can be rationalized by noting that $x = 0$ is the location of potential energy minimum.

P29.35   The Poisson distribution is a widely used discrete probability distribution in science:

$$P(x) = \frac{e^{-\lambda}\lambda^x}{x!}$$

This distribution describes the number of events (x) occurring in a fixed period of time. The events occur with a known average rate corresponding to λ, and event occurrence does not depend on when other events occur. This distribution can be applied to describe the statistics of photon arrival at a detector as illustrated by the following:

a.   Assume that you are measuring a light source with an average output of 5 photons per second. What is the probability of measuring 5 photons in any 1-second interval?

b.   For this same source, what is the probability of observing 8 photons in any 1-second interval?

c.   Assume a brighter photon source is employed with an average output of 50 photons per second. What is the probability of observing 50 photons in any 1-second interval?

(a)   For this question $x = 5$ and $\lambda = 5$ such that:

$$P(5) = \frac{e^{-5}5^5}{5!} = 0.175$$

(b)   For this question $x = 8$, but λ is still equal to 5:

$$P(8) = \frac{e^{-5}5^8}{8!} = 0.0653$$

(c)   A different source is employed corresponding to $\lambda = 50$. With $x = 50$ the probability is:

$$P(50) = \frac{e^{-50}50^{50}}{50!} = 0.0563$$

30 The Boltzmann Distribution

Numerical Problems

P30.1   a.   What is the possible number of microstates associated with tossing a coin N times and having it come up H times heads and T times tails?

         b.   For a series of 1000 tosses, what is the total number of microstates associated with 50% heads and 50% tails?

         c.   How much less probable is the outcome that the coin will land 40% heads and 60% tails?

       (a)   In this case, the number of coin tosses is equal to the number of units (N), and each unit can exist in one of two states: heads (H) or tails (T). Since the number of microstates is equal to the weight (W):

$$W = \frac{N!}{H!T!}$$

       (b)   For this case, $N = 1000$, $H = 500$, and $T = 500$. Substituting into the above expression for W:

$$W = \frac{N!}{H!T!} = \frac{1000!}{(500!)^2}$$

The factorials that require evaluation are generally too large to determine on a calculator; therefore, evaluation of W is performed using Sterling's approximation:

$$\begin{aligned}
\ln W &= \ln(1000!) - 2\ln(500!) \\
&= 1000\ln(1000) - 1000 - 2\big[500\ln(500) - 500\big] \\
&= 1000\ln(1000) - 1000\ln(500) = 1000\ln(2) \\
&= 693
\end{aligned}$$

Therefore, $W = 1.26 \times 10^{14}$.

       (c)   Proceeding as in part (b), but with $H = 400$ and $T = 600$:

$$\begin{aligned}
\ln W &= \ln(1000!) - \ln(400!) - \ln(600!) \\
&= 1000\ln(1000) - 1000 - 400\ln(400) + 400 - 600\ln(600) + 600 \\
&= 1000\ln(1000) - 400\ln(400) - 600\ln(600) = 673
\end{aligned}$$

Therefore, $W = 4.71 \times 10^{13}$. Comparing the answers to parts (b) and (c), the $H = 25, T = 25$ outcome is approximately 2.68 times more likely than the $H = 20, T = 30$ outcome.

P30.3   a.   Realizing that the most probable outcome from a series of N coin tosses is $N/2$ heads and $N/2$ tails, what is the expression for W_{max} corresponding to this outcome?

$$W = \frac{N!}{a_H!a_T!}$$

         b.   Given your answer for part (a), derive the following relationship between the weight for an outcome other than the most probable and W_{max}:

$$\log\left(\frac{W}{W_{max}}\right) = -H\log\left(\frac{H}{N/2}\right) - T\log\left(\frac{T}{N/2}\right)$$

c.   We can define the deviation of a given outcome from the most probable outcome using a "deviation index," $\alpha = \dfrac{H - T}{N}$. Show that the number of heads or tails can be expressed as $H = \dfrac{N}{2}(1 + \alpha)$ and $T = \dfrac{N}{2}(1 - \alpha)$.

d.   Finally, demonstrate that $\dfrac{W}{W_{max}} = e^{-N\alpha^2}$.

(a) $W = \dfrac{N!}{H!T!} = \dfrac{N!}{(N/2)!(N/2)!} = \dfrac{N!}{\left[(N/2)!\right]^2}$

(b) $\ln\left(\dfrac{W}{W_{max}}\right) = \ln W - \ln W_{max} = \ln\left(\dfrac{N!}{H!T!}\right) - \ln\left(\dfrac{N!}{\left[(N/2)!\right]^2}\right)$

$\qquad = \ln(N!) - \ln(H!) - \ln(T!) - \ln(N!) + 2\ln((N/2)!)$

$\qquad = -\ln(H!) - \ln(T!) + 2\ln((N/2)!)$

$\qquad = -H\ln H + H - T\ln T + T + N\ln(N/2) - N$

$\qquad = -H\ln H - T\ln T + N\ln(N/2)$

$\qquad = -H\ln H - T\ln T + (H + T)\ln(N/2)$

$\qquad = -H\ln\left(\dfrac{H}{N/2}\right) - T\ln\left(\dfrac{T}{N/2}\right)$

(c)   Substituting the definition of part (a) into the expressions for H and T:

$$H = \frac{N}{2}(1 + \alpha) = \frac{N}{2}\left(1 + \frac{H - T}{N}\right) = \frac{N}{2} + \frac{H - T}{2} = \frac{H + T}{2} + \frac{H - T}{2} = H$$

$$T = \frac{N}{2}(1 - \alpha) = \frac{N}{2}\left(1 - \frac{H - T}{N}\right) = \frac{N}{2} - \frac{H - T}{2} = \frac{H + T}{2} - \frac{H - T}{2} = T$$

(d)   Substituting in the result of part (c) into the final equation of part (b):

$$\ln\left(\frac{W}{W_{max}}\right) = -\frac{N}{2}(1 + \alpha)\ln(1 + \alpha) - \frac{N}{2}(1 - \alpha)\ln(1 - \alpha)$$

If $|\alpha| \ll 1$, then $\ln(1 \pm \alpha) = \pm\alpha$, therefore:

$$\ln\left(\frac{W}{W_{max}}\right) = -\frac{N}{2}(1 + \alpha)\ln(1 + \alpha) - \frac{N}{2}(1 - \alpha)\ln(1 - \alpha)$$

$$\approx -\frac{N}{2}(1 + \alpha)(\alpha) - \frac{N}{2}(1 - \alpha)(-\alpha) = -N\alpha^2$$

$$\frac{W}{W_{max}} = e^{-N\alpha^2}$$

P30.5   Determine the weight associated with the following card hands:

a.   Having any five cards

b.   Having five cards of the same suit (known as a "flush")

(a)   The problem can be solved by recognizing that there are 52 total cards ($N = 52$), with five cards in the hand ($a_1 = 5$), and 47 out of the hand ($a_0 = 47$):

$$W = \frac{N!}{a_1! a_0!} = \frac{52!}{5!47!} \approx 2.60 \times 10^6$$

(b)   For an individual suit, there are 13 total cards ($N = 13$), five of which must be in the hand ($a_1 = 5$) while the other eight remain in the deck ($a_0 = 47$). Finally, there are four total suits:

$$W = 4\left(\frac{N!}{a_1! a_0!}\right) = 4\left(\frac{13!}{5!8!}\right) = 5148$$

P30.8 Barometric pressure can be understood using the Boltzmann distribution. The potential energy associated with being a given height above the Earth's surface is *mgh*, where *m* is the mass of the particle of interest, *g* is the acceleration due to gravity, and *h* is height. Using this definition of the potential energy, derive the following expression for pressure:

$$P = P_o e^{-mgh/kT}$$

Assuming that the temperature remains at 298 K, what would you expect the relative pressures of N_2 and O_2 to be at the tropopause, the boundary between the troposphere and stratosphere roughly 11 km above the Earth's surface? At the Earth's surface, the composition of air is roughly 78% N_2, 21% O_2, and the remaining 1% is other gases.

At the Earth's surface, $h = 0$ meters and the total pressure is 1 atm. Using the mole fractions of N_2 and O_2, the partial pressures at the Earth's surface are 0.78 and 0.21 atm, respectively. Given this information, the pressure of N_2 at 11 km is given by:

$$P_{11\,km}(N_2) = P_{0\,km}(N_2)e^{-mgh/kT} = (0.78\,atm)e^{-(0.028\,kg\,mol^{-1}\times N_A^{-1})(9.8\,m\,s^{-1})(1.1\times10^4\,m)/(1.38\times10^{-23}\,J\,K^{-1})(298\,K)}$$

$$= 0.23\,atm$$

Performing the identical calculation for O_2 yields:

$$P_{11\,km}(O_2) = P_{0\,km}(O_2)e^{-mgh/kT} = (0.21\,atm)e^{-(0.032\,kg\,mol^{-1}\times N_A^{-1})(9.8\,m\,s^{-1})(1.1\times10^4\,m)/(1.38\times10^{-23}\,J\,K^{-1})(298\,K)}$$

$$= 0.052\,atm$$

P30.10 Consider the energy-level diagrams, modified from Problem P30.9 by the addition of another excited state with energy of $600.\,cm^{-1}$.

a. At what temperature will the probability of occupying the second energy level be 0.15 for the states depicted in part (a) of the figure?

b. Perform the corresponding calculation for the states depicted in part (b) of the figure.

(a) $$0.15 = p_1 = \frac{e^{-\beta\varepsilon_1}}{q} = \frac{e^{-\beta(300.\,cm^{-1})}}{1+e^{-\beta(300.\,cm^{-1})} + e^{-\beta(600.\,cm^{-1})}}$$

$$0.15 + 0.15(e^{-\beta(300.\,cm^{-1})}) + 0.15(e^{-\beta(600.\,cm^{-1})}) = e^{-\beta(300.\,cm^{-1})}$$

$$0.15 - 0.85(e^{-\beta(300.\,cm^{-1})}) + 0.15(e^{-\beta(600.\,cm^{-1})}) = 0$$

The last expression is a quadratic equation with $x = \exp(-\beta(300.\,cm^{-1}))$. This equation has two roots equal to 0.183 and 5.48. Only the 0.183 root will provide temperature greater than zero, therefore:

$$0.183 = e^{-\beta(300.\,cm^{-1})}$$

$$1.70 = \frac{300.\,cm^{-1}}{(0.695\,cm^{-1}\,K^{-1})(T)}$$

$$T = 250\,K$$

(b) $$0.15 = p_1 = \frac{2e^{-\beta\varepsilon_1}}{q} = \frac{2e^{-\beta(300.\,cm^{-1})}}{1 + 2e^{-\beta(300.\,cm^{-1})} + e^{-\beta(600.\,cm^{-1})}}$$

$$0.15 + 0.30(e^{-\beta(300.\,cm^{-1})}) + 0.15(e^{-\beta(600.\,cm^{-1})}) = 2e^{-\beta(300.\,cm^{-1})}$$

$$0.15 - 1.70(e^{-\beta(300.\,cm^{-1})}) + 0.15(e^{-\beta(600.\,cm^{-1})}) = 0$$

The last expression is a quadratic equation with $x = \exp(-\beta(300.\,cm^{-1}))$. This equation has two roots equal to 0.0889 and 11.2. Only the 0.0889 root will provide temperature greater than zero, therefore:

$$0.0899 = e^{-\beta(300.\,cm^{-1})}$$

$$2.42 = \frac{300.\,cm^{-1}}{(0.695\,cm^{-1}\,K^{-1})(T)}$$

$$T = 180\,K$$

P30.12   A set of 13 particles occupies states with energies of 0, 100., and $200.\,\text{cm}^{-1}$. Calculate the total energy and number of microstates for the following configurations of energy:

a. $a_0 = 8$, $a_1 = 5$, and $a_2 = 0$

b. $a_0 = 9$, $a_1 = 3$, and $a_2 = 1$

c. $a_0 = 10$, $a_1 = 1$, and $a_2 = 2$

Do any of these configurations correspond to the Boltzmann distribution?

The total energy is equal to the sum of energy associated with each level times the number of particles in that level. For the occupation numbers in configuration (a):

$$E = \sum_n \varepsilon_n a_n = \varepsilon_0 a_0 + \varepsilon_1 a_1 + \varepsilon_2 a_2$$

$$= (0\,\text{cm}^{-1})(8) + (100.\,\text{cm}^{-1})(5) + (200.\,\text{cm}^{-1})(0) = 500.\,\text{cm}^{-1}$$

Repeating the calculation for the occupation numbers in (b) and (c) yields the same energy of $500.\,\text{cm}^{-1}$. The number of microstates associated with each distribution is given by the weight:

$$W_a = \frac{N!}{\prod_n a_n!} = \frac{N!}{a_0!\,a_1!\,a_2!} = \frac{13!}{(8!)(5!)(0!)} = 1287$$

$$W_b = \frac{13!}{(9!)(3!)(1!)} = 2860$$

$$W_c = \frac{13!}{(10!)(1!)(2!)} = 858$$

The ratio of any two occupation numbers for a set of non-degenerate energy levels is given by:

$$\frac{a_i}{a_j} = e^{-\beta(\varepsilon_i - \varepsilon_j)} = e^{-\left(\frac{\varepsilon_i - \varepsilon_j}{k}\right)\frac{1}{T}}$$

The above expression suggests that the ratio of occupation numbers can be used to determine the temperature. For set (b), comparing the occupation numbers for level 2 and level 0 results in:

$$\frac{a_2}{a_0} = e^{-\left(\frac{\varepsilon_2 - \varepsilon_0}{k}\right)\frac{1}{T}}$$

$$\frac{1}{9} = e^{-\left(\frac{200.\,\text{cm}^{-1} - 0\,\text{cm}^{-1}}{0.695\,\text{cm}^{-1}\,\text{K}^{-1}}\right)\frac{1}{T}}$$

$$T = 131\,\text{K}$$

Repeating the same calculation for level 1 and level 0:

$$\frac{a_1}{a_0} = e^{-\left(\frac{\varepsilon_2 - \varepsilon_0}{k}\right)\frac{1}{T}}$$

$$\frac{3}{9} = e^{-\left(\frac{100.\,\text{cm}^{-1} - 0\,\text{cm}^{-1}}{0.695\,\text{cm}^{-1}\,\text{K}^{-1}}\right)\frac{1}{T}}$$

$$T = 131\,\text{K}$$

The distribution of energy in (b) is in accord with the Boltzmann distribution. Performing a similar calculation for (a) and (c) will demonstrate that the temperatures are not equivalent, that these are not in accord with the Boltzmann distribution.

P30.14 For two non-degenerate energy levels separated by an amount of energy $\varepsilon/k = 500.\,\text{K}$, at what temperature will the population in the higher-energy state be $1/2$ that of the lower-energy state? What temperature is required to make the populations equal?

　　a. The populations are directly related to the probability of occupying the energy levels, and the ratio of energy-level probabilities is related to the energy difference between these levels as follows:

$$\frac{p_1}{p_0} = \frac{1}{2} = e^{-\left(\frac{\varepsilon_1 - \varepsilon_0}{kT}\right)} = e^{-\left(\frac{500.\,\text{K}}{T}\right)}$$

$$\ln 2 = \frac{500.\,\text{K}}{T}$$

$$T = 721\,\text{K}$$

　　b. For equal energy-level probabilities, the ratio of probabilities will equal 1. This is only achieved when $T = \infty$:

$$\frac{p_1}{p_0} = 1 = e^{-\left(\frac{\varepsilon_1 - \varepsilon_0}{kT}\right)} = e^{-\left(\frac{500.\,\text{K}}{T}\right)}$$

$$\ln 1 = 0 = \frac{500\,\text{K}}{T}$$

$$T = \infty$$

P30.16 Consider a molecule having three energy levels as follows:

States	Energy (cm^{-1})	Degeneracy
1	0	1
2	500.	3
3	1500.	5

What is the value of the partition function when $T = 300.$ and $3000.\,\text{K}$?

$$q = \sum_n g_n e^{-\beta \varepsilon_n} = \sum_n g_n e^{-\varepsilon_n/kT}$$

$$= 1 + 3e^{-(6.626 \times 10^{-34}\,\text{J s})(2.998 \times 10^{10}\,\text{cm s}^{-1})(500.\,\text{cm}^{-1})/(1.38 \times 10^{-23}\,\text{J K}^{-1})(300.\,\text{K})}$$

$$+ 5e^{-(6.626 \times 10^{-34}\,\text{J s})(2.998 \times 10^{10}\,\text{cm s}^{-1})(1500.\,\text{cm}^{-1})/(1.38 \times 10^{-23}\,\text{J K}^{-1})(300.\,\text{K})}$$

$$= 1 + 0.272 + 0.004 = 1.28$$

$$q = \sum_n g_n e^{-\beta \varepsilon_n} = \sum_n g_n e^{-\varepsilon_n/kT}$$

$$= 1 + 3e^{-(6.626 \times 10^{-34}\,\text{J s})(2.998 \times 10^{10}\,\text{cm s}^{-1})(500.\,\text{cm}^{-1})/(1.38 \times 10^{-23}\,\text{J K}^{-1})(3000.\,\text{K})}$$

$$+ 5e^{-(6.626 \times 10^{-34}\,\text{J s})(2.998 \times 10^{10}\,\text{cm s}^{-1})(1500.\,\text{cm}^{-1})/(1.38 \times 10^{-23}\,\text{J K}^{-1})(3000.\,\text{K})}$$

$$= 1 + 2.36 + 2.44 = 5.80$$

Notice that the value of the partition function increased with temperature, consistent with an increase in the population of higher-energy states as the temperature increases.

P30.19 The ^{13}C nucleus is a spin $1/2$ particle as is a proton. However, the energy splitting for a given field strength is roughly $1/4$ of that for a proton. Using a 1.45-T magnet as in Example Problem 30.6, what is the ratio of populations in the excited and ground spin states for ^{13}C at 298 K?

Using the information provided in the example problem, the separation in energy is given by:

$$\Delta E = \frac{1}{4}(2.82 \times 10^{-26}\,\text{J T}^{-1})B = \frac{1}{4}(2.82 \times 10^{-26}\,\text{J T}^{-1})(1.45\,\text{T}) = 1.02 \times 10^{-26}\,\text{J}$$

Using this separation in energy, the ratio in spin-state occupation numbers is:

$$\frac{a_+}{a_-} = e^{-\left(\frac{\varepsilon_+ - \varepsilon_-}{kT}\right)} = e^{-\left(\frac{\Delta E}{kT}\right)}$$

$$\frac{a_+}{a_-} = e^{-\left(\frac{1.02\times10^{-26}\ \text{J}}{(1.38\times10^{-23}\ \text{J K}^{-1})(298\ \text{K})}\right)}$$

$$\frac{a_+}{a_-} = 0.999998$$

P30.22   The vibrational frequency of I_2 is $208\ \text{cm}^{-1}$. At what temperature will the population in the first excited state be half that of the ground state?

$$\frac{a_1}{a_0} = \frac{1}{2} = e^{-\beta(\varepsilon_1 - \varepsilon_0)} = e^{-\beta(208\ \text{cm}^{-1})}$$

$$0.5 = e^{\left(\frac{-208\ \text{cm}^{-1}}{(0.695\ \text{cm}^{-1}\ \text{K})(T)}\right)}$$

$$0.693 = \frac{208\ \text{cm}^{-1}}{(0.695\ \text{cm}^{-1}\ \text{K})(T)}$$

$$T = 432\ \text{K}$$

P30.24   Determine the partition function for the vibrational degrees of freedom of $Cl_2(\tilde{v} = 525\ \text{cm}^{-1})$ and calculate the probability of occupying the first excited vibrational level at 300 and 1000 k. Determine the temperature at which identical probabilities will be observed for $F_2(\tilde{v} = 917\ \text{cm}^{-1})$.

For Cl_2 at 300. K:

$$q = \frac{1}{1 - e^{-\beta hc\tilde{v}}} = \frac{1}{1 - e^{\left(\frac{-(6.626\times10^{-34}\ \text{J s})(3.00\times10^{10}\ \text{cm s}^{-1})(525\ \text{cm}^{-1})}{(1.38\times10^{-23}\ \text{J K}^{-1})(300.\ \text{K})}\right)}} = 1.088$$

$$p_1 = \frac{e^{-\beta\varepsilon_1}}{q} = \frac{e^{-\beta hc\tilde{v}}}{q} = \frac{e^{\left(\frac{-(6.626\times10^{-34}\ \text{J s})(3.00\times10^{10}\ \text{cm s}^{-1})(525\ \text{cm}^{-1})}{(1.38\times10^{-23}\ \text{J K}^{-1})(300.\ \text{K})}\right)}}{1.088} = 0.074$$

For Cl_2 at 1000. K:

$$q = \frac{1}{1 - e^{-\beta hc\tilde{v}}} = \frac{1}{1 - e^{\left(\frac{-(6.626\times10^{-34}\ \text{J s})(3.00\times10^{10}\ \text{cm s}^{-1})(525\ \text{cm}^{-1})}{(1.38\times10^{-23}\ \text{J K}^{-1})(1000\ \text{K})}\right)}} = 1.887$$

$$p_1 = \frac{e^{-\beta\varepsilon_1}}{q} = \frac{e^{-\beta hc\tilde{v}}}{q} = \frac{e^{\left(\frac{-(6.626\times10^{-34}\ \text{J s})(3.00\times10^{10}\ \text{cm s}^{-1})(525\ \text{cm}^{-1})}{(1.38\times10^{-23}\ \text{J K}^{-1})(1000\ \text{K})}\right)}}{1.887} = 0.249$$

To determine the temperatures at which F_2 has equivalent populations with Cl_2 in the first vibrational excited state, we first reduce the expression for p_1 as follows:

$$p_1 = \frac{e^{-\beta\varepsilon_1}}{q} = \frac{e^{-\beta hc\tilde{v}}}{(1/1 - e^{-\beta hc\tilde{v}})} = e^{-\beta hc\tilde{v}}(1 - e^{-\beta hc\tilde{v}})$$

Substituting in the values for F_2 and solving as a quadratic equation:

$$p_1 = 0.074 = e^{-\beta hc(917\,\text{cm}^{-1})} - e^{-2\beta hc(917\,\text{cm}^{-1})} = e^{-1319\,\text{K}/T} - e^{-2(1319\,\text{K})/T}$$

$$0 = -0.074 + e^{-1319\,\text{K}/T} - e^{-2(1319\,\text{K})/T}$$

$$T = 523\,\text{K}$$

Repeating this calculation to determine at which temperature p_1 will be equal to the Cl_2 case at 1000. K:

$$p_1 = 0.249 = e^{-\beta hc(917\,\text{cm}^{-1})} - e^{-2\beta hc(917\,\text{cm}^{-1})} = e^{-1319\,\text{K}/T} - e^{-2(1319\,\text{K})/T}$$

$$0 = -0.249 + e^{-1319\,\text{K}/T} - e^{-2(1319\,\text{K})/T}$$

$$T = 1740\,\text{K}$$

P30.26 Calculate the partition function at 298 K for the vibrational energetic degree of freedom for $^1H_2$. where $\tilde{v} = 4401\,\text{cm}^{-1}$. Perform this same calculation for D_2 (or $^2H_2$) assuming the force constant for the bond is the same as in $^1H_2$.

Referring to Example Problem 30.5:

$$q = \frac{1}{1 - e^{-\beta hc\tilde{v}}} = \frac{1}{1 - e^{-hc\tilde{v}/kT}}$$

$$= \frac{1}{1 - e^{-(6.626\times10^{-34}\,\text{J s})(2.998\times10^{10}\,\text{cm s}^{-1})(4401\,\text{cm}^{-1})/(1.38\times10^{-23}\,\text{J K}^{-1})(298\,\text{K})}}$$

$$= 1$$

Using the harmonic oscillator formalism, the vibrational frequency is dependent on the force constant for the bond (κ) and reduced mass (μ) as follows:

$$\frac{\tilde{v}_{D_2}}{\tilde{v}_{H_2}} = \sqrt{\frac{\kappa/\mu_{D_2}}{\kappa/\mu_{H_2}}} = \sqrt{\frac{\mu_{H_2}}{\mu_{D_2}}} = \sqrt{\frac{\dfrac{(1.008\,\text{g mol}^{-1})^2}{2(1.008\,\text{g mol}^{-1})}}{\dfrac{(2.016\,\text{g mol}^{-1})^2}{2(2.016\,\text{g mol}^{-1})}}} = \frac{1}{\sqrt{2}}$$

$$\tilde{v}_{D_2} = \frac{\tilde{v}_{H_2}}{\sqrt{2}} = \frac{4401\,\text{cm}^{-1}}{\sqrt{2}} = 3112\,\text{cm}^{-1}$$

Using this vibrational frequency for D_2, the partition function is:

$$q = \frac{1}{1 - e^{-\beta hc\tilde{v}}} = \frac{1}{1 - e^{-hc\tilde{v}/kT}}$$

$$= \frac{1}{1 - e^{-(6.626\times10^{-34}\,\text{J s})(2.998\times10^{10}\,\text{cm s}^{-1})(3112\,\text{cm}^{-1})/(1.38\times10^{-23}\,\text{J K}^{-1})(298\,\text{K})}}$$

$$= 1$$

Even with the reduction in vibrational energy level spacings for D_2, the spacings are still significantly greater than the amount of thermal energy available (kT). Therefore, only the ground state is populated to a significant extent as evidenced by the partition function having a value of 1.

P30.29 The lowest two electronic energy levels of the molecule NO are illustrated in the text. Determine the probability of occupying one of the higher-energy states at 100., 500., and 2000. K.

Both the lower- and higher-energy states are two-fold degenerate, with an energy spacing of $121.1\,\text{cm}^{-1}$. At 100. K, the partition function is:

$$q = \sum_n g_n e^{-\beta \varepsilon_n} = 2 + 2e^{-\beta(121.1\,\text{cm}^{-1})} = 2 + 2e^{-121.1\,\text{cm}^{-1}/(0.695\,\text{cm}^{-1}\,\text{K}^{-1})(100.\,\text{K})} = 2.35$$

With the partition function evaluated, the probability of occupying the excited energy level is readily determined:

$$p_1 = \frac{g_1 e^{-\beta\varepsilon_1}}{q} = \frac{2e^{-121.1\,\text{cm}^{-1}/(0.695\,\text{cm}^{-1}\,\text{K}^{-1})(100.\,\text{K})}}{2.35} = 0.149$$

At 500. K:

$$q = 2 + 2e^{-121.1\,\text{cm}^{-1}/(0.695\,\text{cm}^{-1}\,\text{K}^{-1})(500.\,\text{K})} = 3.41$$

$$p_1 = \frac{g_1 e^{-\beta\varepsilon_1}}{q} = \frac{2e^{-121.1\,\text{cm}^{-1}/(0.695\,\text{cm}^{-1}\,\text{K}^{-1})(500.\,\text{K})}}{3.41} = 0.414$$

Finally, at 2000. K:

$$q = 2 + 2e^{-121.1\,\text{cm}^{-1}/(0.695\,\text{cm}^{-1}\,\text{K}^{-1})(2000.\,\text{K})} = 3.83$$

$$p_1 = \frac{g_1 e^{-\beta\varepsilon_1}}{q} = \frac{2e^{-121.1\,\text{cm}^{-1}/(0.695\,\text{cm}^{-1}\,\text{K}^{-1})(2000.\,\text{K})}}{3.83} = 0.479$$

Since there are two states per energy level, the probability of occupying an individual excited state is 1/2 of the above probabilities.

31 Ensemble and Molecular Partition Functions

Numerical Problems

P31.2 Evaluate the translational partition function for $^{35}Cl_2$ confined to a volume of 1.00 L at 298 K. How does your answer change if the gas is $^{37}Cl_2$? (*Hint:* Can you reduce the ratio of translational partition functions to an expression involving mass only?)

The translational partition function for $^{35}Cl_2$ is calculated as follows:

$$q_T(^{35}Cl_2) = \frac{V}{\Lambda^3}$$

$$\Lambda = \left(\frac{h^2}{2\pi mkT}\right)^{1/2} = \left(\frac{(6.626 \times 10^{-34}\ J\ s)^2}{2\pi\left(\frac{0.0700\ kg\ mol^{-1}}{N_A}\right)(1.38 \times 10^{-23}\ J\ K^{-1})(298\ K)}\right)^{1/2} = 1.21 \times 10^{-11}\ m$$

$$q_T(^{35}Cl_2) = \frac{V}{(1.21 \times 10^{-11}\ m)^3} = \frac{(1000\ cm^3)(10^{-6}\ m^3\ cm^{-3})}{(1.21 \times 10^{-11}\ m)^3} = 5.66 \times 10^{29}$$

Taking the ratio of translational partition functions for $^{35}Cl_2$ and $^{37}Cl_2$ and cancelling out common terms yields:

$$\frac{q_T(^{37}Cl_2)}{q_T(^{35}Cl_2)} = \frac{\dfrac{V}{\Lambda^3(^{37}Cl_2)}}{\dfrac{V}{\Lambda^3(^{35}Cl_2)}} = \frac{\Lambda^3(^{35}Cl_2)}{\Lambda^3(^{37}Cl_2)} = \left(\frac{m(^{37}Cl_2)}{m(^{35}Cl_2)}\right)^{3/2} = \left(\frac{74.0\ g\ mol^{-1}}{70.0\ g\ mol^{-1}}\right)^{3/2} = 1.087$$

$$q_T(^{37}Cl_2) = 1.087(q_T(^{35}Cl_2))$$

P31.4 Evaluate the translational partition function for Ar confined to a volume of 1000. cm^3 at 298 K. At what temperature will the translational partition function of Ne be identical to that of Ar at 298 K confined to the same volume?

$$q_T(Ar) = \frac{V}{\Lambda^3}$$

$$\Lambda = \left(\frac{h^2}{2\pi mkT}\right)^{1/2} = \left(\frac{(6.626 \times 10^{-34}\ J\ s)^2}{2\pi\left(\frac{0.0399\ kg\ mol^{-1}}{N_A}\right)(1.38 \times 10^{-23}\ J\ K^{-1})(298\ K)}\right)^{1/2} = 1.60 \times 10^{-11}\ m$$

$$q_T(Ar) = \frac{V}{(1.60 \times 10^{-11}\ m)^3} = \frac{(1000.\ cm^3)(10^{-6}\ m^3\ cm^{-3})}{(1.60 \times 10^{-11}\ m)^3} = 2.44 \times 10^{29}$$

If the gases are confined to the same volume, then the partition functions will be equal when the thermal wavelengths are equal:

$$\Lambda(Ne) = \Lambda(Ar) = 1.60 \times 10^{-11} \text{ m}$$

$$\left(\frac{h^2}{2\pi mkT}\right)^{1/2} = 1.60 \times 10^{-11} \text{ m}$$

$$T = \frac{h^2}{2\pi mk(1.60 \times 10^{-11} \text{ m})^2} = \frac{(6.626 \times 10^{-34} \text{ J s})^2}{2\pi\left(\dfrac{0.0202 \text{ kg mol}^{-1}}{N_A}\right)(1.38 \times 10^{-23} \text{ J K}^{-1})(1.60 \times 10^{-11} \text{ m})^2}$$

$$T = 590. \text{ K}$$

P31.8   For N_2 at 77.3 K, 1.00 atm, in a 1.00-cm^3 container, calculate the translational partition function and the ratio of this partition function to the number of N_2 molecules present under these conditions.

$$q_T(N_2) = \frac{V}{\Lambda^3}$$

$$\Lambda = \left(\frac{h^2}{2\pi mkT}\right)^{1/2} = \left(\frac{(6.626 \times 10^{-34} \text{ J s})^2}{2\pi\left(\dfrac{0.028 \text{ kg mol}^{-1}}{N_A}\right)(1.38 \times 10^{-23} \text{ J K}^{-1})(77.3 \text{ K})}\right)^{1/2} = 3.75 \times 10^{-11} \text{ m}$$

$$q_T(N_2) = \frac{V}{(3.75 \times 10^{-11} \text{ m})^3} = \frac{(1 \text{ cm}^3)(10^{-6} \text{ m}^3 \text{ cm}^{-3})}{(3.75 \times 10^{-11} \text{ m})^3} = 1.89 \times 10^{25}$$

Next, the number of molecules (N) present at this temperature is determined using the ideal gas law:

$$n = \frac{PV}{RT} = \frac{(1 \text{ atm})(1.00 \times 10^{-3} \text{ L})}{(0.0821 \text{ L atm mol}^{-1} \text{ K}^{-1})(77.3 \text{ K})} = 1.58 \times 10^{-4} \text{ mol}$$

$$N = n \times N_A = 9.49 \times 10^{19} \text{ molecules}$$

With N, the ratio is readily determined:

$$\frac{q_T}{N} = \frac{1.89 \times 10^{25}}{9.49 \times 10^{19}} = 1.99 \times 10^5$$

P31.12   Consider *para*-$H_2(B = 60.853 \text{ cm}^{-1})$ for which only even J levels are available. Evaluate the rotational partition function for this species at 50. K. Perform this same calculation for HD ($B = 45.655 \text{ cm}^{-1}$).

For *para*-H_2, only even J levels are allowed; therefore, the rotational partition function is:

$$q_R = \sum_{J=0,2,4,6,\dots} (2J + 1)e^{-\beta hcBJ(J+1)} = 1 + 5e^{-\frac{(6.626\times10^{-34} \text{ J s})(3.00\times10^{10} \text{ cm s}^{-1})(60.853 \text{ cm}^{-1})(6)}{(1.38\times10^{-23} \text{ J K}^{-1})(50. \text{ K})}}$$

$$+ 9e^{-\frac{(6.626\times10^{-34} \text{ J s})(3.00\times10^{10} \text{ cm s}^{-1})(60.853 \text{ cm}^{-1})(20)}{(1.38\times10^{-23} \text{ J K}^{-1})(50. \text{ K})}} + \cdots$$

$$= 1 + 1.36 \times 10^{-4} + \cdots$$

$$\approx 1.00$$

Performing this same calculation for HD where both even and odd J states are allowed:

$$q_R = \sum_{J=0,1,2,3,\dots} (2J + 1)e^{-\beta hcBJ(J+1)} = 1 + 3e^{-\frac{(6.626\times10^{-34} \text{ J s})(3.00\times10^{10} \text{ cm s}^{-1})(45.655 \text{ cm}^{-1})(2)}{(1.38\times10^{-23} \text{ J K}^{-1})(50. \text{ K})}}$$

$$+ 5e^{-\frac{(6.626\times10^{-34} \text{ J s})(3.00\times10^{10} \text{ cm s}^{-1})(45.655 \text{ cm}^{-1})(6)}{(1.38\times10^{-23} \text{ J K}^{-1})(50. \text{ K})}} + \cdots$$

$$= 1 + 0.217 + 1.88 \times 10^{-3} + \cdots$$

$$\approx 1.22$$

P31.13 Calculate the rotational partition function for the interhalogen compound $F^{35}Cl$ ($B = 0.516 \text{ cm}^{-1}$) at 298 K.

$$q_R = \frac{1}{\sigma \beta B} = \frac{kT}{B} = \frac{(0.695 \text{ cm}^{-1} \text{ K}^{-1})(298 \text{ K})}{(0.516 \text{ cm}^{-1})} = 401$$

P31.16 Calculate the rotational partition function for SO_2 at 298 K, where $B_A = 2.03 \text{ cm}^{-1}$, $B_B = 0.344 \text{ cm}^{-1}$, and $B_C = 0.293 \text{ cm}^{-1}$.

$$q_R = \frac{\sqrt{\pi}}{\sigma}\left(\frac{1}{\beta B_A}\right)^{1/2}\left(\frac{1}{\beta B_B}\right)^{1/2}\left(\frac{1}{\beta B_C}\right)^{1/2}$$

$$= \frac{\sqrt{\pi}}{2}\left(\frac{(0.695 \text{ cm}^{-1})(298 \text{ K})}{2.03 \text{ cm}^{-1}}\right)^{1/2}\left(\frac{(0.695 \text{ cm}^{-1})(298 \text{ K})}{0.344 \text{ cm}^{-1}}\right)^{1/2}\left(\frac{(0.695 \text{ cm}^{-1})(298 \text{ K})}{0.293 \text{ cm}^{-1}}\right)^{1/2} \cong 5840$$

P31.19 What transition in the rotational spectrum of IF ($B = 0.280 \text{ cm}^{-1}$) is expected to be the most intense at 298 K?

This problem can be solved using the expression relating temperature to J for the maximum transition in the rotational spectrum and solving for J:

$$T = \frac{(2J+1)^2 hcB}{2k} \rightarrow (2J+1)^2 = \frac{2kT}{hcB}$$

$$(2J+1)^2 = \frac{2(1.38 \times 10^{-23} \text{ J K}^{-1})(298 \text{ K})}{(6.626 \times 10^{-34} \text{ J s})(3.00 \times 10^{10} \text{ cm s}^{-1})(0.280 \text{ cm}^{-1})}$$

$$(2J+1)^2 = 1479$$

$$2J+1 = 38.5$$

$$J = 18.5$$

Rounding this answer for J up to 19 results in transitions involving $J = 19$ to $J = 20$ as being predicted to be most intense.

P31.23 a. Calculate the percent population of the first 10 rotational energy levels for HBr ($B = 8.46 \text{ cm}^{-1}$) at 298 K.

b. Repeat this calculation for HF assuming that the bond length of this molecule is identical to that of HBr.

(a) Since $T \gg \Theta_R$, the high-temperature limit is valid. In this limit, the probability of occupying a specific rotational state (p_J) is:

$$p_J = \frac{(2J+1)e^{-\beta hcBJ(J+1)}}{q} = \frac{(2J+1)e^{-\beta hcBJ(J+1)}}{\left(\dfrac{1}{\sigma \beta hcB}\right)}$$

Evaluating the above expression for $J = 0$:

$$p_J = \frac{(2J+1)e^{-\beta hcBJ(J+1)}}{\left(\dfrac{1}{\sigma \beta hcB}\right)} = \frac{1}{\left(\dfrac{1}{\sigma \beta hcB}\right)}$$

$$= \frac{\sigma hcB}{kT} = \frac{(1)(6.626 \times 10^{-34} \text{ J s})(3.00 \times 10^{10} \text{ cm s}^{-1})(8.46 \text{ cm}^{-1})}{(1.38 \times 10^{-23} \text{ J K}^{-1})(298 \text{ K})} = 0.0409$$

Performing similar calculations for $J = 1$ to 9:

J	p_J	J	p_J
0	0.0409	5	0.132
1	0.113	6	0.0955
2	0.160	7	0.0622
3	0.175	8	0.0367
4	0.167	9	0.0196

(b) The rotational constant of HF must be determined before the corresponding level probabilities can be evaluated. The ratio of rotational constants for HBr versus HF yields:

$$\frac{B_{HBr}}{B_{HF}} = \frac{\left(\dfrac{h}{8\pi^2 c I_{HBr}}\right)}{\left(\dfrac{h}{8\pi^2 c I_{HF}}\right)} = \frac{I_{HF}}{I_{HBr}} = \frac{\mu_{HF} r^2}{\mu_{HBr} r^2} = \frac{\mu_{HF}}{\mu_{HBr}} = \frac{\dfrac{m_H m_F}{m_H + m_F}}{\dfrac{m_H m_{Br}}{m_H + m_{Br}}} = 0.962$$

$$B_{HF} = \frac{B_{HBr}}{0.962} = 8.79 \text{ cm}^{-1}$$

With this rotational constant, the p_J values for $J = 0$ to 9 are:

J	p_J	J	p_J
0	0.0425	5	0.131
1	0.117	6	0.0927
2	0.165	7	0.0590
3	0.179	8	0.0338
4	0.163	9	0.0176

P31.25 When ^{4}He is cooled below 2.17 K it becomes a "superfluid" with unique properties such as a viscosity approaching zero. One way to learn about the superfluid environment is to measure the rotational–vibrational spectrum of molecules embedded in the fluid. For example, the spectrum of OCS in a low-temperature ^{4}He droplet has been reported (*Journal of Chemical Physics* 112 [2000]: 4485). For OC^{32}S the authors measured a rotational constant of 0.203 cm^{-1} and found that the intensity of the $J = 0$ to 1 transition was roughly 1.35 times greater than that of the $J = 1$ to 2 transition. Using this information, provide a rough estimate of the temperature of the droplet.

Since the transition intensities are proportional to the occupation numbers for the corresponding initial J states in the transition:

$$\frac{a_{J=0}}{a_{J=1}} = 1.35 = \frac{[(2J+1)e^{-hc\beta BJ(J+1)}]_{J=0}}{[(2J+1)e^{-hc\beta BJ(J+1)}]_{J=1}} = \frac{1}{3e^{-2hc\beta B}}$$

$$e^{-2hc\beta B} = \frac{1}{4.05} = 0.247$$

$$\frac{2hcB}{kT} = 1.399$$

$$T = \frac{2hcB}{(1.399) \text{ K}} = \frac{2(6.626 \times 10^{-34} \text{ J s})(3.00 \times 10^{10} \text{ cm s}^{-1})(0.203 \text{ cm}^{-1})}{(1.399)(1.38 \times 10^{-23} \text{ J K}^{-1})} = 0.418 \text{ K}$$

P31.27 Determine the rotational partition function for I^{35}Cl ($B = 0.114$ cm^{-1}) at 298 K.

$$q_R = \frac{1}{\sigma\beta hcB} = \frac{kT}{\sigma hcB}$$

$$= \frac{(1.38 \times 10^{-23} \text{ J K}^{-1})(298 \text{ K})}{(1)(6.626 \times 10^{-34} \text{ J s})(3.00 \times 10^{10} \text{ cm s}^{-1})(0.114 \text{ cm}^{-1})}$$

$$= 1820$$

P31.29   Evaluate the vibrational partition function for H_2O at 2000. K, where the vibrational frequencies are 1615, 3694, and $3802 \, cm^{-1}$.

The total vibrational partition function is the product of partition functions for each vibrational degree of freedom:

$$q_{V, total} = (q_{V,1})(q_{V,2})(q_{V,3})$$

$$= \left(\frac{1}{1 - e^{\beta hc\tilde{v}_1}}\right)\left(\frac{1}{1 - e^{\beta hc\tilde{v}_2}}\right)\left(\frac{1}{1 - e^{\beta hc\tilde{v}_3}}\right)$$

$$= \left(\frac{1}{1 - e^{\frac{(6.626 \times 10^{-34} \, J \, s)(3.00 \times 10^{10} \, cm \, s^{-1})(1615 \, cm^{-1})}{(1.38 \times 10^{-23} \, J \, K^{-1})(2000. \, K)}}}\right)\left(\frac{1}{1 - e^{\frac{(6.626 \times 10^{-34} \, J \, s)(3.00 \times 10^{10} \, cm \, s^{-1})(3694 \, cm^{-1})}{(1.38 \times 10^{-23} \, J \, K^{-1})(2000. \, K)}}}\right)$$

$$\times \left(\frac{1}{1 - e^{\frac{(6.626 \times 10^{-34} \, J \, s)(3.00 \times 10^{10} \, cm \, s^{-1})(3802 \, cm^{-1})}{(1.38 \times 10^{-23} \, J \, K^{-1})(2000. \, K)}}}\right)$$

$$= 1.67$$

P31.31   Evaluate the vibrational partition function for NH_3 at 1000. K for which the vibrational frequencies are 950., 1627.5 (doubly degenerate), 3335, and $3414 \, cm^{-1}$ (doubly degenerate). Are there any modes that you can disregard in this calculation? Why or why not?

The total vibrational partition function is the product of partition functions for each vibrational degree of freedom, with the partition function for the mode with degeneracy raised to the power equal to the degeneracy:

$$q_{V, total} = (q_{V,1})(q_{V,2})^2(q_{V,3})(q_{V,4})^2$$

$$= \left(\frac{1}{1 - e^{\beta hc\tilde{v}_1}}\right)\left(\frac{1}{1 - e^{\beta hc\tilde{v}_2}}\right)^2\left(\frac{1}{1 - e^{\beta hc\tilde{v}_3}}\right)\left(\frac{1}{1 - e^{\beta hc\tilde{v}_4}}\right)^2$$

$$= \left(\frac{1}{1 - e^{\frac{(6.626 \times 10^{-34} \, J \, s)(3.00 \times 10^{10} \, cm \, s^{-1})(950. \, cm^{-1})}{(1.38 \times 10^{-23} \, J \, K^{-1})(1000. \, K)}}}\right)\left(\frac{1}{1 - e^{\frac{(6.626 \times 10^{-34} \, J \, s)(3.00 \times 10^{10} \, cm \, s^{-1})(1627.5 \, cm^{-1})}{(1.38 \times 10^{-23} \, J \, K^{-1})(1000. \, K)}}}\right)^2$$

$$\times \left(\frac{1}{1 - e^{\frac{(6.626 \times 10^{-34} \, J \, s)(3.00 \times 10^{10} \, cm \, s^{-1})(3335 \, cm^{-1})}{(1.38 \times 10^{-23} \, J \, K^{-1})(1000. \, K)}}}\right)\left(\frac{1}{1 - e^{\frac{(6.626 \times 10^{-34} \, J \, s)(3.00 \times 10^{10} \, cm \, s^{-1})(3414 \, cm^{-1})}{(1.38 \times 10^{-23} \, J \, K^{-1})(1000. \, K)}}}\right)^2$$

$$= (1.34)(1.11)^2(1.01)(1.01)^2$$

$$= 1.69$$

Notice that the two highest-frequency vibrational degrees of freedom have partition functions that are near unity; therefore, their contribution to the total partition function is modest and could be ignored to a reasonable approximation in evaluating the total partition function.

P31.34   Isotopic substitution is employed to isolate features in a vibrational spectrum. For example, the $C{=}O$ stretch of individual carbonyl groups in the backbone of a polypeptide can be studied by substituting $^{13}C^{18}O$ for $^{12}C^{16}O$.

a.   From quantum mechanics the vibrational frequency of a diatomic molecules depends on the bond force constant (κ) and reduced mass (μ) as follows:

$$\tilde{v} = \sqrt{\frac{\kappa}{\mu}}$$

If the vibrational frequency of $^{12}C^{16}O$ is $1680 \, cm^{-1}$, what is the expected frequency for $^{13}C^{18}O$?

$$\frac{\tilde{v}_{13,18}}{\tilde{v}_{12,16}} = \sqrt{\frac{\kappa}{\mu_{13,18}}} \bigg/ \sqrt{\frac{\kappa}{\mu_{12,16}}} = \sqrt{\frac{\mu_{12,16}}{\mu_{13,18}}} = \sqrt{\frac{\dfrac{(12.0 \text{ g mol}^{-1})(16.0 \text{ g mol}^{-1})}{(12.0 \text{ g mol}^{-1}) + (16.0 \text{ g mol}^{-1})}}{\dfrac{(13.0 \text{ g mol}^{-1})(18.0 \text{ g mol}^{-1})}{(13.0 \text{ g mol}^{-1}) + (18.0 \text{ g mol}^{-1})}}} = 0.953$$

$$\tilde{v}_{13,18} = \tilde{v}_{12,16}(0.953) = 1680 \text{ cm}^{-1}(0.953) = 1601 \text{ cm}^{-1}$$

b.   Using the vibrational frequencies for $^{12}C^{16}O$ and $^{13}C^{18}O$, determine the value of the corresponding vibrational partition functions at 298 K. Does this isotopic substitution have a dramatic effect on q_V?

$$q_V(12,16) = \frac{1}{1 - e^{-\beta hc\tilde{v}_{12,16}}} = \frac{1}{1 - e^{-(6.626\times10^{-34} \text{ J s})(3.00\times10^{10} \text{ cm s}^{-1})(1680 \text{ cm}^{-1})/(1.38\times10^{-23} \text{ J K}^{-1})(298 \text{ K})}}$$

$$= \frac{1}{1 - e^{-8.12}} = 1.0003 \approx 1$$

$$q_V(13,18) = \frac{1}{1 - e^{-\beta hc\tilde{v}_{12,16}}} = \frac{1}{1 - e^{-(6.626\times10^{-34} \text{ J s})(3.00\times10^{10} \text{ cm s}^{-1})(1601 \text{ cm}^{-1})/(1.38\times10^{-23} \text{ J K}^{-1})(298 \text{ K})}}$$

$$= \frac{1}{1 - e^{-7.74}} = 1.0004 \approx 1$$

Both partition functions have values extremely close to 1; therefore, this isotopic substitution has little effect on the value of q_V.

P31.39   Consider a particle free to translate in one dimension. The classical Hamiltonian is $H = \dfrac{p^2}{2m}$.

a.   Determine $q_{classical}$ for this system. To what quantum system should you compare it in order to determine the equivalence of the classical and quantum statistical mechanical treatments?

b.   Derive $q_{classical}$ for a system with translational motion in three dimensions for which:

$$H = (p_x^2 + p_y^2 + p_z^2)/2m$$

(a)   The particle in a one-dimensional box model is the appropriate quantum mechanical model for comparison. Integrating

$$q_{class} = \frac{1}{h} \int_{-\infty}^{\infty} \int_0^L e^{\frac{-\beta p^2}{2m}} \, dx \, dp = \frac{2L}{h} \int_0^{\infty} e^{\frac{-\beta p^2}{2m}} \, dp = \frac{2L}{h}\left(\frac{1}{2}\sqrt{\frac{\pi}{\beta/2m}}\right) = L\left(\frac{h^2}{2\pi mkT}\right)^{-1/2} = \frac{L}{\Lambda}$$

(b) $q_{class} = \dfrac{1}{h^3} \int_{-\infty}^{\infty}\int_{-\infty}^{\infty}\int_{-\infty}^{\infty} \int_0^{L_z}\int_0^{L_y}\int_0^{L_x} e^{\frac{-\beta(p_x^2+p_y^2+p_z^2)}{2m}} \, dx \, dy \, dz \, dp_x \, dp_y \, dp_z$

$$= \frac{8(L_x L_y L_z)}{h^3} \int_0^{\infty}\int_0^{\infty}\int_0^{\infty} e^{\frac{-\beta(p_x^2+p_y^2+p_z^2)}{2m}} \, dp_x \, dp_y \, dp_z$$

$$= \frac{8V}{h^3}\left(\frac{1}{2}\sqrt{\frac{\pi}{\beta/2m}}\right)^3 = V\left(\frac{h^2}{2\pi mkT}\right)^{-3/2} = \frac{V}{\Lambda^3}$$

P31.43   a.   Evaluate the electronic partition function for atomic Si at 298 K given the following energy levels:

Level (n)	Energy (cm^{-1})	Degeneracy
0	0	1
1	77.1	3
2	223.2	5
3	6298	5

b. At what temperature will the $n = 3$ energy level contribute 0.100 to the electronic partition function?

(a) $$q_E = \sum_n g_n e^{-\beta \varepsilon_n} = 1e^{-0} + 3e^{-\beta(77.1\,\text{cm}^{-1})} + 5e^{-\beta(223.2\,\text{cm}^{-1})} + 5e^{-\beta(6298\,\text{cm}^{-1})}$$

$$= 1 + 3e^{-\frac{77.1\,\text{cm}^{-1}}{(0.695\,\text{cm}^{-1}\,\text{K}^{-1})(298\,\text{K})}} + 5e^{-\frac{223.2\,\text{cm}^{-1}}{(0.695\,\text{cm}^{-1}\,\text{K}^{-1})(298\,\text{K})}} + 3e^{-\frac{6298\,\text{cm}^{-1}}{(0.695\,\text{cm}^{-1}\,\text{K}^{-1})(298\,\text{K})}}$$

$$= 1 + 3(0.689) + 5(0.340) + 5(6.22 \times 10^{-14})$$

$$q_E = 4.77$$

(b) Focusing on the contribution to q_E from the $n = 3$ level:

$$0.1 = g_3 e^{-\beta \varepsilon_3} = 5e^{-\frac{6298\,\text{cm}^{-1}}{(0.695\,\text{cm}^{-1}\,\text{K}^{-1})(T)}}$$

$$3.91 = \frac{6298\,\text{cm}^{-1}}{(0.695\,\text{cm}^{-1}\,\text{K}^{-1})(T)}$$

$$T = \frac{6298\,\text{cm}^{-1}}{(0.695\,\text{cm}^{-1}\,\text{K}^{-1})(3.91)}$$

$$T = 2320\,\text{K}$$

P31.46 Determine the total molecular partition function for I_2, confined to a volume of $1000.\,\text{cm}^3$ at 298 K. Other information you will find useful: $B = 0.0374\,\text{cm}^{-1}$, $\tilde{\nu} = 208\,\text{cm}^{-1}$, and the ground electronic state is non degenerate.

Since q_{Total} is the product of partition functions for each energetic degree of freedom (translational, rotational, vibrational, and electronic), it is more straightforward to calculate the partition function for each of these degrees of freedom separately, and then take the product of these functions:

$$q_T = \frac{V}{\Lambda^3}$$

$$\Lambda = \left(\frac{h^2}{2\pi m k T}\right)^{1/2} = \left(\frac{(6.626 \times 10^{-34}\,\text{J s})^2}{2\pi \left(\frac{0.254\,\text{kg mol}^{-1}}{N_A}\right)(1.38 \times 10^{-23}\,\text{J K}^{-1})(298\,\text{K})}\right)^{1/2} = 6.35 \times 10^{-12}\,\text{m}$$

$$q_T = \frac{V}{(6.35 \times 10^{-12}\,\text{m})^3} = \frac{(1000.\,\text{cm}^3)(10^{-6}\,\text{m}^3\,\text{cm}^{-3})}{(6.35 \times 10^{-12}\,\text{m})^3} = 3.91 \times 10^{30}$$

$$q_R = \left(\frac{1}{\sigma \beta h c B}\right) = \left(\frac{(1.38 \times 10^{-23}\,\text{J K}^{-1})(298\,\text{K})}{(2)(6.626 \times 10^{-34}\,\text{J s})(3.00 \times 10^{10}\,\text{cm s}^{-1})(0.0374\,\text{cm}^{-1})}\right)$$

$$= 2.77 \times 10^3$$

$$q_V = \frac{1}{1 - e^{-\beta h c \tilde{\nu}}} = \frac{1}{1 - e^{-\frac{(6.626 \times 10^{-34}\,\text{J s})(3.00 \times 10^{10}\,\text{cm s}^{-1})(208\,\text{cm}^{-1})}{(1.38 \times 10^{-23}\,\text{J K}^{-1})(298\,\text{K})}}}$$

$$= 1.58$$

$$q_E = 1$$

$$q_{total} = q_T q_R q_V q_E = (3.91 \times 10^{30})(2.77 \times 10^3)(1.58)(1) = 1.71 \times 10^{34}$$

32 Statistical Thermodynamics

Numerical Problems

P32.3   Consider two separate molar ensembles of particles characterized by the energy-level diagram provided in the text. Derive expressions for the internal energy for each ensemble. At 298 K, which ensemble is expected to have the greatest internal energy?

$$U = -\left(\frac{\partial \ln Q}{\partial \beta}\right)_V = -N\left(\frac{\partial \ln q}{\partial \beta}\right)_V = \frac{-N}{q}\left(\frac{\partial q}{\partial \beta}\right)_V$$

$$q_A = \sum_n g_n e^{-\beta \varepsilon_n} = 1 + e^{-\beta(300.\,\mathrm{cm}^{-1})} + e^{-\beta(600.\,\mathrm{cm}^{-1})}$$

$$U_A = \frac{-N}{1 + e^{-\beta(300.\,\mathrm{cm}^{-1})} + e^{-\beta(600.\,\mathrm{cm}^{-1})}}\left(\frac{\partial}{\partial \beta}(1 + e^{-\beta(300.\,\mathrm{cm}^{-1})} + e^{-\beta(600.\,\mathrm{cm}^{-1})})\right)$$

$$= \frac{-N}{1 + e^{-\beta(300.\,\mathrm{cm}^{-1})} + e^{-\beta(600.\,\mathrm{cm}^{-1})}}(-((300.\,\mathrm{cm}^{-1})e^{-\beta(300\,\mathrm{cm}^{-1})} + (600.\,\mathrm{cm}^{-1})e^{-\beta(600.\,\mathrm{cm}^{-1})}))$$

$$= \frac{N((300.\,\mathrm{cm}^{-1})e^{-\beta(300.\,\mathrm{cm}^{-1})} + (600.\,\mathrm{cm}^{-1})e^{-\beta(600.\,\mathrm{cm}^{-1})})}{1 + e^{-\beta(300.\,\mathrm{cm}^{-1})} + e^{-\beta(600.\,\mathrm{cm}^{-1})}}$$

$$q_B = \sum_n g_n e^{-\beta \varepsilon_n} = 1 + 2e^{-\beta(300.\,\mathrm{cm}^{-1})} + e^{-\beta(600.\,\mathrm{cm}^{-1})}$$

$$U_B = \frac{-N}{1 + 2e^{-\beta(300.\,\mathrm{cm}^{-1})} + e^{-\beta(600.\,\mathrm{cm}^{-1})}}\left(\frac{\partial}{\partial \beta}(1 + 2e^{-\beta(300.\,\mathrm{cm}^{-1})} + e^{-\beta(600.\,\mathrm{cm}^{-1})})\right)$$

$$= \frac{-N}{1 + 2e^{-\beta(300.\,\mathrm{cm}^{-1})} + e^{-\beta(600.\,\mathrm{cm}^{-1})}}(-((600.\,\mathrm{cm}^{-1})e^{-\beta(300.\,\mathrm{cm}^{-1})} + (600.\,\mathrm{cm}^{-1})e^{-\beta(600.\,\mathrm{cm}^{-1})}))$$

$$= \frac{N(600.\,\mathrm{cm}^{-1})(e^{-\beta(300.\,\mathrm{cm}^{-1})} + e^{-\beta(600.\,\mathrm{cm}^{-1})})}{1 + 2e^{-\beta(300.\,\mathrm{cm}^{-1})} + e^{-\beta(600.\,\mathrm{cm}^{-1})}}$$

Evaluating the expressions for U_A and U_B at 298 K:

$$U_A = \frac{N\left((300.\,\mathrm{cm}^{-1})e^{-\frac{(300.\,\mathrm{cm}^{-1})}{(0.695\,\mathrm{cm}^{-1}\,\mathrm{K}^{-1})(298\,\mathrm{K})}} + (600.\,\mathrm{cm}^{-1})e^{-\frac{(600.\,\mathrm{cm}^{-1})}{(0.695\,\mathrm{cm}^{-1}\,\mathrm{K}^{-1})(298\,\mathrm{K})}}\right)}{1 + e^{-\frac{(300.\,\mathrm{cm}^{-1})}{(0.695\,\mathrm{cm}^{-1}\,\mathrm{K}^{-1})(298\,\mathrm{K})}} + e^{-\frac{(600.\,\mathrm{cm}^{-1})}{(0.695\,\mathrm{cm}^{-1}\,\mathrm{K}^{-1})(298\,\mathrm{K})}}}$$

$$= N(80.3\,\mathrm{cm}^{-1})$$

$$U_B = \frac{N(600.\,\mathrm{cm}^{-1})\left(e^{-\frac{(300.\,\mathrm{cm}^{-1})}{(0.695\,\mathrm{cm}^{-1}\,\mathrm{K}^{-1})(298\,\mathrm{K})}} + e^{-\frac{(600.\,\mathrm{cm}^{-1})}{(0.695\,\mathrm{cm}^{-1}\,\mathrm{K}^{-1})(298\,\mathrm{K})}}\right)}{1 + 2e^{-\frac{(300.\,\mathrm{cm}^{-1})}{(0.695\,\mathrm{cm}^{-1}\,\mathrm{K}^{-1})(298\,\mathrm{K})}} + e^{-\frac{(600.\,\mathrm{cm}^{-1})}{(0.695\,\mathrm{cm}^{-1}\,\mathrm{K}^{-1})(298\,\mathrm{K})}}}$$

$$= N(114\,\mathrm{cm}^{-1})$$

Ensemble B will have the larger internal energy.

P32.9 The lowest four energy levels for atomic vanadium (V) have the following energies and degeneracies:

Level (n)	Energy (cm^{-1})	Degeneracy
0	0	4
1	137.38	6
2	323.46	8
3	552.96	10

What is the contribution to the average energy from electronic degrees of freedom for V when $T = 298$ K?

$$q = \sum_n g_n e^{-\beta \varepsilon_n} = 4 + 6e^{-\beta(137.38 \text{ cm}^{-1})} + 8e^{-\beta(323.46 \text{ cm}^{-1})} + 10e^{-\beta(552.96 \text{ cm}^{-1})}$$

$$\left(\frac{\partial q}{\partial \beta}\right)_V = -((824.28 \text{ cm}^{-1})e^{-\beta(137.38 \text{ cm}^{-1})} + (2587.68 \text{ cm}^{-1})e^{-\beta(323.46 \text{ cm}^{-1})} + (5529.60)e^{-\beta(552.96 \text{ cm}^{-1})})$$

$$U = \frac{-N}{q}\left(\frac{\partial q}{\partial \beta}\right)_V = N\frac{(824.28 \text{ cm}^{-1})e^{-\beta(137.38 \text{ cm}^{-1})} + (2587.68 \text{ cm}^{-1})e^{-\beta(323.46 \text{ cm}^{-1})} + (5529.30 \text{ cm}^{-1})e^{-\beta(552.96 \text{ cm}^{-1})}}{4 + 6e^{-\beta(137.38 \text{ cm}^{-1})} + 8e^{-\beta(323.46 \text{ cm}^{-1})} + 10e^{-\beta(552.96 \text{ cm}^{-1})}}$$

Using $\beta = (kT)^{-1} = (0.695 \text{ cm}^{-1}\text{K}^{-1} \times 298 \text{ K})^{-1}$

$$U = N(143 \text{ cm}^{-1})$$

Converting to J:

$$U = N(143 \text{ cm}^{-1})hc = nN_A(143 \text{ cm}^{-1})(6.626 \times 10^{-34} \text{ J s})(3.00 \times 10^{10} \text{ cm s}^{-1}) = n(1.71 \text{ kJ mol}^{-1})$$

$$U_m = 1.71 \text{ kJ mol}^{-1}$$

P32.11 Consider an ensemble of units in which the first excited electronic state at energy ε_1 is m_1-fold degenerate, and the energy of the ground state is m_0-fold degenerate with energy ε_0.

a. Demonstrate that if $\varepsilon_0 = 0$, the expression for the electronic partition function is

$$q_E = m_0\left(1 + \frac{m_1}{m_0}e^{-\varepsilon_1/kT}\right)$$

b. Determine the expression for the internal energy U of an ensemble of N such units. What is the limiting value of U as the temperature approaches zero and infinity?

(a) $q = m_0 e^{-\beta \varepsilon_0} + m_1 e^{-\beta \varepsilon_1} = m_0 + m_1 e^{-\beta \varepsilon_1}$

$$= m_0 + m_0\left(\frac{m_1}{m_0}\right)e^{-\beta \varepsilon_1}$$

$$= m_0\left(1 + \left(\frac{m_1}{m_0}\right)e^{-\beta \varepsilon_1}\right)$$

$$= m_0\left(1 + \left(\frac{m_1}{m_0}\right)e^{-\frac{\varepsilon_1}{kT}}\right)$$

(b) $U = \frac{-N}{q}\left(\frac{\partial q}{\partial \beta}\right)_V = \frac{-N}{q}\left(\frac{\partial}{\partial \beta}\left(m_0\left(1 + \left(\frac{m_1}{m_0}\right)e^{-\beta \varepsilon_1}\right)\right)\right)_V$

$$= \frac{-N}{q}(-m_1 \varepsilon_1 e^{-\beta \varepsilon_1})$$

$$= \frac{Nm_1\varepsilon_1 e^{-\beta \varepsilon_1}}{m_0\left(1 + \left(\frac{m_1}{m_0}\right)e^{-\beta \varepsilon_1}\right)} = \frac{Nm_1\varepsilon_1 e^{-\frac{\varepsilon_1}{kT}}}{m_0\left(1 + \left(\frac{m_1}{m_0}\right)e^{-\frac{\varepsilon_1}{kT}}\right)}$$

Looking at the limiting behavior with temperature:

$$\lim_{T \to 0} U = \lim_{T \to 0} \frac{N m_1 \varepsilon_1 e^{-\frac{\varepsilon_1}{kT}}}{m_0 \left(1 + \left(\frac{m_1}{m_0}\right) e^{-\frac{\varepsilon_1}{kT}}\right)} = \lim_{T \to 0} \frac{N m_1 \varepsilon_1}{m_0 \left(e^{\frac{\varepsilon_1}{kT}} + \left(\frac{m_1}{m_0}\right)\right)} = 0$$

$$\lim_{T \to \infty} U = \lim_{T \to 0} \frac{N m_1 \varepsilon_1 e^{-\frac{\varepsilon_1}{kT}}}{m_0 \left(1 + \left(\frac{m_1}{m_0}\right) e^{-\frac{\varepsilon_1}{kT}}\right)} = \frac{N m_1 \varepsilon_1}{m_0 + m_1}$$

P32.15   Determine the vibrational contribution to C_V for a mole of HCl ($\tilde{\nu} = 2886 \text{ cm}^{-1}$) over a temperature range from 500. to 5000. K in 500.-K intervals and plot your result. At what temperature do you expect to reach the high-temperature limit for the vibrational contribution to C_V?

The problem requires evaluation of the following expression versus temperature:

$$C_V = \frac{N}{kT^2} (hc\tilde{\nu})^2 \frac{e^{\frac{hc\tilde{\nu}}{kT}}}{\left(e^{\frac{hc\tilde{\nu}}{kT}} - 1\right)^2}$$

Using Excel or a similar program, the following plot of the molar heat capacity versus temperature can be constructed.

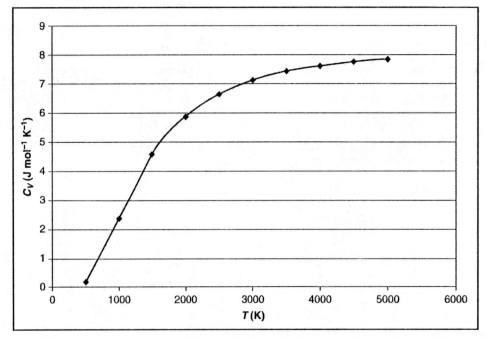

The high-temperature limit value for the molar heat capacity is $(1 \text{ mol}) \times R = 8.314 \text{ J K}^{-1}$. Comparison of this value to the graph illustrates that the high-temperature limit will be valid for temperatures well above 6000. Specifically, the high-temperature limit is applicable when $T > 10\Theta_V$ or $\sim 40,000$ K for HCl.

P32.16   Determine the vibrational contribution to C_V for HCN, where $\tilde{\nu}_1 = 2041 \text{ cm}^{-1}$, $\tilde{\nu}_2 = 712 \text{ cm}^{-1}$ (doubly degenerate), and $\tilde{\nu}_3 = 3369 \text{ cm}^{-1}$ at $T = 298, 500.$, and $1000.$ K.

The total vibrational heat capacity will be equal to the sum of heat capacity contributions from each vibrational degree of freedom. Keeping track of the degeneracy of the 712 cm^{-1} mode, the total heat capacity can be written as:

$$C_{V,total} = C_{V,\tilde{\nu}_1} + 2 C_{V,\tilde{\nu}_2} + C_{V,\tilde{\nu}_3}$$

where the heat capacity for a specific mode is determined using:

$$C_V = \frac{N}{kT^2}(hc\tilde{v})^2 \frac{e^{\frac{hc\tilde{v}}{kT}}}{\left(e^{\frac{hc\tilde{v}}{kT}} - 1\right)^2}$$

Evaluating this expression for the $2041\,cm^{-1}$ mode at $500.\,K$ yields:

$$C_V = \frac{N}{kT^2}(hc\tilde{v})^2 \frac{e^{\frac{hc\tilde{v}}{kT}}}{\left(e^{\frac{hc\tilde{v}}{kT}} - 1\right)^2}$$

$$= \frac{N}{(1.38 \times 10^{-23}\,J\,K^{-1})(500.\,K)^2}((6.626 \times 10^{-34}\,J\,s)(3.00 \times 10^{10}\,cm\,s^{-1})(2041\,cm^{-1}))^2$$

$$\times \frac{e^{\frac{(6.626 \times 10^{-34}\,J\,s)(3.00 \times 10^{10}\,cm\,s^{-1})(2041\,cm^{-1})}{(1.38 \times 10^{-23}\,J\,K^{-1})(500.\,K)}}}{\left(e^{\frac{(6.626 \times 10^{-34}\,J\,s)(3.00 \times 10^{10}\,cm\,s^{-1})(2041\,cm^{-1})}{(1.38 \times 10^{-23}\,J\,K^{-1})(500.\,K)}} - 1\right)^2}$$

$$= N(4.76 \times 10^{-22}\,J\,K^{-1})(2.82 \times 10^{-3})$$

$$= nN_A(1.34 \times 10^{-24}\,J\,K^{-1})$$

$$= n(0.808\,J\,mol^{-1}\,K^{-1})$$

Similar calculations for the other vibrational degrees of freedom and temperatures of interest yield the following table of molar constant volume heat capacities (units of $J\,mol^{-1}\,K^{-1}$):

	298 K	500 K	1000 K
$2041\,cm^{-1}$	0.042	0.808	4.24
$712\,cm^{-1}$	3.37	5.93	7.62
$3369\,cm^{-1}$	0.000	0.048	1.56
Total	6.78	12.72	21.04

P32.20 The speed of sound is given by the relationship

$$c_{sound} = \left(\frac{\frac{C_P}{C_V}RT}{M}\right)^{1/2}$$

where C_P is the constant pressure heat capacity (equal to $C_V + R$), R is the ideal gas constant, T is temperature, and M is molar mass.

a. What is the expression for the speed of sound for an ideal monatomic gas?

b. What is the expression for the speed of sound of an ideal diatomic gas?

c. What is the speed of sound in air at 298 K, assuming that air is mostly made up of nitrogen ($B = 2.00\,cm^{-1}$ and $\tilde{v} = 2359\,cm^{-1}$)?

(a)   For a monatomic gas, only translational degrees of freedom contribute to the C_V:

$$C_V = \frac{3}{2}Nk = \frac{3}{2}nR$$

$$\frac{C_P}{C_V} = \frac{\frac{5}{2}nR}{\frac{3}{2}nR} = \frac{5}{3}$$

$$c_{sound} = \left(\frac{\frac{C_P}{C_V}RT}{M}\right)^{1/2} = \left(\frac{\frac{5}{3}RT}{M}\right)^{1/2}$$

(b)   In addition to translations, rotational and vibrational degrees of freedom will also contribute to C_V:

$$C_V = C_{V,trans} + C_{V,rot} + C_{V,vib} = \frac{3}{2}nR + nR + nR\beta^2(hc\tilde{\nu})^2\frac{e^{\beta hc\tilde{\nu}}}{(e^{\beta hc\tilde{\nu}}-1)^2}$$

$$= nR\left(\frac{5}{2} + \beta^2(hc\tilde{\nu})^2\frac{e^{\beta hc\tilde{\nu}}}{(e^{\beta hc\tilde{\nu}}-1)^2}\right)$$

$$\frac{C_P}{C_V} = \frac{\left(\frac{7}{2} + \beta^2(hc\tilde{\nu})^2\dfrac{e^{\beta hc\tilde{\nu}}}{(e^{\beta hc\tilde{\nu}}-1)^2}\right)}{\left(\frac{5}{2} + \beta^2(hc\tilde{\nu})^2\dfrac{e^{\beta hc\tilde{\nu}}}{(e^{\beta hc\tilde{\nu}}-1)^2}\right)}$$

$$c_{sound} = \left(\frac{\frac{C_P}{C_V}RT}{M}\right)^{1/2} = \left(\frac{\left(\dfrac{\left(\frac{7}{2} + \beta^2(hc\tilde{\nu})^2\dfrac{e^{\beta hc\tilde{\nu}}}{(e^{\beta hc\tilde{\nu}}-1)^2}\right)}{\left(\frac{5}{2} + \beta^2(hc\tilde{\nu})^2\dfrac{e^{\beta hc\tilde{\nu}}}{(e^{\beta hc\tilde{\nu}}-1)^2}\right)}\right)RT}{M}\right)^{1/2}$$

(c)   First, evaluation of the vibrational contribution to C_V demonstrates that this contribution is small relative to the contribution from translational and rotational degrees of freedom, and can be neglected to good approximation:

$$\frac{C_{V,vib}}{nR} = \beta^2(hc\tilde{\nu})^2\frac{e^{\beta hc\tilde{\nu}}}{(e^{\beta hc\tilde{\nu}}-1)^2}$$

$$= \left(\frac{(6.626\times10^{-34}\text{ J s})(3.00\times10^{10}\text{ cm s}^{-1})(2359\text{ cm}^{-1})}{(1.38\times10^{-23}\text{ J K}^{-1})(298\text{ K})}\right)^2$$

$$\times \frac{e^{\frac{(6.626\times10^{-34}\text{ J s})(3.00\times10^{10}\text{ cm s}^{-1})(2359\text{ cm}^{-1})}{(1.38\times10^{-23}\text{ J K}^{-1})(298\text{ K})}}}{\left(e^{\frac{(6.626\times10^{-34}\text{ J s})(3.00\times10^{10}\text{ cm s}^{-1})(2359\text{ cm}^{-1})}{(1.38\times10^{-23}\text{ J K}^{-1})(298\text{ K})}} - 1\right)^2}$$

$$= 1.46\times10^{-3} \ll \frac{5}{2}$$

Therefore, keeping only the translational and rotational contributions to C_V yields the following value for the speed of sound in N_2:

$$c_{sound} = \left(\frac{\frac{C_p}{C_V}RT}{M}\right)^{1/2} \cong \left(\frac{\frac{7}{5}RT}{M}\right)^{1/2} = \left(\frac{\frac{7}{5}(8.314 \text{ J mol}^{-1} \text{ K}^{-1})(298 \text{ K})}{(0.028 \text{ kg mol}^{-1})}\right)^{1/2}$$

$$= 352 \text{ m s}^{-1}$$

P32.22 The molar constant volume heat capacity for $I_2(g)$ is $28.6 \text{ J mol}^{-1} \text{ K}^{-1}$. What is the vibrational contribution to the heat capacity? You can assume that the contribution from the electronic degrees of freedom is negligible.

The translational and rotational degrees of freedom contribute $3/2 \, R$ and R to the constant volume heat capacity, respectively. Therefore, the remainder must be the contribution from vibrational degrees of freedom:

$$C_{V,total} = C_{V,trans} + C_{V,rot} + C_{V,vib}$$

$$28.6 \text{ J mol}^{-1} \text{ K}^{-1} = \frac{3}{2}R + R + C_{V,vib}$$

$$28.6 \text{ J mol}^{-1} \text{ K}^{-1} - \frac{5}{2}R = C_{V,vib}$$

$$7.82 \text{ J mol}^{-1} \text{ K}^{-1} = C_{V,vib}$$

P32.25 Determine the molar entropy for 1 mol of gaseous Ar at 200., 300., and 500. K and $V = 1000. \text{ cm}^3$ assuming that Ar can be treated as an ideal gas. How does the result of this calculation change if the gas is Kr instead of Ar?

Determining the molar entropy for gaseous Ar at 200. K yields:

$$S = \frac{U}{T} + k \ln Q = \frac{3}{2}Nk + k \ln\left(\frac{q^N}{N!}\right) = \frac{3}{2}Nk + Nk \ln q - k \ln(N!)$$

$$= \frac{3}{2}Nk + Nk \ln q - k(N \ln N - N)$$

$$= \frac{5}{2}nR + nR \ln q - nR \ln(nN_A)$$

$$S_m = \frac{5}{2}R + R \ln\left(\frac{V}{\Lambda^3}\right) - R \ln(6.022 \times 10^{23})$$

$$= 20.79 \text{ J mol}^{-1} \text{ K}^{-1} + R \ln\left(\frac{V}{\left(\frac{h^2}{2\pi mkT}\right)^{3/2}}\right) - 456 \text{ J mol}^{-1} \text{ K}^{-1}$$

$$= 20.79 \text{ J mol}^{-1} \text{ K}^{-1}$$

$$+ R \ln\left(\frac{1.00 \times 10^{-3} \text{ m}^3}{\left(\frac{(6.626 \times 10^{-34} \text{ J s})^2}{2\pi\left(\frac{0.040 \text{ kg mol}^{-1}}{N_A}\right)(1.38 \times 10^{-23} \text{ J K}^{-1})(200. \text{ K})}\right)^{3/2}}\right) - 456 \text{ J mol}^{-1} \text{ K}^{-1}$$

$$= 122 \text{ J mol}^{-1} \text{ K}^{-1}$$

Repeating the calculation for $T = 300$ and 500 K, the molar entropy is found to be $128 \, \text{J mol}^{-1} \text{K}^{-1}$ and $135 \, \text{J mol}^{-1} \text{K}^{-1}$, respectively. Kr is heavier than Ar; therefore, the thermal wavelength will be shorter, and the translational partition function will correspondingly be larger. Since the molar entropy is linear related to $\ln(q)$, we would expect the molar entropy for Kr to be greater than that of Ar. This expectation can be confirmed by repeating the preceding calculation for Kr, or by simply looking at the difference in entropy between Kr and Ar:

$$S_{Kr} - S_{Ar} = \left(\frac{U}{T} + k\ln Q\right)_{Kr} - \left(\frac{U}{T} + k\ln Q\right)_{Ar}$$

$$= k\ln\left(\frac{Q_{Kr}}{Q_{Ar}}\right) = Nk\ln\left(\frac{q_{Kr}}{q_{Ar}}\right) = Nk\ln\left(\frac{\Lambda_{Ar}^3}{\Lambda_{Kr}^3}\right) = Nk\ln\left(\frac{m_{kr}}{m_{Ar}}\right)^{3/2}$$

$$S_{m,Kr} - S_{m,Ar} = \frac{3}{2}R\ln\left(\frac{m_{kr}}{m_{Ar}}\right) = \frac{3}{2}R \, \ln(2.10)$$

$$S_{m,Kr} = 9.24 \, \text{J mol}^{-1} \text{K}^{-1} + S_{m,Ar}$$

P32.27   Determine the standard molar entropy of N_2O, a linear triatomic molecule at 298 K and $P = 1.00$ atm. For this molecule, $B = 0.419 \, \text{cm}^{-1}$ and $\tilde{v}_1 = 1285 \, \text{cm}^{-1}$, $\tilde{v}_2 = 589 \, \text{cm}^{-1}$ (doubly degenerate), and $\tilde{v}_3 = 2224 \, \text{cm}^{-1}$.

At 298 K both the translational and rotational degrees of freedom will be in the high-temperature limit, but the vibrational contributions must be calculated. Using the standard volume of 24.5 L, the entropy is determined as follows:

$$U_m^\circ = U_{T,m}^\circ + U_{R,m}^\circ + U_{V,m}^\circ + U_{E,m}^\circ$$

$$= \frac{3}{2}RT + RT + N_A hc\left[\left(\frac{\tilde{v}_1}{e^{\beta hc\tilde{v}_1} - 1}\right) + 2\left(\frac{\tilde{v}_2}{e^{\beta hc\tilde{v}_2} - 1}\right) + \left(\frac{\tilde{v}_3}{e^{\beta hc\tilde{v}_3} - 1}\right)\right]$$

$$= \frac{5}{2}R(298 \, \text{K}) + 906 \, \text{J mol}^{-1} = 7.10 \, \text{kJ mol}^{-1}$$

$$S_m^\circ = \frac{U_m^\circ}{T} + k\ln Q = \frac{U_m^\circ}{T} + k\ln\left(\frac{q_{total}^N}{N!}\right) = \frac{U_m^\circ}{T} + Nk\ln(q_{total}) - k\ln(N!)$$

$$= \frac{7.10 \, \text{kJ mol}^{-1}}{298 \, \text{K}} + R\ln(q_T q_R q_V q_E) - R\ln((1 \, \text{mol})N_A) + R$$

$$= -424 \, \text{J mol}^{-1} \text{K}^{-1} + R\ln(q_T q_R q_V q_E)$$

$$q_T = \left(\frac{V}{\Lambda^3}\right) = \frac{0.0245 \, \text{m}^3}{3.55 \times 10^{-33} \, \text{m}^3} = 6.91 \times 10^{30}$$

$$q_R = \left(\frac{kT}{\sigma B}\right) = \frac{(0.695 \, \text{cm}^{-1} \text{K}^{-1})(298.15 \, \text{K})}{0.42 \, \text{cm}^{-1}} = 493$$

$$q_V = \left(\frac{1}{1 - e^{-\beta \tilde{v}_1}}\right)\left(\frac{1}{1 - e^{-\beta \tilde{v}_2}}\right)^2\left(\frac{1}{1 - e^{-\beta \tilde{v}_3}}\right)$$

$$= \left(\frac{1}{1 - e^{-\frac{1285 \, \text{cm}^{-1}}{(0.695 \, \text{cm}^{-1})(298 \, \text{K})}}}\right)\left(\frac{1}{1 - e^{-\frac{589 \, \text{cm}^{-1}}{(0.695 \, \text{cm}^{-1})(298 \, \text{K})}}}\right)^2\left(\frac{1}{1 - e^{-\frac{2224 \, \text{cm}^{-1}}{(0.695 \, \text{cm}^{-1})(298 \, \text{K})}}}\right)$$

$$= 1.00$$

$$q_E = 1.00$$

$$S_m^\circ = -424 \, \text{J mol}^{-1} \text{K}^{-1} + R \, \ln(q_T q_R q_V q_E) = -424 \, \text{J mol}^{-1} \text{K}^{-1} + 643 \, \text{J mol}^{-1} \text{K}^{-1}$$

$$S_m^\circ = 219 \, \text{J mol}^{-1} \text{K}^{-1}$$

P32.33 The standard molar entropy of CO is $197.7 \ \text{J mol}^{-1} \ \text{K}^{-1}$. How much of this value is due to rotational and vibrational motion of CO?

The contribution of rotational and vibrational energetic degrees of freedom is equal to the difference between the standard molar entropy and the contribution from translational degrees of freedom. The translational contribution to the molar entropy can be calculated using the Sackur–Tetrode equation as follows:

$$S = nR \ln \left[\frac{RTe^{5/2}}{\Lambda^3 N_A P} \right]$$

$$S_m = R \ln \left[\frac{RTe^{5/2}}{\Lambda^3 N_A P} \right]$$

$$\Lambda^3 = \left(\frac{h^2}{2\pi m k T} \right)^{3/2} = \left(\frac{(6.626 \times 10^{-34} \ \text{J s})^2}{2\pi (4.65 \times 10^{-26} \ \text{kg})(1.38 \times 10^{-23} \ \text{J K}^{-1})(298.15 \ \text{K})} \right)^{3/2} = 6.99 \times 10^{-33} \ \text{m}^3$$

$$S_m = R \ln \left[\frac{RTe^{5/2}}{\Lambda^3 N_A P} \right] = nR \ln \left[\frac{(8.21 \times 10^{-5} \ \text{m}^3 \ \text{atm mol}^{-1} \ \text{K}^{-1})(298.15 \ \text{K})e^{5/2}}{(6.99 \times 10^{-33} \ \text{m}^3)N_A (1 \ \text{atm})} \right] = 148.6 \ \text{J mol}^{-1} \ \text{K}^{-1}$$

$$S_{rot,vib} = S_{total} - S_{trans} = 197.7 \ \text{J mol}^{-1} \ \text{K}^{-1} - 148.6 \ \text{J mol}^{-1} \ \text{K}^{-1} = 49.1 \ \text{J mol}^{-1} \ \text{K}^{-1}$$

P32.36 Consider the molecule NNO, which has a rotational constant nearly identical to CO_2. Would you expect the standard molar entropy for NNO to be greater or less than CO_2? If greater, can you provide a rough estimation of how much greater?

Since NNO has almost exactly the same mass as CO_2, we would expect the translational contribution to the standard molar entropy to be the same. However, the rotational partition function of NNO should be two-fold greater due to the reduction in the symmetry number. Therefore, we would expect the standard molar entropy of NNO to be greater than that of CO_2. Since $q_{trans,NNO} \approx q_{trans,CO_2}$ and $q_{rot,NNO} \approx 2q_{rot,CO_2}$:

$$S_{NNO}^{\circ} - S_{CO_2}^{\circ} \approx \left(\frac{7}{2}R + R \ln q_{trans} + R \ln q_{rot} - R \ln N_A \right)_{NNO} - \left(\frac{7}{2}R + R \ln q_{trans} + R \ln q_{rot} - R \ln N_A \right)_{CO_2}$$

$$= R \ln \left(\frac{q_{trans,NNO}}{q_{trans,CO_2}} \right) + R \ln \left(\frac{q_{rot,NNO}}{q_{rot,CO_2}} \right)$$

$$= R \ln(1) + R \ln(2)$$

$$= R \ln 2 = 5.76 \ \text{J mol}^{-1} \ \text{K}^{-1}$$

Our estimate is consistent with the experimental value for the standard molar entropy of NNO being $220. \ \text{J mol}^{-1} \ \text{K}^{-1}$, compared to a value of $213.8 \ \text{J mol}^{-1} \ \text{K}^{-1}$.

P32.38 The molecule NO has a ground electronic level that is doubly degenerate, and a first excited level at $121.1 \ \text{cm}^{-1}$ that is also two-fold degenerate. Determine the contribution of electronic degrees of freedom to the standard molar entropy of NO. Compare your result to $R \ln(4)$. What is the significance of this comparison?

$$q_E = g_0 + g_1 e^{-\beta \varepsilon_1} = 2 + 2e^{-\beta hc(121.1 \ \text{cm}^{-1})} = 3.11$$

$$U_E = \frac{-N}{q_E} \left(\frac{\partial q_E}{\partial \beta} \right)_V = \frac{2Nhc(121.1 \ \text{cm}^{-1})e^{-\beta hc(121.1 \ \text{cm}^{-1})}}{2 + 2e^{-\beta hc(121.1 \ \text{cm}^{-1})}}$$

$$= \frac{Nhc(121.1 \ \text{cm}^{-1})e^{-\beta hc(121.1 \ \text{cm}^{-1})}}{1 + e^{-\beta hc(121.1 \ \text{cm}^{-1})}}$$

$$U_{E,m} = \frac{N_A hc(121.1 \ \text{cm}^{-1})e^{-\beta hc(121.1 \ \text{cm}^{-1})}}{1 + e^{-\beta hc(121.1 \ \text{cm}^{-1})}} = 518 \ \text{J mol}^{-1}$$

$$S_{E,m} = \frac{U_{E,m}}{T} + R\ln(q_E) = \frac{518\,\text{J mol}^{-1}}{298.15\,\text{K}} + R\ln(3.11)$$

$$= 1.73\,\text{J mol}^{-1}\,\text{K}^{-1} + 9.43\,\text{J mol}^{-1}\,\text{K}^{-1}$$

$$= 11.2\,\text{J mol}^{-1}\,\text{K}^{-1}$$

$$R\ln(4) = 11.5\,\text{J mol}^{-1}\,\text{K}^{-1}$$

At sufficiently high temperatures, $q_E = 4$, and the contribution of the two lowest electronic energy levels to the molar entropy would equal $R\ln(4)$. The similarity between the calculated and limiting values demonstrates that this limiting behavior is being approached at this temperature.

P32.43   Calculate the standard Helmholtz energy for molar ensembles of Ne and Kr at 298 K.

First, performing the calculation for Ne ($M = 0.020\,\text{kg mol}^{-1}$):

$$A = -kT\ln Q = -kT\ln\left(\frac{q^N}{N!}\right)$$

$$= -NkT\ln q + kT\ln(N!)$$

$$= -NkT\ln q + kT(N\ln N - N)$$

$$= -NkT(\ln q - \ln N + 1)$$

$$q = q_T = \frac{V}{\Lambda^3} = \frac{0.0245\,\text{m}^3}{\left(\dfrac{h^2}{2\pi mkT}\right)^{3/2}} = \frac{0.0245\,\text{m}^3}{1.08\times10^{-32}\,\text{m}^3} = 2.27\times10^{30}$$

$$N = n\times N_A = 6.022\times10^{23}$$

$$A = -nRT(\ln q - \ln N + 1)$$

$$A_m^\circ = -(8.314\,\text{J mol}^{-1}\,\text{K}^{-1})(298.15\,\text{K})(\ln(2.27\times10^{30}) - \ln(6.022\times10^{23}) + 1)$$

$$A_m^\circ = -40.0\,\text{kJ mol}^{-1}$$

This calculation can be repeated for Kr ($M = 0.083\,\text{kg mol}^{-1}$), or the difference between the Helmholtz energy of Kr and Ne can be determined:

$$A_{Kr} - A_{Ar} = -kT(\ln Q_{Kr} - \ln Q_{Ar})$$

$$= -kT\ln\left(\frac{Q_{Kr}}{Q_{Ar}}\right)$$

$$= -NkT\ln\left(\frac{q_{Kr}}{q_{Ar}}\right)$$

$$= -NkT\ln\left(\left(\frac{\Lambda_{Ar}}{\Lambda_{Kr}}\right)^3\right)$$

$$= -NkT\ln\left(\left(\frac{m_{Kr}}{m_{Ar}}\right)^{3/2}\right)$$

$$= -\frac{3}{2}NkT\ln\left(\frac{m_{Kr}}{m_{Ar}}\right)$$

$$= -\frac{3}{2}nRT\ln\left(\frac{0.083}{0.020}\right)$$

$$A_{Kr,m}^\circ = -\frac{3}{2}RT\ln(4.15) + A_{Ar,m}^\circ = -45.3\,\text{kJ mol}^{-1}$$

P32.47   Determine the equilibrium constant for the dissociation of sodium at 298 K:

$$Na_2(g) \rightleftharpoons 2Na(g)$$

For Na_2, $B = 0.155 \, \text{cm}^{-1}$, $\tilde{v} = 159 \, \text{cm}^{-1}$, the dissociation energy is $70.4 \, \text{kJ/mol}$, and the ground-state electronic degeneracy for Na is 2.

$$K = \frac{\left(\dfrac{q}{N_A}\right)^2_{Na}}{\left(\dfrac{q}{N_A}\right)_{Na_2}} e^{-\beta\varepsilon_D}$$

$$q_{Na} = q_T q_E = \left(\frac{V}{\Lambda^3}\right)(2) = \left(\frac{0.0245 \, \text{m}^3}{9.38 \times 10^{-33} \, \text{m}^3}\right)(2) = 5.22 \times 10^{30}$$

$$q_{Na_2} = q_T q_R q_V q_E = \left(\frac{V}{\Lambda^3}\right)\left(\frac{kT}{\sigma B}\right)\left(\frac{1}{1 - e^{-\beta hc\tilde{v}}}\right)(1)$$

$$= \left(\frac{0.0245 \, \text{m}^3}{3.32 \times 10^{-33} \, \text{m}^3}\right)\left(\frac{(0.695 \, \text{cm}^{-1}\,\text{K}^{-1})(298.15 \, \text{K})}{(2)(0.155 \, \text{cm}^{-1})}\right)$$

$$\times \left(\frac{1}{1 - e^{-\dfrac{(6.626\times10^{-34} \, \text{J s})(3.00\times10^{10} \, \text{cm s}^{-1})(159 \, \text{cm}^{-1})}{(1.38\times10^{-23} \, \text{J K}^{-1})(298.15 \, \text{K})}}}\right)(1)$$

$$= 9.22 \times 10^{33}$$

$$K = \frac{\left(\dfrac{5.22 \times 10^{30}}{N_A}\right)^2_{Na}}{\left(\dfrac{9.22 \times 10^{33}}{N_A}\right)_{Na_2}} e^{-\dfrac{70,400 \, \text{J mol}^{-1}/N_A}{(1.38\times10^{-23} \, \text{J K}^{-1})(298.15 \, \text{K})}}$$

$$= 2.25 \times 10^{-9}$$

33 Kinetic Theory of Gases

Numerical Problems

P33.1 Consider a collection of gas particles confined to translate in two dimensions (for example, a gas molecule on a surface). Derive the Maxwell speed distribution for such a gas.

Beginning with the Maxwell–Boltzmann velocity distribution in one dimension

$$f(v_j) = \left(\frac{m}{2\pi kT}\right)^{1/2} e^{-\frac{m}{2kT}v_j^2}$$

and the definition of speed in two dimensions

$$v = (v_x^2 + v_y^2)^{1/2}$$

the speed distribution in two dimensions is given by:

$$F\,dv = f(v_x)f(v_y)\,dv_x\,dv_y$$

where dv_j is the differential of velocity in the jth direction. Thus

$$F\,dv = \left(\frac{m}{2\pi kT}\right)^{1/2} e^{-\frac{m}{2kT}v_x^2}\left(\frac{m}{2\pi kT}\right)^{1/2} e^{-\frac{m}{2kT}v_y^2}\,dv_x\,dv_y$$

$$= \left(\frac{m}{2\pi kT}\right) e^{-\frac{m}{2kT}(v_x^2+v_y^2)}\,dv_x\,dv_y$$

$$F\,dv = \left(\frac{m}{2\pi kT}\right) e^{-\frac{m}{2kT}v^2}\,dv_x\,dv_y$$

The differential is defined as:

$$dv_x\,dv_y = 2\pi v\,dv$$

Substituting this into the expression for Fdv:

$$F\,dv = \left(\frac{m}{2\pi kT}\right) e^{-\frac{m}{2kT}v^2}(2\pi v)\,dv = \frac{m}{kT} e^{-\frac{m}{2kT}v^2} v\,dv$$

P33.3 Compute v_{mp}, v_{ave}, and v_{rms} for O_2 at 300. and 500. K. How would your answers change for H_2?

O_2: $M = 0.0320\ \text{kg mol}^{-1}$ H_2: $M = 0.00202\ \text{Kg mol}^{-1}$

$O_2 @ 300.\ \text{K}$:

$$v_{mp} = \sqrt{\frac{2RT}{M}} = \sqrt{\frac{2(8.314\ \text{J mol}^{-1}\ \text{K}^{-1})(300.\ \text{K})}{0.0320\ \text{kg mol}^{-1}}} = 395\ \text{m s}^{-1}$$

$$v_{ave} = \sqrt{\frac{8RT}{\pi M}} = \sqrt{\frac{8(8.314\ \text{J mol}^{-1}\ \text{K}^{-1})(300.\ \text{K})}{\pi(0.0320\ \text{kg mol}^{-1})}} = 446\ \text{m s}^{-1}$$

$$v_{rms} = \sqrt{\frac{3RT}{M}} = \sqrt{\frac{3(8.314\ \text{J mol}^{-1}\ \text{K}^{-1})(300.\ \text{K})}{0.0320\ \text{kg mol}^{-1}}} = 484\ \text{m s}^{-1}$$

$O_2 @ 500.\,K$

$$v_{mp} = \sqrt{\frac{2RT}{M}} = \sqrt{\frac{2(8.314 \text{ J mol}^{-1} \text{ K}^{-1})(500.\,K)}{0.0320 \text{ kg mol}^{-1}}} = 510 \text{ m s}^{-1}$$

$$v_{ave} = \sqrt{\frac{8RT}{\pi M}} = \sqrt{\frac{8(8.314 \text{ J mol}^{-1} \text{ K}^{-1})(500.\,K)}{\pi(0.0320 \text{ kg mol}^{-1})}} = 575 \text{ m s}^{-1}$$

$$v_{rms} = \sqrt{\frac{3RT}{M}} = \sqrt{\frac{3(8.314 \text{ J mol}^{-1} \text{ K}^{-1})(500.\,K)}{0.0320 \text{ kg mol}^{-1}}} = 624 \text{ m s}^{-1}$$

All of these values depend on the square root of the mass of the particle, so that:

$$v_{H_2} = \left(\frac{M_{O_2}}{M_{H_2}}\right)^{1/2} v_{O_2}$$

$$v_{H_2} = \left(\frac{0.0320 \text{ kg mol}^{-1}}{0.00202 \text{ kg mol}^{-1}}\right)^{1/2} v_{O_2}$$

$$v_{H_2} = (3.98)\, v_{O_2} \text{ for each velocity and } T.$$

P33.5 Compare the average speed and average kinetic energy of O_2 with that of CCl_4 at 298 K.

$$M_{O_2} = 0.0320 \text{ kg mol}^{-1} \quad M_{CCl_4} = 0.154 \text{ kg mol}^{-1}$$

At the same temperature, the speed for two particles of different mass is related by the square root of the mass ratios. For this case of O_2 and CCl_4:

$$v_{CCl_4} = \left(\frac{M_{O_2}}{M_{CCl_4}}\right)^{1/2} v_{O_2}$$

$$v_{CCl_4} = \left(\frac{0.0320 \text{ kg mol}^{-1}}{0.154 \text{ kg mol}^{-1}}\right)^{1/2} v_{O_2} = (0.456) v_{O_2}$$

The average speed for O_2 at 298 K is:

$$v_{ave} = \sqrt{\frac{8RT}{\pi M}} = \sqrt{\frac{8(8.314 \text{ J mol}^{-1} \text{ K}^{-1})(298 \text{ K})}{\pi \cdot 0.0320 \text{ kg mol}^{-1}}} = 444 \text{ m s}^{-1}$$

Using this result, the average speed for CCl_4 at this same temperature is:

$$v_{ave,\,CCl_4} = 0.456 \times v_{ave,\,O_2}$$

$$v_{ave,\,CCl_4} = (0.456)(444 \text{ m s}^{-1})$$

$$v_{ave,\,CCl_4} = 202 \text{ m s}^{-1}$$

The average kinetic energy is mass independent; thus for a given temperature, all gases have the same kinetic energy. The average translational kinetic energy per gas particle is therefore:

$$\langle KE \rangle = \frac{3}{2} kT = \frac{3}{2}(1.38 \times 10^{-23} \text{ J K}^{-1})(298 \text{ K})$$

$$\langle KE \rangle = 6.17 \times 10^{-21} \text{ J}$$

P33.9   At what temperature is the v_{rms} of Ar equal to that of SF$_6$ at 298 K? Perform the same calculation for v_{mp}.

$$M_{Ar} = 0.0400 \text{ kg mol}^{-1} \text{ and } M_{SF_6} = 0.146 \text{ kg mol}^{-1}$$

Setting the rms speeds equal and reducing yields:

$$\sqrt{\frac{3RT_{Ar}}{M_{Ar}}} = \sqrt{\frac{3RT_{SF_6}}{M_{SF_6}}}$$

$$T_{Ar} = \frac{M_{Ar}}{M_{SF_6}} \cdot T_{SF_6}$$

Therefore:

$$T_{Ar} = \left(\frac{0.0400 \text{ kg mol}^{-1}}{0.146 \text{ kg mol}^{-1}}\right) 298 \text{ K}$$

$$T_{Ar} = 81.5 \text{ K}$$

Since v_{rms} and v_{mp} are related by non-gas-dependent factors, the temperature relation for v_{mp} is the same as the temperature relation for v_{rms}.

P33.11   The probability that a particle will have a velocity in the x direction in the range of $-v_{x0}$ and v_{x0} is given by

$$f(-v_{x_0} \leq v_x \leq v_{x_0}) = \left(\frac{m}{2\pi kT}\right)^{1/2} \int_{-v_{x_0}}^{v_{x_0}} e^{-mv_x^2/2kT} dv_x$$

$$= \left(\frac{2m}{\pi kT}\right)^{1/2} \int_{0}^{v_{x_0}} e^{-mv_x^2/2kT} dv_x$$

The preceding integral can be rewritten using the following substitution: $\xi^2 = mv_x^2/2kT$, resulting in $f(-v_{x0} \leq v_x \leq v_{x0}) = 2/\sqrt{\pi}\left(\int_0^{\xi_0} e^{-\xi^2} d\xi\right)$, which can be evaluated using the error function defined as $\text{erf}(z) = 2/\sqrt{\pi}\left(\int_0^z e^{-x^2} dx\right)$. The complementary error function is defined as $\text{erfc}(z) = 1 - \text{erf}(z)$. Finally, a plot of both $\text{erf}(z)$ and $\text{erfc}(z)$ as a function of z is shown here:

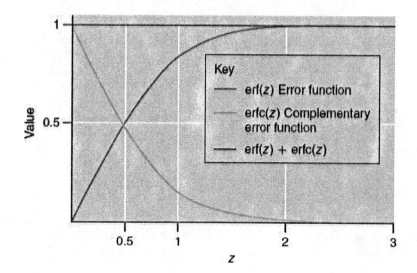

Using this graph of $\text{erf}(z)$, determine the probability that $|v_x| \leq (2kT/m)^{1/2}$. What is the probability that $|v_x| > (2kT/m)^{1/2}$?

The probability that a particle will have a velocity in the x direction in the range of $-v_{x_0}$ and v_{x_0} is given by

$$f(-v_{x_0} \leq v_x \leq v_{x_0}) = \left(\frac{m}{2\pi kT}\right)^{1/2} \int_{-v_{x_0}}^{v_{x_0}} e^{\frac{-mv_x^2}{2kT}} dv_x = \left(\frac{2m}{\pi kT}\right)^{1/2} \int_0^{v_{x_0}} e^{\frac{-mv_x^2}{2kT}} dv_x$$

The preceding integral can be rewritten using the following substitution $\xi^2 = \dfrac{mv_x^2}{2kT}$, resulting in

$$f(-v_{x_0} \leq v_x \leq v_{x_0}) = \frac{2}{\sqrt{\pi}} \int_0^{\xi_0} e^{-\xi^2} d\xi,$$ which can be evaluated using the error function defined as

$\text{erf}(z) = \dfrac{2}{\sqrt{\pi}} \int_0^z e^{-x^2} dx.$ The complementary error function is defined as $\text{erfc}(z) = 1 - \text{erf}(z)$. Finally, a plot of both erf(z) and erfc(z) as a function of z is shown in the text (tabulated values are available in the Math Supplement, Appendix A): Using this graph of erf(z), determine the probability that $|v_x| \leq (2kT/m)^{1/2}$. What is the probability that $|v_x| > (2kT/m)^{1/2}$?

If $|v_x| \leq \left(\dfrac{2kT}{m}\right)^{1/2}$, then

$$\xi^2 \leq \frac{m}{2kT}\left(\frac{2kT}{m}\right) = 1$$

and

$$\xi_0 \leq \sqrt{1} \leq 1$$

The probability that the particle has $|v_x| \leq \left(\dfrac{2kT}{m}\right)^{1/2}$ is given by:

$$f_\leq = \frac{2}{\sqrt{\pi}} \int_0^1 e^{-\xi^2} d\xi$$
$$f_\leq = \text{erf}(1)$$
$$f_\leq = 0.8427$$

The probability that the particle would have a velocity $|v_x| > \left(\dfrac{2kT}{m}\right)^{1/2}$ is found from the previous part, since the total probability must be 1:

$$f_> = 1 - 0.8427$$
$$f_> = 0.1573$$

P33.12   The speed of sound is given by $v_{sound} = \sqrt{\dfrac{\gamma kT}{m}} = \sqrt{\dfrac{\gamma RT}{M}}$, where $\gamma = C_P/C_V$.

a.   What is the speed of sound in Ne, Kr, and Ar at 1000. K?

b.   At what temperature will the speed of sound in Kr equal the speed of sound in Ar at 1000. K?

x. $M_{Ne} = 0.0202 \text{ kg mol}^{-1}$ $M_{Kr} = 0.0840 \text{ kg mol}^{-1}$ $M_{Ar} = 0.0400 \text{ kg mol}^{-1}$

The heat capacities for the three gases are the same:

$$C_V = 12.5 \text{ J K}^{-1} \text{ mol}^{-1} \text{ and } C_p = 20.8 \text{ J K}^{-1} \text{ mol}^{-1}$$

Thus, $\gamma = 1.6666$.

(a)   Ne:

$$v_{sound} = \sqrt{\frac{1.67(8.314 \text{ J mol}^{-1}\text{K})(1000.\text{ K})}{0.0202 \text{ kg mol}^{-1}}} = 828 \text{ m s}^{-1}$$

Kr:

$$v_{sound} = \sqrt{\frac{1.67(8.314 \text{ J mol}^{-1}\text{ K}^{-1})(1000.\text{ K})}{0.0840 \text{ kg mol}^{-1}}} = 406 \text{ m s}^{-1}$$

Ar:

$$v_{sound} = \sqrt{\frac{1.67(8.314 \text{ J mol}^{-1}\text{ K}^{-1})(1000.\text{ K})}{0.0400 \text{ kg mol}}} = 589 \text{ m s}^{-1}$$

(b)   Setting the speed of sound equal for Kr and Ar:

$$\sqrt{\frac{\gamma R T_{Kr}}{M_{Kr}}} = \sqrt{\frac{\gamma R T_{Ar}}{M_{Ar}}}$$

$$T_{Kr} = \frac{M_{Kr}}{M_{Ar}} \cdot T_{Ar}$$

Therefore:

$$T_{Kr} = \left(\frac{0.0840 \text{ kg mol}^{-1}}{0.0400 \text{ kg mol}^{-1}}\right) 1000.\text{ K}$$

$$T_{Kr} = (2.10)1000.\text{ K}$$

$$T_{Kr} = 2100 \text{ K}$$

P33.19 Starting with the Maxwell speed distribution, demonstrate that the probability distribution for translational kinetic energy for $\varepsilon_{Tr} \gg kT$ is given by:

$$f(\varepsilon_{Tr}) d\varepsilon_{Tr} = 2\pi\left(\frac{1}{\pi kT}\right)^{3/2} e^{-\varepsilon_{Tr}/kT} \varepsilon_{Tr}^{1/2} d\varepsilon_{Tr}$$

The translational kinetic energy of a particle can be related to the velocity of the particle by the expressions

$$\varepsilon_{Tr} = \frac{1}{2}mv^2$$

$$v = \sqrt{\frac{2\varepsilon_{T}}{m}}$$

$$dv = \sqrt{\frac{2}{m}} \cdot \frac{1}{2}\sqrt{\frac{1}{\varepsilon_{Tr}}} d\varepsilon_{Tr} = \frac{1}{2}\sqrt{\frac{2}{m\varepsilon_{Tr}}} d\varepsilon_{Tr}$$

Substituting this result into the Maxwell speed distribution:

$$f(\varepsilon_{Tr}) d\varepsilon_{Tr} = 4\pi\left(\frac{m}{2\pi kT}\right)^{3/2}\left(\frac{2\varepsilon_{Tr}}{m}\right) e^{-\frac{m}{2kT}\left(\frac{2\varepsilon_{Tr}}{m}\right)}\left(\frac{1}{2}\sqrt{\frac{2}{m\varepsilon_{Tr}}} d\varepsilon_{Tr}\right)$$

$$= 2\pi\left(\frac{1}{\pi kT}\right)^{3/2} e^{-\frac{\varepsilon_{Tv}}{kT}} \varepsilon_{Tr}^{1/2} d\varepsilon_{Tr}$$

P33.23   Imagine a cubic container with sides 1 cm in length that contains 1 atm of Ar at 298 K. How many gas–wall collisions are there per second?

The collisional rate is given by:

$$\frac{dN_c}{dt} = \frac{PAv_{ave}}{4kT} = \frac{PAN_A v_{ave}}{4RT}$$

With $M_{Ar} = 0.0400 \text{ kg mol}^{-1}$ and $T = 298 \text{ K}$ the average speed is:

$$v_{ave} = \sqrt{\frac{8RT}{\pi M}} = \sqrt{\frac{8(8.314 \text{ J mol}^{-1} \text{ K})(298 \text{ K})}{\pi(0.0400 \text{ kg mol}^{-1})}} = 398 \text{ m s}^{-1}$$

Substituting into the expression for the collisional rate:

$$\frac{dN_c}{dt} = \frac{\left(1 \text{ atm} \cdot \dfrac{101.325 \times 10^3 \text{ N m}^{-2}}{1 \text{ atm}}\right)\left(1 \text{ cm}^2 \left(\dfrac{1 \text{ m}}{100 \text{ cm}}\right)^2\right)(6.022 \times 10^{23} \text{ mol}^{-1})(398 \text{ m s}^{-1})}{4(8.314 \text{ J mol}^{-1} \text{ K}^{-1})(298 \text{ K})}$$

$$\frac{dN_c}{dt} = 2.45 \times 10^{23} \text{ coll. per sec. per wall}$$

Taking into account the six walls that comprise the container, the total collisional rate is:

$$\frac{dN_c}{dt} = 6 \text{ walls} \cdot 2.45 \times 10^{23} \text{ coll.s}^{-1} \text{ wall}^{-1}$$

$$\frac{dN_c}{dt} = 1.47 \times 10^{24} \text{ coll.s}^{-1}$$

P33.26   a.   How many molecules strike a 1.00-cm^2 surface during 1 min if the surface is exposed to O_2 at 1.00 atm and 298 K?

b.   Ultra-high-vacuum studies typically employ pressures on the order of 10^{-10} Torr. How many collisions will occur at this pressure at 298 K?

For O_2, $M = 0.0320 \text{ kg mol}^{-1}$ and $Z_c = \dfrac{PN_A}{(2\pi MRT)^{1/2}}$.

(a)   @ 1 atm: $Z_c = \dfrac{(1.00 \text{ atm})\left(\dfrac{101.325 \times 10^3 \text{ Pa}}{1 \text{ atm}}\right)(6.022 \times 10^{23} \text{ mol}^{-1})}{(2\pi(0.032 \text{ kg mol}^{-1})(8.314 \text{ J mol}^{-1} \text{ K}^{-1})(298 \text{ K}))^{1/2}}$

$Z_c = 2.73 \times 10^{27} \text{ m}^{-2} \text{ s}^{-1}$

$\dfrac{dN_c}{dt} = Z_c \times A = (2.73 \times 10^{27} \text{ m}^{-2} \text{ s}^{-1})(1.00 \text{ cm}^2)\left(\dfrac{1 \text{ m}}{100 \text{ cm}}\right)^2$

$\dfrac{dN_c}{dt} = 2.73 \times 10^{23} \text{ coll.s}^{-1}$

(b)   @ 10^{-10} torr: $Z_c = \dfrac{(10^{-10} \text{ torr})\left(\dfrac{133.32 \text{ Pa}}{1 \text{ torr}}\right)(6.022 \times 10^{23} \text{ mol}^{-1})}{(2\pi(0.032 \text{ kg mol}^{-1})(8.314 \text{ J mol}^{-1} \text{ K}^{-1})(298 \text{ K}))^{1/2}}$

$Z_c = 3.60 \times 10^{14} \text{ m}^{-2} \text{ s}^{-1}$

$\dfrac{dN_c}{dt} = Z_c \times A = (3.60 \times 10^{14} \text{ m}^{-2} \text{ s}^{-1})(1.00 \text{ cm}^2)\left(\dfrac{1 \text{ m}}{100 \text{ cm}}\right)^2$

$\dfrac{dN_c}{dt} = 3.60 \times 10^{14} \text{ coll.s}^{-1}$

P33.28   Many of the concepts developed in this chapter can be applied to understanding the atmosphere. Because atmospheric air is comprised primarily of N_2 (roughly 78% by volume), approximate the atmosphere as consisting only of N_2 in answering the following questions:

   a.   What is the single-particle collisional frequency at sea level, with $T = 298$ K and $P = 1.0$ atm? The corresponding single-particle collisional frequency is reported as 10^{10} s^{-1} in the *CRC Handbook of Chemistry and Physics* (62nd ed., p. F-171).

   b.   At the tropopause (11 km in altitude), the collisional frequency decreases to 3.16×10^9 s^{-1}, primarily due to a reduction in temperature and barometric pressure (i.e., fewer particles). The temperature at the tropopause is ~220 K. What is the pressure of N_2 at this altitude?

   c.   At the tropopause, what is the mean free path for N_2?

   The collisional cross section of N_2 is $\sigma = 4.3 \times 10^{-19}$ m^2 (see Table 33.1) and $M = 0.0280$ kg mol^{-1}.

   (a) $$z_{11} = \sqrt{2}\sigma \frac{PN_A}{RT}\left(\frac{8RT}{\pi M}\right)^{1/2}$$

   $$= \sqrt{2}(4.30 \times 10^{-19}\ \text{m}^2)\frac{(1.00\ \text{atm})(6.022 \times 10^{23}\ \text{mol}^{-1})}{(8.21 \times 10^{-2}\ \text{L atm mol}^{-1}\ \text{K}^{-1})(298\ \text{K})}\left(\frac{8(8.314\ \text{J mol}^{-1}\ \text{K}^{-1})(298\ \text{K})}{\pi(0.028\ \text{kg mol}^{-1})}\right)^{1/2}$$

   $$= \sqrt{2}(2.46 \times 10^{22}\ \text{L}^{-1})(4.3 \times 10^{-19}\ \text{m}^2)(475\ \text{m s}^{-1})\left(\frac{1000\ \text{L}}{\text{m}^3}\right)$$

   $$z_{11} = 7 \times 10^9\ \text{s}^{-1}$$

   (b) $$3.16 \times 10^9\ \text{s}^{-1} = \frac{\sqrt{2}(P)(6.022 \times 10^{23}\ \text{mol}^{-1})}{(8.21 \times 10^{-2}\ \text{L atm mol}^{-1}\ \text{K}^{-1})(220.\ \text{K})}(4.30 \times 10^{-19}\ \text{m}^2)\left(\frac{8(8.314\ \text{J mol}^{-1}\ \text{K}^{-1})(220.\ \text{K})}{\pi(0.0280\ \text{kg mol}^{-1})}\right)^{1/2}$$

   $$P = \frac{(3.16 \times 10^9\ \text{s}^{-1})}{\sqrt{2}(3.34 \times 10^{22}\ \text{L}^{-1}\ \text{atm}^{-1})(4.3 \times 10^{-19}\ \text{m}^2)(408\ \text{m s}^{-1})}\left(\frac{1\ \text{m}^3}{1000\ \text{L}}\right)$$

   $$P = 0.38\ \text{atm}$$

   (c) $$\lambda = \left(\frac{RT}{PN_A}\right)\frac{1}{\sqrt{2}\sigma} = \left(\frac{(8.21 \times 10^{-2}\ \text{L atm mol}^{-1}\ \text{s}^{-1})(220.\ \text{K})}{(0.38\ \text{atm})(6.022 \times 10^{23}\ \text{mol}^{-1})}\right)\frac{1}{\sqrt{2}(4.30 \times 10^{-19}\ \text{m}^2)}$$

   $$\lambda = \left(\frac{18.1\ \text{L atm mol}^{-1}}{1.40 \times 10^5\ \text{atm mol}^{-1}\ \text{m}^2}\right)\left(\frac{1\ \text{m}^3}{1000\ \text{L}}\right)$$

   $$\lambda = 1.3 \times 10^{-7}\ \text{m}$$

P33.32   Determine the mean free path for Ar at 298 K at the following pressures:

   a.   0.500 atm

   b.   0.00500 atm

   c. 5.00×10^{-6} atm

   For Ar, $\sigma = 3.6 \times 10^{-19}$ m^2 (see Table 33.1) and $M = 0.040$ kg mol^{-1}.

   (a) $$\lambda = \left(\frac{RT}{PN_A}\right)\frac{1}{\sqrt{2}\sigma} = \left(\frac{(8.21 \times 10^{-2}\ \text{L atm mol}^{-1}\ \text{K}^{-1})(298\ \text{K})}{(0.500\ \text{atm})(6.022 \times 10^{23}\ \text{mol}^{-1})}\right)\frac{1}{\sqrt{2}\ (3.6 \times 10^{-19}\ \text{m}^2)}\left(\frac{1\ \text{m}^3}{1000\ \text{L}}\right)$$

   $$\lambda = 1.6 \times 10^{-7}\ \text{m}$$

(b)   Since the mean free path is inversely proportional to pressure, the result from part (a) can be used to determine the mean free path at pressures specified in parts (b) and (c) as follows:

$$\lambda_{0.005} = \lambda_{0.5}\left(\frac{0.500 \text{ atm}}{0.00500 \text{ atm}}\right) = 1.6 \times 10^{-7} \text{ m } (100)$$

$$\lambda_{0.005} = 1.6 \times 10^{-5} \text{ m}$$

(c) $\lambda_{5\times 10^{-6}} = \lambda_{0.5}\left(\dfrac{0.500 \text{ atm}}{5.00 \times 10^{-6} \text{ atm}}\right) = 1.60 \times 10^{-7} \text{ m } (10^5)$

$$\lambda_{5\times 10^{-6}} = 1.6 \times 10^{-2} \text{ m}$$

P33.35   A comparison of v_{ave}, v_{mp}, and v_{rms} for the Maxwell speed distribution reveals that these three quantities are not equal. Is the same true for the one-dimensional velocity distributions?

$$v_{avg} = \langle v \rangle = \int_{-\infty}^{\infty} v_x \left(\frac{M}{2\pi RT}\right)^{1/2} e^{-\frac{M}{2RT}v_x^2}\, dv_x$$

$$= \left(\frac{M}{2\pi RT}\right)^{1/2} \int_{-\infty}^{\infty} v_x e^{-\frac{M}{2RT}v_x^2}\, dv_x$$

$$v_{avg} = 0$$

$$v_{mp} \Rightarrow 0 = \frac{\partial}{\partial v_x}\left[\left(\frac{M}{2\pi RT}\right)^{1/2} e^{-\frac{M}{2RT}v_x^2}\right]$$

$$0 = \left(\frac{M}{2\pi RT}\right)^{1/2}\left(\frac{M}{2\pi RT}v_x\right)e^{-\frac{M}{2RT}v_x^2}$$

The above equality will be true when $v_x = 0$; therefore, $v_{mp} = 0$.

$$v_{rms} = \langle v^2 \rangle^{1/2} = \left[\int_{-\infty}^{\infty} v_x^2 \left(\frac{M}{2\pi RT}\right)^{1/2} e^{-\frac{M}{2RT}v_x^2}\, dv_x\right]^{1/2}$$

$$= \left[\left(\frac{\beta}{\pi}\right)^{1/2} \int_{-\infty}^{\infty} v_x^2 e^{-\beta v_x^2}\, dv_x\right]^{1/2} \quad \text{for } \beta = \frac{M}{2RT}$$

$$= \left[\left(\frac{\beta}{\pi}\right)^{1/2}\left(2\int_{0}^{\infty} v_x^2 e^{-\beta v_x^2}\, dv_x\right)\right]^{1/2}$$

$$= \left[\left(\frac{\beta}{\pi}\right)^{1/2} \cdot \frac{1}{2}\left(\frac{\pi}{\beta^3}\right)^{1/2}\right]^{1/2}$$

$$= \left[\frac{1}{2}\beta^{-1}\right]^{1/2}$$

$$v_{rms} = \sqrt{\frac{RT}{M}}$$

34 Transport Phenomena

Numerical Problems

P34.3 a. The diffusion coefficient for Xe at 273 K and 1 atm is 0.5×10^{-5} m s^{-1}. What is the collisional cross section of Xe?

  b. The diffusion coefficient of N_2 is three-fold greater than that of Xe under the same pressure and temperature conditions. What is the collisional cross section of N_2?

(a) $D_{Xe} = 0.5 \times 10^{-5}$ m^2 s^{-1} @ 273 K and 1 atm

$$\sigma = \frac{1}{3}\sqrt{\frac{8kT}{\pi M}}\left(\frac{RT}{PN_A}\right)\frac{1}{\sqrt{2}D}$$

$$= \frac{1}{3\sqrt{2}}\sqrt{\frac{8(8.314 \text{ J mol}^{-1}\text{ K}^{-1})(273 \text{ K})}{\pi(0.131 \text{ kg mol}^{-1})}}\left(\frac{(8.21 \times 10^{-2}\text{ L atm mol}^{-1}\text{ K})\left(\frac{1 \text{ m}^3}{1000 \text{ L}}\right)(273 \text{ K})}{(1 \text{ atm})(6.022 \times 10^{23}\text{ mol}^{-1})(0.5 \times 10^{-5}\text{ m}^2\text{ s}^{-1})}\right)$$

$$\sigma = 0.368 \text{ nm}^2 \approx 0.4 \text{ nm}^2$$

(b) The ratio of collisional cross sections is given by:

$$\frac{\sigma_{N_2}}{\sigma_{Xe}} = \frac{D_{Xe}}{D_{N_2}}\sqrt{\frac{M_{Xe}}{M_{N_2}}}$$

$$\sigma_{N_2} = \sigma_{Xe}\left(\frac{D_{Xe}}{D_{N_2}}\right)\sqrt{\frac{M_{Xe}}{M_{N_2}}}$$

$$= (0.368 \text{ nm}^2)\left(\frac{1}{3}\right)\sqrt{\frac{0.131 \text{ kg mol}^{-1}}{0.028 \text{ kg mol}^{-1}}}$$

$$\sigma_{N_2} = 0.265 \text{ nm}^2 \approx 0.3 \text{ nm}^2$$

P34.10 Determine the thermal conductivity of the following species at 273 K and 1.00 atm:

  a. Ar ($\sigma = 0.36$ nm^2)

  b. $Cl_2(\sigma = 0.93$ nm^2)

  c. $SO_2(\sigma = 0.58$ nm^2, geometry: bent)

  You will need to determine $C_{V,m}$ for the species listed. You can assume that the translational and rotational degrees of freedom are in the high-temperature limit, and that the vibrational contribution to $C_{V,m}$ can be ignored at this temperature.

$$\kappa = \frac{1}{3}\frac{C_{V,m}}{N_A}\cdot\left(\frac{8RT}{\pi M}\right)^{1/2}\frac{1}{\sqrt{2}\sigma}$$

164

(a) $\quad C_{V,m}^{Av} = \dfrac{3}{2}R$

$$\kappa = \frac{1}{3}\left(\frac{3}{2}\frac{R}{N_A}\right)\cdot\left(\frac{8RT}{\pi M}\right)^{1/2}\frac{1}{\sqrt{2}\sigma}$$

$$= \frac{8.314\ \text{J mol}^{-1}\ \text{K}^{-1}}{2(6.022\times10^{23}\ \text{mol}^{-1})}\cdot\left(\frac{8(8.314\ \text{J mol}^{-1}\ \text{K}^{-1})(273\ \text{K})}{\pi(0.040\ \text{kg mol}^{-1})}\right)^{1/2}\frac{1}{\sqrt{2}(0.36\ \text{nm}^2)}\cdot\left(\frac{10^9\ \text{nm}}{1\ \text{m}}\right)^2$$

$$= 0.0052\ \text{J K}^{-1}\ \text{m}^{-1}\ \text{s}^{-1}$$

(b) $\quad C_{V,m}^{Cl_2} = C_V^T + C_V^R = \dfrac{3}{2}R + R = \dfrac{5}{2}R$

$$\kappa = \frac{1}{3}\left(\frac{5}{2}\frac{R}{N_A}\right)\left(\frac{8RT}{\pi M}\right)^{1/2}\frac{1}{\sqrt{2}\sigma}$$

$$= \frac{5(8.314\ \text{J mol}^{-1}\ \text{K}^{-1})}{6(6.022\times10^{23}\ \text{mol}^{-1})}\left(\frac{8(8.314\ \text{J mol}^{-1}\ \text{K}^{-1})(273\ \text{K})}{\pi(0.071\ \text{kg mol}^{-1})}\right)^{1/2}\frac{1}{\sqrt{2}(0.93\ \text{nm}^2)}\left(\frac{10^9\ \text{nm}}{1\ \text{m}}\right)^2$$

$$= 0.0025\ \text{J K}^{-1}\ \text{m}^{-1}\ \text{s}^{-1}$$

(c) $\quad C_{V,m}^{SO_2} = C_V^T + C_V^R = \dfrac{3}{2}R + \dfrac{3}{2}R = 3R$

$$\kappa = \frac{1}{3}\left(\frac{3R}{N_A}\right)\left(\frac{8RT}{\pi M}\right)^{1/2}\frac{1}{\sqrt{2}\sigma}$$

$$= \frac{(8.314\ \text{J mol}^{-1}\ \text{K}^{-1})}{(6.022\times10^{23}\ \text{mol}^{-1})}\left(\frac{8(8.314\ \text{J mol}^{-1}\ \text{K}^{-1})(273\ \text{K})}{\pi(0.064\ \text{kg mol}^{-1})}\right)^{1/2}\frac{1}{\sqrt{2}(0.58\ \text{nm}^2)}\left(\frac{10^9\ \text{nm}}{1\ \text{m}}\right)^2$$

$$= 0.0051\ \text{J K}^{-1}\ \text{m}^{-1}\ \text{s}^{-1}$$

P34.11   The thermal conductivity of Kr is $0.0087\ \text{J k}^{-1}\ \text{m}^{-1}\ \text{s}^{-1}$ at 273 K and 1 atm. Estimate the collisional cross section of Kr.

Treating Kr as an ideal monatomic gas, $C_{V,m} = \dfrac{3}{2}R$, and the thermal conductivity is:

$$\kappa = \frac{1}{3}C_{V,m}v_{ave}\frac{1}{\sqrt{2}\sigma} = \frac{1}{3}\left(\frac{3}{2}\frac{R}{N_A}\right)\left(\frac{8RT}{\pi M}\right)^{1/2}\frac{1}{\sqrt{2}\sigma}$$

Rearranging to isolate the collisional cross section:

$$\kappa = \frac{1}{3}C_{V,m}v_{ave}\frac{1}{\sqrt{2}\sigma} = \frac{1}{3}\left(\frac{3}{2}\frac{R}{N_A}\right)\left(\frac{8RT}{\pi M}\right)^{1/2}\frac{1}{\sqrt{2}\sigma}$$

$$\sigma = \frac{1}{3}\left(\frac{3}{2}\frac{R}{N_A}\right)\left(\frac{8RT}{\pi M}\right)^{1/2}\frac{1}{\sqrt{2}\kappa}$$

$$= \frac{1}{2}\left(\frac{8.314\ \text{J mol}^{-1}\ \text{K}^{-1}}{6.022\times10^{23}\ \text{mol}^{-1}}\right)\left(\frac{8(8.314\ \text{J mol}^{-1}\ \text{K}^{-1})(273\ \text{K})}{\pi(0.0838\ \text{kg mol}^{-1})}\right)^{1/2}\frac{1}{\sqrt{2}(0.0087\ \text{J K}^{-1}\ \text{m}^{-1}\ \text{s}^{-1})} = 1.5\times10^{-19}\ \text{m}^2$$

P34.15   The thermal conductivities of acetylene (C_2H_2) and N_2 at 273 K and 1 atm are 0.01866 and 0.0240 $\text{J m}^{-1}\ \text{K}^{-1}$, respectively. Based on these data, what is the ratio of the collisional cross section of acetylene relative to N_2?

$$\frac{\kappa_{C_2H_2}}{\kappa_{N_2}} = \frac{\dfrac{C_{V,m}^{C_2H_2}}{3N_A}\left(\dfrac{8RT}{\pi M_{C_2H_2}}\right)^{1/2}\dfrac{1}{\sqrt{2}\sigma_{C_2H_2}}}{\dfrac{C_{V,m}^{N_2}}{3N_A}\left(\dfrac{8RT}{\pi M_{N_2}}\right)^{1/2}\dfrac{1}{\sqrt{2}\sigma_{N_2}}}$$

$$= \frac{C_{V,m}^{C_2H_2}}{C_{V,m}^{N_2}} \left(\frac{M_{N_2}}{M_{C_2H_2}}\right)^{1/2} \frac{\sigma_{N_2}}{\sigma_{C_2H_2}}$$

Rearranging to isolate the ratio of collisional cross sections:

$$\frac{\sigma_{C_2H_2}}{\sigma_{N_2}} = \frac{C_{V,m}^{C_2H_2}}{C_{V,m}^{N_2}} \left(\frac{M_{N_2}}{M_{C_2H_2}}\right)^{1/2} \frac{\kappa_{N_2}}{\kappa_{C_2H_2}}$$

Both C_2H_2 and N_2 are linear molecules, and will therefore have the same heat capacity (ignoring vibrational degrees of freedom) so that the collision cross section ratio depends only on the mass and thermal conductivity ratios:

$$\frac{\sigma_{C_2H_2}}{\sigma_{N_2}} = \left(\frac{0.0280 \text{ kg mol}^{-1}}{0.0260 \text{ kg mol}^{-1}}\right)^{1/2} \left(\frac{0.0240 \text{ J m}^{-1}\text{ s}^{-1}\text{ K}^{-1}}{0.01866 \text{ J m}^{-1}\text{ s}^{-1}\text{ K}^{-1}}\right)$$

$$\frac{\sigma_{C_2H_2}}{\sigma_{N_2}} = 1.33$$

P34.19   The viscosity of H_2 at 273 K at 1 atm is 84.0 μP. Determine the viscosities of D_2 and HD.

The expression for viscosity is

$$\eta = \frac{1}{3}\left(\frac{8RT}{\pi M}\right)^{1/2} \frac{1}{\sqrt{2}\sigma} \frac{M}{N_A}$$

Taking the ratio of viscosities for two species (denoted as 1 and 2) yields

$$\frac{\eta_2}{\eta_1} = \sqrt{\frac{M_2}{M_1}}\left(\frac{\sigma_1}{\sigma_2}\right)$$

Assuming that the collisional cross sections for the species are the same, the ratio of velocities reduces to:

$$\frac{\eta_2}{\eta_1} = \sqrt{\frac{M_2}{M_1}}$$

Substituting the molecular weights into the above expression yields the following viscosities for D_2 and HD:

$$\eta_{D_2} = \eta_{H_2}\sqrt{\frac{M_{D_2}}{M_{H_2}}}$$

$$= (84.0 \ \mu\text{P})\sqrt{\frac{4.04 \text{ g mol}^{-1}}{2.02 \text{ g mol}^{-1}}}$$

$$\eta_{D_2} = 119 \ \mu\text{P}$$

$$\eta_{HD} = 84.0 \ \mu\text{P}\sqrt{\frac{3.03 \text{ g mol}^{-1}}{2.02 \text{ g mol}^{-1}}}$$

$$= 103 \ \mu\text{P}$$

P34.21   How long will it take to pass 200. mL of H_2 at 273 K through a 10.0-cm-long capillary tube of 0.250 mm if the gas input and output pressures are 1.05 and 1.00 atm, respectively?

The flow rate is given as

$$\frac{\Delta V}{\Delta t} = \frac{\pi r^4}{8\eta}\left(\frac{P_2 - P_1}{x_2 - x_1}\right)$$

Substituting into this expression and solving for Δt yields:

$$\frac{(0.2\,\text{L})}{\Delta t} = \frac{\pi(2.5\times10^{-4}\,\text{m})^4}{8(84\times10^{-6}\,\text{P})\left(\dfrac{0.1\,\text{kg m}^{-1}\,\text{s}^{-1}}{1\,\text{P}}\right)}\left(\frac{1.05\,\text{atm} - 1.00\,\text{atm}}{0.1\,\text{m}}\right)$$

$$\frac{(0.2\,\text{L})}{\Delta t} = 9.13\times10^{-11}\,\text{m}^4\,\text{kg}^{-1}\,\text{s} \cdot \frac{101{,}325\,\text{N m}^{-2}}{1\,\text{atm}} \cdot$$

$$\frac{(0.2\,\text{L})}{\Delta t} = 9.13\times10^{-6}\,\text{m}^3\,\text{s}^{-1}$$

$$\frac{(0.2\,\text{L})\left(\dfrac{1\,\text{m}^3}{1000\,\text{L}}\right)}{9.13\times10^{-5}\,\text{m}^3\,\text{s}^{-1}} = 21.9\,\text{s} = \Delta t$$

P34.25 Poiseuille's Law can be used to describe the flow of blood through blood vessels. Using Poiseuille's Law, determine the pressure drop accompanying the flow of blood through 5.00 cm of the aorta ($r = 1.00$ cm). The rate of blood flow through the body is 0.080 L s^{-1} and the viscosity of blood is approximately 4 cP at 310 K.

$$\frac{\Delta V}{\Delta t} = \frac{\pi r^4}{8\eta}\left(\frac{P_2 - P_1}{x_2 - x_1}\right) = \frac{\pi r^4}{8\eta}\left(\frac{\Delta P}{\Delta x}\right)$$

$$\Delta P = \left(\frac{\Delta V}{\Delta t}\right)\frac{8\eta\,\Delta x}{\pi r^4} = (8.00\times10^{-5}\,\text{m}^3\,\text{s}^{-1})\frac{8(0.00400\,\text{kg m}^{-1}\,\text{s}^{-1})(0.050\,\text{m})}{\pi(0.0100\,\text{m})^4} = 4.07\,\text{Pa}$$

P34.27 Myoglobin is a protein that participates in oxygen transport. For myoglobin in water at 20°C, $\bar{s} = 2.04\times10^{-13}$ s, $D = 1.13\times10^{-10}$ m^2 s^{-1}, and $\bar{V} = 0.740$ cm^3 g^{-1}. The density of water is 0.998 g cm^3 and the viscosity is 1.002 cP at this temperature.

    a.   Using the information provided, estimate the size of myoglobin.

    b.   What is the molecular weight of myoglobin?

    (a)   Using the Stokes–Einstein equation, the radius of myoglobin is:

$$r = \frac{kT}{6\pi\eta D}$$

$$= \frac{(1.38\times10^{-23}\,\text{J K}^{-1})(293\,\text{K})}{6\pi(0.01002\,\text{P})\left(\dfrac{0.1\,\text{kg m}^{-1}\,\text{s}^{-1}}{1\,\text{P}}\right)(1.13\times10^{-10}\,\text{m}^2\,\text{s}^{-1})}$$

$$= 1.89\times10^{-9}\,\text{m}$$

$$= 1.89\,\text{nm}$$

    (b)   The molecular weight of myoglobin can be found as follows:

$$M = \frac{RT\bar{s}}{D(1 - \bar{V}\rho)}$$

$$= \frac{(8.314\,\text{J mol}^{-1}\,\text{K}^{-1})(293\,\text{K})(2.04\times10^{-13}\,\text{s})}{(1.13\times10^{-10}\,\text{m}^2\,\text{s}^{-1})(1 - (0.740\,\text{cm}\,\text{g}^{-1})(0.998\,\text{g cm}^{-3}))}$$

$$= 16.8\,\text{kg mol}^{-1}$$

P34.30 Boundary centrifugation is performed at an angular velocity of 40,000. rpm to determine the sedimentation coefficient of cytochrome c ($M = 13{,}400$ g mol^{-1}) in water at 20°C ($\rho = 0.998$ g cm^{-3}, $\eta = 1.002$ cP). The following data are obtained on the position of the boundary layer as a function of time:

Time (h)	x_b(cm)
0	4.00
2.5	4.11
5.2	4.23
12.3	4.57
19.1	4.91

a.   What is the sedimentation coefficient for cytochrome c under these conditions?

b.   The specific volume of cytochrome c is 0.728 cm g^{-1}. Estimate the size of cytochrome c.

(a)   Using the data from the table, a plot of $\ln\left(\dfrac{x_b}{x_{b,t=0}}\right)$ versus t can be constructed, the slope of which is equal to $\omega^2 \bar{s}$:

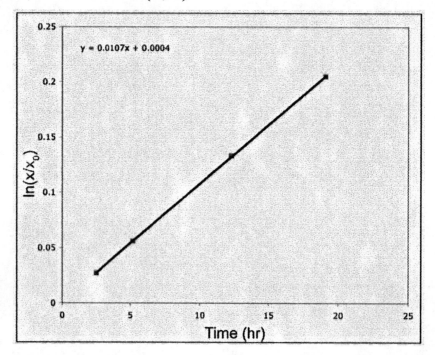

The slope from the best fit to the line is 0.0107 hr^{-1}. Using this slope, the sedimentation coefficient is determined as follows:

$$\omega^2 \bar{s} = 0.0107 \text{ hr}^{-1}$$

$$\bar{s} = \frac{0.0107 \text{ hr}^{-1}}{\omega^2}\left(\frac{1 \text{ hr}}{3600 \text{ s}}\right) = \frac{2.97 \times 10^{-6} \text{ s}^{-1}}{\omega^2}$$

$$= \frac{2.97 \times 10^{-6} \text{ s}^{-1}}{(4.00 \times 10^4 \text{ rev min}^{-1})^2 \left(\dfrac{2\pi \text{ rad}}{1 \text{ rev}}\right)^2 \left(\dfrac{1 \text{ min}}{60 \text{ s}}\right)^2}$$

$$= 1.69 \times 10^{-13} \text{ s}$$

(b)   First, the frictional force is calculated as follows:

$$f = \frac{m(1 - \bar{V}\rho)}{\bar{s}}$$

$$= \frac{\left(\dfrac{13.4 \text{ kg mol}^{-1}}{6.022 \times 10^{23} \text{ mol}^{-1}}\right)(1 - (0.728 \text{ cm}^3 \text{ g}^{-1})(0.998 \text{ g cm}^{-3}))}{1.70 \times 10^{-13} \text{ s}}$$

$$f = 3.59 \times 10^{-11} \text{ kg s}^{-1}$$

With the frictional force, the particle radius can be determined:

$$6\pi\eta r = f$$

$$r = \frac{f}{6\pi\eta} = \frac{3.58 \times 10^{-11} \text{ kg s}^{-1}}{6\pi(0.01002 \text{ P})\left(\dfrac{0.1 \text{ kg m}^{-1}\text{s}^{-1}}{\text{P}}\right)}$$

$$= 1.89 \times 10^{-9} \text{ m}$$

$$= 1.90 \text{ nm}$$

P34.39 In the determination of molar conductivities, it is convenient to define the cell constant, K, as $K = \dfrac{l}{A}$, where l is the separation between the electrodes in the conductivity cell, and A is the area of the electrodes.

a. A standard solution of KCl (conductivity or $\kappa = 1.06296 \times 10^{-6} \text{ S m}^{-1}$ at 298 K) is employed to standardize the cell, and a resistance of $4.2156 \ \Omega$ is measured. What is the cell constant?

b. The same cell is filled with a solution of HCl and a resistance of $1.0326 \ \Omega$ is measured. What is the conductivity of the HCl solution?

(a) The conductivity is defined as

$$\kappa = \frac{\ell}{R \cdot A} = \frac{K}{R}$$

where R is the resistance and K is the cell constant. Using this relationship, the cell constant is determined as follows:

$$\kappa = 1.06296 \times 10^{-6} \text{ s m}^{-1} \quad R = 4.2156 \, \Omega$$

$$K = \kappa R = (1.06296 \times 10^{-6} \text{ S m}^{-1})(4.2156 \, \Omega)$$

$$K = 4.4810 \times 10^{-6} \text{ S m}^{-1} \, \Omega$$

(b) $\kappa = \dfrac{K}{R} = \dfrac{4.48 \times 10^{-6} \text{ S m}^{-1} \, \Omega}{1.0326 \, \Omega}$

$\qquad = 4.3395 \times 10^{-6} \text{ S m}^{-1}$

35 Elementary Chemical Kinetics

Numerical Problems

P35.2   Consider the first-order decomposition of cyclobutane at $438\,^{\circ}C$ at constant volume: $C_4H_8(g) \rightarrow 2C_2H_4(g)$.

   a.   Express the rate of the reaction in terms of the change in total pressure as a function of time.

   b.   The rate constant for the reaction is $2.48 \times 10^{-4}\ s^{-1}$. What is the half-life?

   c.   After initiation of the reaction, how long will it take for the initial pressure of C_4H_8 to drop to 90.0% of its initial value?

   (a)   The rate is given as

$$R = -\frac{1}{RT}\frac{d\,P_{C_4H_8}}{dt} = \frac{1}{2\,RT}\frac{d\,P_{C_2H_4}}{dt}$$

The pressure at time t is given as $P_t = P_{t=0} - P_{C_4H_8} + P_{C_2H_4}$, where $P_{t=0}$ is the initial pressure, and $P_{C_4H_8}$ and $P_{C_2H_4}$ represent the pressures of the individual gases at a specific time. By stoichiometry, $P_{C_4H_8} = \frac{1}{2}P_{C_2H_4}$ so that:

$$P_t = P_{t=0} - \frac{1}{2}P_{C_2H_4} + P_{C_2H_4} = P_{t=0} + \frac{1}{2}P_{C_2H_4}$$

Since $P_{C_2H_4}$ is dependent on time and the initial pressure is time independent resulting in the following:

$$\frac{dP_t}{dt} = \frac{1}{2}\frac{d\,P_{C_2H_4}}{dt}$$

and

$$R = \frac{1}{RT}\frac{dP_t}{dt}$$

   (b)   The reaction is first order with respect to C_4H_8; therefore:

$$t_{1/2} = \frac{\ln 2}{k}$$

so

$$t_{1/2} = \frac{0.693}{2.48 \times 10^{-4}\ s^{-1}}$$

$$= 2.79 \times 10^3\ s$$

   (c)   Using the integrated rate law

$$-kt = \ln\left(\frac{[A]}{[A]_0}\right)$$

where

$$\frac{[A]}{[A]_0} = 0.900$$

one finds

$$t = -\frac{\ln(0.900)}{2.48 \times 10^{-4}\ s^{-1}}$$

$$= \frac{0.105}{2.48 \times 10^{-4}\ s^{-1}}$$

$$= 425\ s$$

P35.7 The reaction rate as a function of initial reactant pressures was investigated for the reaction $2NO(g) + 2H_2(g) \rightarrow N_2(g) + 2H_2O(g)$, and the following data were obtained:

Run	P_0H_2 (kPa)	P_0NO (kPa))	Rate (kPa s^{-1})
1	53.3	40.0	0.137
2	53.3	20.3	0.033
3	38.5	53.3	0.213
4	19.6	53.3	0.105

What is the rate law expression for this reaction?

The rate law expression is:

$$Rate = kP_{H_2}^{\alpha}\, P_{NO}^{\beta}$$

We first find the order of the reaction. Using the method of initial rates for the first two runs, the order of the reaction with respect to NO is:

$$\frac{Rate_1}{Rate_2} = \frac{P_{NO_1}^{\beta}}{P_{NO_2}^{\beta}} = \left(\frac{P_{NO_1}}{P_{NO_2}}\right)^{\beta}$$

$$\ln\left(\frac{Rate_1}{Rate_2}\right) = \beta \ln\left(\frac{P_{NO_1}}{P_{NO_2}}\right)$$

$$\frac{\ln\left(\dfrac{Rate_1}{Rate_2}\right)}{\ln\left(\dfrac{P_{NO_1}}{P_{NO_2}}\right)} = \beta$$

$$\frac{\ln\left(\dfrac{0.137}{0.033}\right)}{\ln\left(\dfrac{40}{20.3}\right)} = 2.09 \approx 2 = \beta$$

Next, for the second two runs, the order of the reaction with respect to H_2 is:

$$\frac{Rate_3}{Rate_4} = \frac{P_{H_2,\ run3}^{\alpha}}{P_{H_2,\ run4}^{\alpha}}$$

$$\ln\left(\frac{Rate_3}{Rate_4}\right) = \alpha\, \ln\left(\frac{P_{H_2,\ run3}}{P_{H_2,\ run4}}\right)$$

$$\frac{\ln\left(\dfrac{Rate_3}{Rate_4}\right)}{\ln\left(\dfrac{P_{H_2\ \text{run3}}}{P_{H_2\ \text{run4}}}\right)} = \alpha$$

$$\frac{\ln\left(\dfrac{0.213}{0.105}\right)}{\ln\left(\dfrac{38.5}{19.6}\right)} = 1.04 \approx 1 = \alpha$$

With the order dependence, the rate constant can be determined using the known concentrations (or pressures) of each species for a given run. Using the first run:

$$Rate_1 = kP_{H_2}P_{NO}^2$$

$$\frac{Rate_1}{P_{H_2}P_{NO}^2} = \frac{0.137 \text{ kPa s}^{-1}}{(53.3 \text{ kPa})(40.0 \text{ kPa})^2} = k$$

$$1.61 \times 10^{-6} \text{ kPa}^2 \text{ s}^{-1}$$

Alternatively, one could determine the rate constant by recognizing that the reaction is first order in H_2 so that a plot of $\ln(P_{H_2})$ versus t employing data where NO is constant will yield a straight line, the slope of which is equal to $-k$. Finally, the rate law expression is:

$$Rate = 1.61 \times 10^{-6} \text{ kPa}^{-2} \text{ s}^{-1}P_{H_2}P_{NO}^2$$

P35.12 Consider the schematic reaction $A \xrightarrow{k} P$.

    a.   If the reaction is one-half order with respect to [A], what is the integrated rate law expression for this reaction?

    b.   What plot would you construct to determine the rate constant k for the reaction?

    c.   What would be the half-life for this reaction? Will it depend on initial concentration of the reactant?

    (a) $\dfrac{d[A]}{dt} = -k[A]^{1/2}$

$$[A]^{-1/2}d[A] = -k\,dt$$

$$\int_{[A]_0}^{[A]} [A]^{-1/2}d[A] = -k\int_0^t dt$$

$$2([A]^{1/2} - [A]_0^{1/2}) = -kt$$

$$[A]^{1/2} = [A]_0^{1/2} - \frac{kt}{2}$$

$$[A] = \left([A]_0^{1/2} - \frac{kt}{2}\right)^2$$

    (b)   A plot of $[A]^{1/2}$ vs. t will be linear, and the rate constant is equal to -2 times the slope of the line.

    (c)   At $t_{1/2}$, $[A] = 0.5[A]_0$:

$$\frac{1}{2}[A]_0 = \left([A]_0^{1/2} - \frac{1}{2}kt_{1/2}\right)^2$$

$$\frac{1}{\sqrt{2}}[A]_0^{1/2} = [A]_0^{1/2} - \frac{1}{2}kt_{1/2}$$

$$\frac{1}{2}kt_{1/2} = \left(1 - \frac{1}{\sqrt{2}}\right)[A]_0^{1/2}$$

$$t_{1/2} = \frac{2}{k}\left(1 - \frac{1}{\sqrt{2}}\right)[A]_0^{1/2}$$

$$t_{1/2} = \frac{1}{k}(2 - \sqrt{2})[A]_0^{1/2}$$

The half-life depends on the square root of the initial concentration.

P35.14 The half-life of ^{238}U is 4.5×10^9 years. How many disintegrations occur in 1 minute for a 10.0-mg sample of this element?

The rate constant can be found from the half-life:

$$t_{1/2} = \frac{\ln 2}{k} \quad or \quad k = \frac{\ln 2}{t_{1/2}}$$

Thus

$$k = \frac{0.693}{4.5 \times 10^9 \text{ yr}}\left(\frac{1 \text{ yr}}{364.25 \text{ day}}\right)\left(\frac{1 \text{ day}}{24 \text{ hr}}\right)\left(\frac{1 \text{ hr}}{60 \text{ min}}\right)$$

$$k = 2.94 \times 10^{-16} \text{ min}^{-1}$$

Converting 10.0 mg into the number of ^{238}U atoms:

$$N_{238U} = (1 \times 10^{-2} \text{ g})(238 \text{ g mol}^{-1})(6.022 \times 10^{23} \text{ mol}^{-1}) = 1.43 \times 10^{24}$$

Employing the rate law, the number of ^{235}U atoms left over after 1 minute is:

$$N = N_o \, e^{-kt}$$

$$= (1.43 \times 10^{24}) \, e^{-2.94 \times 10^{-16} \text{ min}^{-1} \cdot 1 \text{ min}}$$

$$= 1.43 \times 10^{24}$$

The rate constant is so small that a negligible number of disintegrations occur in 1 minute.

P35.19 A technique for radioactively labeling proteins is electrophilic radioiodination, in which an aromatic substitution of ^{131}I onto a tyrosine residue is performed as follows:

Using the activity of ^{131}I, one can measure protein lifetimes in a variety of biological processes. ^{131}I undergoes beta decay with a half-life of 8.02 days. Initially a protein labeled with ^{131}I has a specific activity of 1.0 μCi, which corresponds to 37,000 decay events every second. The protein is suspended in aqueous solution and exposed to oxygen for 5 days. After isolating the protein from solution, the protein sample is found to have a specific activity of 0.32 μCi. Is oxygen reacting with the tyrosine residues of the protein resulting in the loss of ^{131}I?

To answer this question, we need to determine what the specific activity of the protein should be due to beta emission of ^{131}I. First, we need to determine the rate constant for beta emission using the half-life information provided in the problem:

$$\ln 2 = kt_{1/2}$$

$$\frac{\ln 2}{t_{1/2}} = k$$

$$\frac{\ln 2}{8.02 \text{ days}} = 0.0864 \text{ days}^{-1} = k$$

With k, we can determine how much ^{131}I should be present after 5 days.

$$[^{131}\text{I}] = [^{131}\text{I}]_0 e^{-kt} = (1.0 \, \mu\text{Ci}) e^{-(0.0864 \text{ days}^{-1} \times 5 \text{ days})} = (1.0 \, \mu\text{Ci})(0.65) = 0.65 \, \mu\text{Ci}$$

Since the amount of labeled protein is less than this amount, we can conclude that there was another decay process resulting in the loss of labeled protein, and that oxygen reacted with the tyrosine residues.

P35.23   For the sequential reaction $A \xrightarrow{k_A} B \xrightarrow{k_B} C$, the rate constants are $k_A = 5 \times 10^6 \text{ s}^{-1}$ and $k_B = 3 \times 10^6 \text{ s}^{-1}$. Determine the time at which [B] is at a maximum.

For a sequential reaction

$$[B] = \frac{k_A}{k_B - k_A}(e^{-k_A t} - e^{-k_B t})[A_0]$$

The maximum occurs when

$$\frac{d[B]}{dt} = 0 = \frac{k_A}{k_B - k_A}[A]_0 \frac{d}{dt}(e^{-k_A t} - e^{-k_B t})$$

$$0 = \frac{k_A}{k_B - k_A}[A]_0(-k_A e^{-k_A t} + k_B e^{-k_B t})$$

The above equality will be true when the term in parentheses equals zero, therefore:

$$k_A e^{-k_A t} = k_B e^{-k_B t}$$

$$\ln k_A - k_A t = \ln k_B - k_B t$$

$$\ln k_A - \ln k_B = (k_A - k_B)t$$

$$\frac{1}{k_A - k_B}\ln\left(\frac{k_A}{k_B}\right) = t$$

Substituting the values of k_A and k_B into the previous expression yields:

$$t = \frac{1}{(5 \times 10^6 \text{ s}^{-1}) - (3 \times 10^6 \text{ s}^{-1})}\ln\left(\frac{5 \times 10^6 \text{ s}^{-1}}{3 \times 10^6 \text{ s}^{-1}}\right)$$

$$= (5 \times 10^{-7} \text{ s})(0.511)$$

$$t = 2.6 \times 10^{-7} \text{ s}$$

P35.26   For a type II second-order reaction, the reaction is 60% complete in 60 seconds when $[A]_0 = 0.1 \text{ M}$ and $[B]_0 = 0.5 \text{ M}$.

a.   What is the rate constant for this reaction?

b.   Will the time for the reaction to reach 60% completion change if the initial reactant concentrations are decreased by a factor of two?

(a)   The integrated rate law expression for a second-order reaction of type II is:

$$kt = \frac{1}{[B]_0 - [A]_0}\ln\left(\frac{[B]/[B]_0}{[A]/[A]_0}\right)$$

at $t = 60$ s, $[A] = 0.04$ M with 1:1 stoichiometry so that $[B] = 0.44$ M. Substituting these values into the above expression and using $t = 60$ s yields:

$$k = \frac{1}{60\,\text{s}(0.5\,\text{M} - 0.1\,\text{M})}\ln\left(\frac{0.44\,\text{M}/0.5\,\text{M}}{0.04\,\text{M}/0.1\,\text{M}}\right)$$

$$= (0.0417\,\text{M}^{-1}\,\text{s}^{-1})(0.788)$$

$$= 0.0329\,\text{M}^{-1}\,\text{s}^{-1} \approx 0.03\,\text{M}^{-1}\,\text{s}^{-1}$$

(b)   The time will double, assuming the k value is the same. Numerically checking this expectation:

$$t = \frac{1}{k([B]_0 - [A]_0)}\ln\left(\frac{[B]/[B]_0}{[A]/[A]_0}\right)$$

$$= \frac{1}{(0.0329\,\text{M}^{-1}\,\text{s}^{-1})\,(0.25\,\text{M} - 0.05\,\text{M})}\ln\left(\frac{0.22\,\text{M}/0.25\,\text{M}}{0.02\,\text{M}/0.05\,\text{M}}\right)$$

$$= (152\,\text{s})(0.788)$$

$$= 120\,\text{s}$$

P35.30   In the stratosphere, the rate constant for the conversion of ozone to molecular oxygen by atomic chlorine is $Cl + O_3 \rightarrow ClO \bullet + O_2$ $[k = (1.7 \times 10^{10}\,\text{M}^{-1}\,\text{s}^{-1})e^{-260K/T}]$.

a.   What is the rate of this reaction at 20 km where $[Cl] = 5 \times 10^{-17}$ M, $[O_3] = 8 \times 10^{-9}$ M, and $T = 220.$ K?

b.   The actual concentrations at 45 km are $[Cl] = 3 \times 10^{-15}$ M and $[O_3] = 8 \times 10^{-11}$ M. What is the rate of the reaction at this altitude where $T = 270.$ K?

c.   (Optional) Given the concentrations in part (a), what would you expect the concentrations at 45 km to be assuming that the gravity represents the operative force defining the potential energy?

(a)   Based on the units of k, the reaction is second order overall so that the rate law expression is:

$$Rate = k[Cl][O_3]$$

For $[Cl] = 5 \times 10^{-17}$ M, $[O_3] = 8 \times 10^{-9}$ M, and $T = 220.$ K

$$k = 1.7 \times 10^{10}\,\text{M}^{-1}\,\text{s}^{-1}e^{-260.\,K/220.\,K}$$

$$k = 5.21 \times 10^9\,\text{M}^{-1}\,\text{s}^{-1}$$

$$Rate = 5.21 \times 10^9\,\text{M}^{-1}\,\text{s}^{-1}(5 \times 10^{-17}\,\text{M})(8 \times 10^{-9}\,\text{M})$$

$$= 2.08 \times 10^{-15}\,\text{M}\,\text{s}^{-1}$$

(b) $[Cl] = 3 \times 10^{-15}$ M $[O_3] = 8 \times 10^{-11}$ M $T = 270.$ K

$$k = 1.7 \times 10^{10}\,\text{M}^{-1}\,\text{s}^{-1}e^{-260.\,K/270.\,K}$$

$$k = 6.49 \times 10^9\,\text{M}^{-1}\,\text{s}^{-1}$$

$$Rate = 6.49 \times 10^9\,\text{M}^{-1}\,\text{s}^{-1}(3 \times 10^{-15}\,\text{M})(8 \times 10^{-11}\,\text{M})$$

$$= 1.56 \times 10^{-15}\,\text{M}\,\text{s}^{-1}$$

(c)   The ratio of pressures at two altitudes is given by:

$$\frac{[P]_1}{[P]_2} = e^{-\frac{mg(h_1 - h_2)}{kT}}$$

Using this expression to determine the difference in concentration for Cl at 45 versus 20 km yields:

$$\frac{[Cl]_{45}}{[Cl]_{20}} = e^{-\frac{(0.035 \text{ kg mol}^{-1})(1/N_A)(9.80 \text{ m s}^{-2})(2.5\times10^4 \text{ m})}{(1.38\times10^{-23} \text{ J K}^{-1})(270 \text{ K})}} = 0.0219$$

$$[Cl]_{45} = (5\times10^{-17} \text{ M})(0.0219) = 1.10\times10^{-18} \text{ M}$$

Performing the same calculation for O_3:

$$\frac{[O_3]_{45}}{[O_3]_{20}} = e^{-\frac{(0.048 \text{ kg mol}^{-1})(1/N_A)(9.80 \text{ m s}^{-2})(2.5\times10^4 \text{ m})}{(1.38\times10^{-23} \text{ J K}^{-1})(270 \text{ K})}} = 0.0053$$

$$[O_3]_{45} = (8\times10^{-9} \text{ M})(0.0053) = 4.24\times10^{-11} \text{ M}$$

Finally, the rate is:

$$Rate = 6.49\times10^9 \text{ M}^{-1} \text{s}^{-1}(1.10\times10^{-18} \text{ M})(4.24\times10^{-11} \text{ M})$$

$$= 3.03\times10^{-19} \text{ M s}^{-1}$$

Notice that since this simple model for the concentration dependence versus altitude significantly underestimates the concentration of Cl, the rate of ozone depletion by reaction with Cl is also significantly underestimated.

P35.36 The conversion of NO_2 to NO and O_2 can occur through the following reaction:

$$NO_2(g) \rightarrow 2NO(g) + O_2(g)$$

The activation energy for this reaction is 111 kJ mol^{-1} and the preexponential factor is $2.0\times10^{-9} \text{ M}^{-1}\text{s}^{-1}$; assume that these quantities are temperature independent.

a.   What is the rate constant for this reaction at 298 K?

$$k = Ae^{-E_a/RT} = (2.0\times10^{-9} \text{ M}^{-1}\text{s}^{-1})e^{-(1.11\times10^5 \text{ J mol}^{-1})/(8.314 \text{ J mol}^{-1} \text{K}^{-1})(298 \text{ K})}$$

$$= (2.0\times10^9 \text{ M}^{-1}\text{s}^{-1})e^{-44.8} = 7.0\times10^{-11} \text{ M}^{-1}\text{s}^{-1}$$

b.   What is the rate constant for this reaction at the tropopause where $T = 225 \text{ K}$?

Assuming that k and E_a are temperature independent:

$$k = Ae^{-E_a/RT} = (2.0\times10^{-9} \text{ M}^{-1}\text{s}^{-1})e^{-(1.11\times10^5 \text{ J mol}^{-1})/(8.314 \text{ J mol}^{-1} \text{K}^{-1})(225 \text{ K})}$$

$$= (2.0\times10^9 \text{ M}^{-1}\text{s}^{-1})e^{-59.3} = 3.5\times10^{-17} \text{ M}^{-1}\text{s}^{-1}$$

Notice that temperature has a pronounced effect on the rate constant for this reaction.

P35.42 Consider the reaction $A + B \underset{k'}{\overset{k}{\rightleftharpoons}} P$. A temperature-jump experiment is performed where the relaxation time constant is measured to be 310 μs, resulting in an equilibrium where $K_{eq} = 0.7$ with $[P]_{eq} = 0.2 \text{ M}$. What are k and k'? (Watch the units!)

$$\tau = 310\times10^{-6} \text{ s} \qquad K_{eq} = 0.7 \qquad [P]_{eq} = 0.2 \text{ M}$$

Assuming the following rate law,

$$\frac{d[A]}{dt} = -k[A][B] + k'[P] = 0,$$

The post-jump equilibrium concentrations with respect to the initial concentrations and concentration shift are:

$$[A] - \xi = [A]_{eq}$$
$$[B] - \xi = [B]_{eq}$$
$$[P] + \xi = [P]_{eq}$$

Therefore, the differential rate expression for the concentration shift, ξ, is:

$$\frac{d\xi}{dt} = -k([A]_{eq} + \xi)([B]_{eq} + \xi) + k'([P]_{eq} - \xi)$$

$$\frac{d\xi}{dt} = -\xi(k[A]_{eq} + k[B]_{eq} + k\xi + k')$$

$$= -\xi(k([A]_{eq} + k[B]_{eq}) + k') + O(\xi^2)$$

Ignoring terms on the order ξ^2, the relaxation time is:

$$\tau = [k([A]_{eq} + k[B]_{eq}) + k']^{-1}$$

Next, the equilibrium constant is given by:

$$K = \frac{k}{k'} = \frac{[P]_{eq}}{[A]_{eq}[B]_{eq}} = 0.7$$

If we assume that $[A]_0 = [B]_0$ then $[A]_{eq} = [B]_{eq}$ and

$$0.7 = \frac{0.2\,M}{x^2} \Rightarrow x = 0.535\,M = [A]_{eq} = [B]_{eq}$$

And using the expression for K, we know that $k^+ = 0.7\,k^{+'}$. Use these last two results in the expression for the relaxation time yields:

$$310 \times 10^{-6}\,s = \frac{1}{k(0.535 + 0.535) + k'}$$

$$310 \times 10^{-6}\,s = \frac{1}{(0.7k')(1.070) + k'}$$

$$310 \times 10^{-6}\,s = \frac{1}{k'(1.749)}$$

$$k' = 1800\,s^{-1}$$

$$k = 0.7k' = 1300\,M^{-1}\,s^{-1}$$

The units of the rate constants are consistent with the forward reaction being second order and the reverse reaction being first order.

P35.45 Imidazole is a common molecular species in biological chemistry. For example, it constitutes the side chain of the amino acid histidine. The rate constant for the protonation reaction is $5.5 \times 10^{10}\,M^{-1}\,s^{-1}$. Assuming that the reaction is diffusion controlled, estimate the diffusion coefficient of imidazole when $D(H^+) = 9.31 \times 10^{-5}\,cm^2\,s^{-1}$, $r(H^+) \sim 1.0\,\text{Å}$, and $r(\text{imidazole}) = 6\,\text{Å}$. Use this information to predict the rate of deprotonation of imidazole by OH^- ($D = 5.30 \times 10^{-5}\,cm^2\,s^{-1}$ and $r = \sim 1.5\,\text{Å}$).

The diffusion constant for imidazole (D_I) can be determined as follows:

$$k_d = 4\pi N_A (r_H + r_I)(D_H + D_I)$$

$$D_I = \frac{k_d}{4\pi(r_H + r_I)} - D_H$$

$$D_I = \frac{(5.5 \times 10^{10}\,M^{-1}\,s^{-1})}{4\pi(6.022 \times 10^{23}\,mol^{-1}) \times (1 \times 10^{-8}\,cm + 6 \times 10^{-8}\,cm)}\left(\frac{1000\,cm^3}{L}\right) - (9.31 \times 10^{-5}\,cm^2\,s^{-1})$$

$$D_I = 1 \times 10^{-5}\,cm^2\,s^{-1}$$

The rate constant for the deprotonation of imidazole by OH^- is given by:

$$k_d = 4\pi N_A (r_{OH} + r_I)(D_{OH} + D_I)$$
$$k_d = 4\pi (6.022 \times 10^{23} \text{ mol}^{-1}) \times (1.5 \times 10^{-8} \text{cm} + 6.0 \times 10^{-8} \text{cm})$$
$$\times \left(\frac{L}{1000 \text{ cm}^3}\right) \times (5.30 \times 10^{-5} \text{ cm}^2 \text{ s}^{-1} + 1.073 \times 10^{-5} \text{ cm}^2 \text{ s}^{-1})$$
$$k_d = 3.62 \times 10^{10} \text{ M}^{-1} \text{ s}^{-1} \approx 3.6 \times 10^{10} \text{ M}^{-1} \text{ s}^{-1}$$

P35.50 Consider the "unimolecular" isomerization of methylcyanide, a reaction that will be discussed in detail in Chapter 36:

$$CH_3NC(g) \rightarrow CH_3CN(g)$$

The Arrhenius parameters for this reaction are $A = 2.5 \times 10^{16} \text{ s}^{-1}$ and $E_a = 272 \text{ kJ mol}^{-1}$. Determine the Eyring parameters $\Delta H^{\ddagger}$ and $\Delta S^{\ddagger}$ for this reaction with $T = 300. \text{ K}$.

For a unimolecular gas-phase reaction:

$$E_a = \Delta H^{\ddagger} + RT$$
$$\Delta H^{\ddagger} = 272 \times 10^3 \text{ J mol}^{-1} - (8.314 \text{ J mol}^{-1} \text{ K}^{-1})(300 \text{ K})$$
$$= 272 \times 10^3 \text{ J mol}^{-1} - 2.49_4 \times 10^3 \text{ J mol}^{-1}$$
$$= 269.5 \times 10^3 \text{ J mol}^{-1}$$

$$A = \frac{ek_BT}{h} e^{\Delta S^{\ddagger}/R}$$

$$\Delta S^{\ddagger} = R\ln\left(\frac{Ah}{ek_BT}\right)$$
$$= 8.314 \text{ J mol}^{-1} \text{K}^{-1} \ln\left(\frac{(2.5 \times 10^{16} \text{ s}^{-1})(6.626 \times 10^{-34} \text{ J s})}{e(1.38 \times 10^{-23} \text{ J K}^{-1})(300. \text{ K})}\right)$$
$$= (8.314 \text{ J mol}^{-1} \text{ K}^{-1})\ln(1472)$$
$$= 60.6 \text{ J mol}^{-1} \text{ K}^{-1}$$

36 Complex Reaction Mechanisms

Numerical Problems

P36.1 A proposed mechanism for the formation of $N_2O_5(g)$ from $NO_2(g)$ and $O_3(g)$ is

$$NO_2(g) + O_3(g) \xrightarrow{k_1} NO_3(g) + O_2(g)$$
$$NO_3(g) + NO_2(g) + M(g) \xrightarrow{k_2} N_2O_5(g) + M(g)$$

Determine the rate law expression for the production of $N_2O_5(g)$ given this mechanism.

$$\frac{d[N_2O_5]}{dt} = k_2[NO_2][NO_3]$$
$$\frac{d[NO_3]}{dt} = k_1[NO_2][O_3][M] - k_2[NO_2][NO_3][M]$$

Applying the steady-state approximation to the intermediate NO_3 and substituting back into the differential rate expression for N_2O_5 yields:

$$\frac{d[NO_3]}{dt} = 0 = k_1[NO_2][O_3][M] - k_2[NO_2][NO_3][M]$$
$$k_2[NO_2][NO_3][M] = k_1[NO_2][O_3][M]$$
$$[NO_3] = \frac{k_1}{k_2}[O_3]$$

$$\frac{d[N_2O_5]}{dt} = k_2[NO_2][NO_3]$$
$$= k_2[NO_2]\left(\frac{k_1}{k_2}[O_3]\right)$$
$$= k_1[NO_2][O_3]$$

The mechanism predicts that the reaction is first order in NO_2 and O_3, second order overall.

P36.3 In the troposphere carbon monoxide and nitrogen dioxide undergo the following reaction:

$$NO_2(g) + CO(g) \rightarrow NO(g) + CO_2(g)$$

Experimentally, the rate law for the reaction is second order in $NO_2(g)$, and $NO_3(g)$ has been identified as an intermediate in this reaction. Construct a reaction mechanism that is consistent with these experimental observations.

A potential mechanism that is consistent with the experimental observations is:

$$NO_2(g) + NO_2(g) \xrightarrow{k_1} NO_3(g) + NO(g)$$
$$\underline{NO_3(g) + CO(g) \xrightarrow{k_2} NO_2(g) + CO_2(g)}$$
$$NO_2(g) + CO(g) \rightarrow CO_2(g) + NO(g)$$

Expressing the reaction rate in terms of the rate for $CO_2(g)$ production yields:

$$R = \frac{d[CO_2]}{dt} = k_2[NO_3][CO]$$

Writing the differential rate expression for $NO_3(g)$ and applying the steady-state approximation yields:

$$\frac{d[NO_3]}{dt} = k_1[NO_2]^2 - k_2[NO_3][CO] = 0$$

$$k_2[NO_3][CO] = k_1[NO_2]^2$$

$$[NO_3] = \frac{k_1}{k_2}\frac{[NO_2]^2}{[CO]}$$

Substituting this result into the rate expression for the reaction results in:

$$R = k_2[NO_3][CO] = k_2\left(\frac{k_1}{k_2}\frac{[NO_2]^2}{[CO]}\right)[CO] = k_1[NO_2]^2$$

P36.4   The Rice–Herzfeld mechanism for the thermal decomposition of acetaldehyde ($CH_3CO(g)$) is

$$CH_3CHO(g) \xrightarrow{k_1} CH_3\cdot(g) + CHO\cdot(g)$$

$$CH_3\cdot(g) + CH_3CHO(g) \xrightarrow{k_2} CH_4(g) + CH_2CHO\cdot(g)$$

$$CH_2CHO\cdot(g) \xrightarrow{k_3} CO(g) + CH_3\cdot(g)$$

$$CH_3\cdot(g) + CH_3\cdot(g) \xrightarrow{k_4} C_2H_6(g)$$

Using the steady-state approximation, determine the rate of methane ($CH_4(g)$) formation.

The differential rate expressions for methane and relevant intermediate species are:

$$\frac{d[CH_4]}{dt} = k_2[CH_3\cdot][CH_3CHO]$$

$$\frac{d[CH_2CHO]}{dt} = k_2[CH_3\cdot][CH_3CHO] - k_3[CH_2CHO\cdot]$$

$$\frac{d[CH_3\cdot]}{dt} = k_1[CH_3CHO] - k_2[CH_3\cdot][CH_3CHO] + k_3[CH_2CHO\cdot] - 2k_4[CH_3\cdot]^2$$

Applying the steady-state approximation to the differential rate expressions for the intermediates:

$$\frac{d[CH_2CHO\cdot]}{dt} = \frac{d[CH_3\cdot]}{dt} = 0$$

or

$$0 = k_2[CH_3\cdot][CH_3CHO] - k_3[CH_2CHO\cdot]$$

$$0 = k_1[CH_3CHO] - k_2[CH_3\cdot][CH_3CHO] + k_3[CH_2CHO\cdot] - 2k_4[CH_3\cdot]^2$$

Adding the above two equations yields the following expression:

$$2k_4[CH_3\cdot]^2 = k_1[CH_3CHO]$$

$$[CH_3\cdot] = \sqrt{\frac{k_1}{2k_4}}[CH_3CHO]^{1/2}$$

Substitution into the rate expression for $[CH_4]$ yields

$$\frac{d[CH_4]}{dt} = k_2\left[\sqrt{\frac{k_1}{2k_4}}[CH_3CHO]^{1/2}\right][CH_3CHO]$$

$$\frac{d[CH_4]}{dt} = k_{eff}[CH_3CHO]^{3/2} \quad \text{where} \quad k_{eff} = k_2\sqrt{\frac{k_1}{2k_4}}$$

P36.6   Consider the formation of double-stranded (DS) DNA from two complementary single strands (S and S′) through the following mechanism involving an intermediate helix (IH):

$$S + S' \underset{k_{-1}}{\overset{k_1}{\rightleftharpoons}} IH$$

$$IH \xrightarrow{k_2} DS$$

a.   Derive the rate law expression for this reaction employing the preequilibrium approximation.

The rate law expression for the reaction is:

$$R = \frac{d[DS]}{dt} = k_2[IH]$$

We apply the steady-state approximation to define [IH] as follows:

$$k_1[S][S'] = k_{-1}[IH]$$

$$[IH] = \frac{k_1}{k_{-1}}[S][S']$$

Substituting this result into the rate law expression yields:

$$R = k_2[IH] = k_2\frac{k_1}{k_{-1}}[S][S']$$

b.   What is the corresponding rate law expression for the reaction employing the steady-state approximation for the intermediate IH?

Writing the differential rate expression for IH and applying the steady-state approximation yields:

$$\frac{d[IH]}{dt} = k_1[S][S'] - k_{-1}[IH] - k_2[IH] = 0$$

$$[IH] = \frac{k_1[S][S']}{k_{-1} + k_2}$$

Substituting this result into the rate law expression for the reaction yields:

$$R = k_2[IH] = \frac{k_2 k_1}{k_{-1} + k_2}[S][S']$$

Notice that if $k_{-1} \gg k_2$, then the steady-state rate law becomes equivalent to the preequilibrium rate law.

P36.8   a.   For the hydrogen–bromine reaction presented in Problem P36.7, imagine initiating the reaction with only Br_2 and H_2 present. Demonstrate that the rate law expression at $t = 0$ reduces to

$$\left(\frac{d[HBr]}{dt}\right)_{t=0} = 2k_2\left(\frac{k_1}{k_5}\right)^{1/2}[H_2]_0[Br_2]_0^{1/2}$$

b.   The activation energies for the rate constants are as follows:

Rate Constant	ΔE_a (kJ/mol)
k_1	192
k_2	0
k_5	74

What is the overall activation energy for this reaction?

c.   How much will the rate of the reaction change if the temperature is increased to 400. K from 298 K?

(a)   At $t = 0$: $[Br_2] = [Br_2]_0$ $[H_2] = [H_2]_0$ $[HBr] = 0$
Thus:

$$\left(\frac{d[HBr]}{dt}\right)_{t=0} = \frac{k[Br_2]_0^{1/2}[H]_0}{1 + m \cdot \dfrac{0}{[Br_2]_0}} = k[Br_2]_0^{1/2}[H_2]_0$$

$$\left(\frac{d[HBr]}{dt}\right)_{t=0} = 2k_2\sqrt{\frac{k_1}{k_{-1}}}[Br_2]_0^{1/2}[H_2]_0$$

(b) $$k = 2k_2\sqrt{\frac{k_1}{k_{-1}}} = 2A_2 e^{-Ea_2/RT}\left(\frac{A_1 e^{-Ea_1/RT}}{A_{-1} e^{-Ea_{-1}/RT}}\right)^{1/2}$$

$$k = 2A_2\left(\frac{A_1}{A_{-1}}\right)^{1/2}\left(e^{-\frac{2Ea_2 + Ea_1 - Ea_{-1}}{RT}}\right)^{1/2} = 2A_2\left(\frac{A_1}{A_{-1}}\right)^{1/2} e^{-\frac{0.5(2Ea_2 + Ea_1 - Ea_{-1})}{RT}}$$

$$k = 2A_2\left(\frac{A_1}{A_{-1}}\right)^{1/2} e^{-\frac{0.5(192 \text{ kJ mol}^{-1} - 74 \text{ kJ mol}^{-1})}{RT}}$$

$$k = 2A_2\left(\frac{A_1}{A_{-1}}\right)^{1/2} e^{-\frac{59 \text{ kJ mol}^{-1}}{RT}}$$

Therefore, $E_a = 59$ kJ mol^{-1}.

(c) $$\frac{\text{rate}_{400 \text{ K}}}{\text{rate}_{298 \text{ K}}} = \frac{e^{-\frac{5.9 \times 10^4 \text{ J mol}^{-1}}{(8.314 \text{ J mol}^{-1}\text{ K}^{-1})(400.\text{ K})}}}{e^{-\frac{5.9 \times 10^4 \text{ J mol}^{-1}}{(8.314 \text{ J mol}^{-1}\text{ K}^{-1})(298 \text{ K})}}} = 434$$

P36.12   Consider the gas-phase isomerization of cyclopropane. Are the following data of the observed rate constant as a function of pressure consistent with the Lindemann mechanism?

P (Torr)	k (10^4s^{-1})	P (Torr)	k (10^4s^{-1})
84.1	2.98	1.37	1.30
34.0	2.82	0.569	0.857
11.0	2.23	0.170	0.486
6.07	2.00	0.120	0.392
2.89	1.54	0.067	0.303

If the data obeyed the Lindemann mechanism, then a plot of the data would fit the equation

$$\frac{1}{k_{uni}} = \frac{k_{-1}}{k_1 k_2} + \frac{1}{k_1}\frac{1}{[M]}$$

$$\frac{1}{k_{uni}} = \frac{k_{-1}}{k_1 k_2} + \frac{1}{k_1}\frac{RT}{P}$$

Therefore, the Lindemann mechanism predicts that a plot of $\dfrac{1}{k_{uni}}$ versus $\dfrac{1}{P}$ should be a straight line. This plot is

provided below, and demonstrates that the plot shows significant curvature so that the Lindemann mechanism does not provide an adequate description of the reaction mechanism.

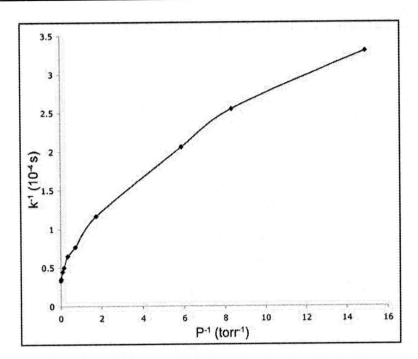

P36.17 The enzyme catalase catalyzes the decomposition of hydrogen peroxide. The following data are obtained regarding the rate of reaction as a function of substrate concentration:

$[H_2O_2]_0$ (M)	0.001	0.002	0.005
Initial Rate (M s^{-1})	1.38×10^{-3}	2.67×10^{-3}	6.00×10^{-3}

The concentration of catalase is 3.5×10^{-9} M. Use these data to determine $rate_{max}$, K_m, and the turnover number for this enzyme.

Analyzing the data using the Lineweaver–Burk plot:

$$\frac{1}{rate_0} = \frac{1}{rate_{max}} + \frac{K_m}{rate_{max}}\frac{1}{[S]_0}$$

The corresponding plot of $rate_0^{-1}$ versus $[H_2O_2]^{-1}$ is as follows:

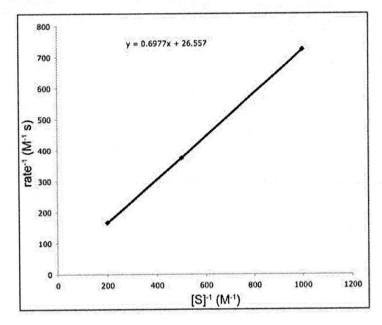

Best fit to the data by a straight line yields the following equation:

$$\frac{1}{rate_0} = (0.698 \text{ s}) \frac{1}{[S]_0} + 26.6 \text{ M}^{-1} \text{ s}$$

The $rate_{max}$ is related to the intercept

$$rate_{max} = \frac{1}{y - int} = \frac{1}{26.6 \text{ M}^{-1} \text{ s}}$$

$$rate_{max} = 3.75 \times 10^{-2} \text{ M s}^{-1}$$

With $rate_{max}$, the Michaelis constant is determined using the slope:

$$K_m = (slope) \times (rate_{max}) = (0.698 \text{ s}) \times (3.75 \times 10^{-2} \text{ M s}^{-1})$$

$$K_m = 2.63 \times 10^{-2} \text{ M}$$

Finally, the turnover number is determined using $rate_{max}$ and $[E]_0$ as follows:

$$rate_{max} = k_2[E]_0$$

$$\frac{rate_{max}}{[E]_0} = k_2$$

$$\frac{3.75 \times 10^{-2} \text{ M s}^{-1}}{3.5 \times 10^{-9} \text{ M}} = k_2$$

$$1.08 \times 10^7 \text{ s}^{-1} = k_2$$

P36.19 Protein tyrosine phosphatases (PTPases) are a general class of enzymes that are involved in a variety of disease processes, including diabetes and obesity. In a study by Z.-Y. Zhang and coworkers [*J. Medicinal Chemistry* 43 (2000): 146], computational techniques were used to identify potential competitive inhibitors of a specific PTPase known as PTP1B. The reaction rate was determined in the presence and absence of inhibitor, I, and revealed the following initial reaction rates as a function of substrate concentration:

[S] (μM)	$Rate_0$ (μM s^{-1}), [I] = 0	$Rate_0$ (μM s^{-1}), [I] = 200 μM
0.299	0.071	0.018
0.500	0.100	0.030
0.820	0.143	0.042
1.22	0.250	0.070
1.75	0.286	0.105
2.85	0.333	0.159
5.00	0.400	0.200
5.88	0.500	0.250

a. Determine K_m and $rate_{max}$ for PTP1B.

b. Demonstrate that the inhibition is competitive, and determine K_i.

(a)   Performing a Lineweaver–Burk plot for the uninhibited reaction yields:

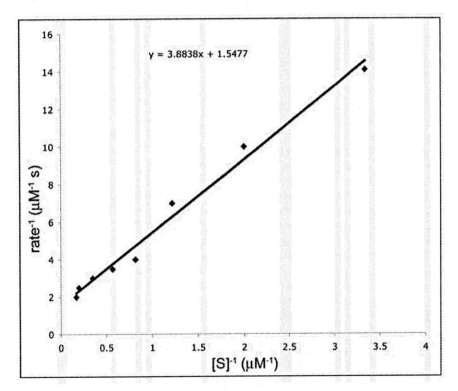

y = 3.8838x + 1.5477

Best fit to the data by a straight line yields the following equation:

$$\frac{1}{rate_0} = 3.88\,\text{s}\,\frac{1}{[S]_0} + 1.55 \times 10^6\,\text{M}^{-1}\,\text{s}$$

The maximum rate is equal to the inverse of the y intercept:

$$rate_{max} = \frac{1}{y\text{-}int} = \frac{1}{1.55 \times 10^6\,\text{M}^{-1}\,\text{s}}$$

$$rate_{max} = 6.5 \times 10^{-7}\,\text{M}\,\text{s}^{-1}$$

With $rate_{max}$, the Michaelis constant is determined from the slope:

$$K_m = (slope) \times (rate_{max}) = (3.88\,\text{s}) \times (6.45 \times 10^{-7}\,\text{M}\,\text{s}^{-1})$$

$$K_m = 2.5 \times 10^{-6}\,\text{M}$$

(b)   With the inhibitor present, the modified Lineweaver–Burk equation is used where:

$$\frac{1}{rate_0} = \frac{1}{rate_{max}} + \frac{K_m^*}{rate_{max}}\frac{1}{[S]_0}$$

Again, a plot of the inverse of the initial rate with the inverse of substrate concentration should yield a straight line. The plot is as follows:

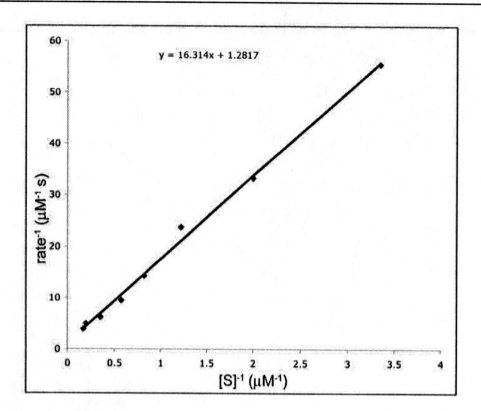

Best fit to the data by a straight line yields the following equation:

$$\frac{1}{rate_0} = 1.63\,s\,\frac{1}{[S]_0} + 1.28 \times 10^6\,M^{-1}\,s$$

The maximum rate is given by the inverse of the *y* intercept:

$$rate_{max} = \frac{1}{y\text{-}int} = \frac{1}{1.28 \times 10^6\,M^{-1}s}$$

$$rate_{max} = 7.9 \times 10^{-7}\,M\,s^{-1}$$

With $rate_{max}$ and the slope of the line the apparent Michaelis constant is:

$$K_m^* = (slope) \times (rate_{max}) = (16.31\,s) \times (7.94 \times 10^{-7}\,M\,s^{-1})$$

$$= 1.3 \times 10^{-5}\,M$$

With K_m, K_m^* and [I], K_I is determined as follows:

$$K_I = \frac{[I]}{\dfrac{K_m^*}{K_m} - 1}$$

$$= \frac{200 \times 10^{-6}\,M}{\left(\dfrac{1.3 \times 10^{-5}\,M}{2.5 \times 10^{-6}\,M}\right) - 1}$$

$$= 4.8 \times 10^{-5}\,M$$

P36.21 The enzyme glycogen synthase kinase (GSK-3β) plays a central role in Alzheimer's disease. The onset of Alzheimer's disease is accompanied by the production of highly phosphorylated forms of a protein referred to as "τ." GSK-3β contributes to the hyperphosphorylation of τ such that inhibiting the activity of this enzyme represents a pathway for the development of an Alzheimer's drug. A compound known as Ro 31-8220 is a competitive inhibitor of GSK-3β. The following data were obtained for the rate of GSK-3β activity in the presence and absence of Ro 31-8220 [A. Martinez *et al.*, *J. Medicinal Chemistry* 45 (2002): 1292]:

$[S]$ (μM)	$Rate_0$ $(\mu M\ s^{-1})$, $[I] = 0$	$Rate_0$ $(\mu M\ s^{-1})$ $[I] = 200\ \mu M$
66.7	4.17×10^{-8}	3.33×10^{-8}
40.0	3.97×10^{-8}	2.98×10^{-8}
20.0	3.62×10^{-8}	2.38×10^{-8}
13.3	3.27×10^{-8}	1.81×10^{-8}
10.0	2.98×10^{-8}	1.39×10^{-8}
6.67	2.31×10^{-8}	1.04×10^{-8}

Determine K_m and $rate_{max}$ for GSK-3β and, using the data with the inhibitor, determine K_m^* and K_I. Analyzing the data without inhibitor using a Lineweaver–Burk plot yields:

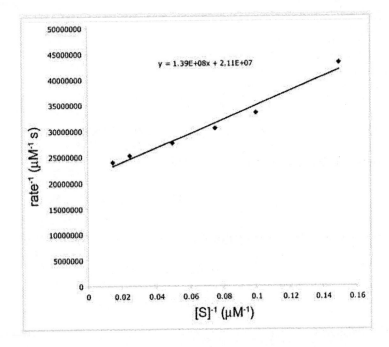

The best-fit straight line to the data yields the following equation:

$$\frac{1}{rate_0} = 1.39 \times 10^8\ s\ \frac{1}{[S]_0} + 2.11 \times 10^6\ \mu M^{-1}\ s$$

The maximum rate is equal to the inverse of the y intercept:

$$rate_{max} = \frac{1}{y\text{-}int} = \frac{1}{2.11 \times 10^7\ \mu M^{-1}\ s}$$

$$rate_{max} = 4.74 \times 10^{-8}\ \mu M\ s^{-1}$$

With the maximum rate and slope of the best-fit line, the Michaelis constant is obtained as follows:

$$K_m = (slope) \times (rate_{max}) = (1.39 \times 10^8\ s) \times 4.74 \times 10^{-8}\ \mu M\ s^{-1}$$

$$K_m = 6.49\ \mu M$$

Using the inhibited data, the Lineweaver–Burk plot is:

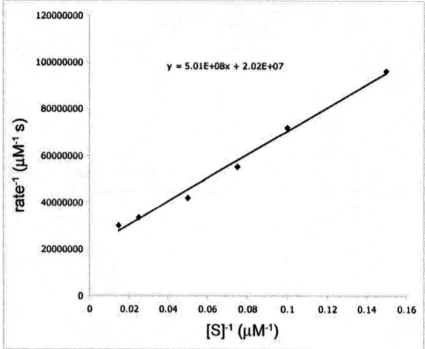

Best fit by a straight line to the data yields the following equation:

$$\frac{1}{rate_0} = 5.01 \times 10^8 \text{ s} \frac{1}{[S]_0} + 2.01 \times 10^7 \ \mu M^{-1} \text{ s}$$

The maximum rate with inhibitor is equal to the inverse of the y intercept:

$$rate_{max} = \frac{1}{2.01 \times 10^7 \ \mu M^{-1} \text{ s}}$$

$$= 4.98 \times 10^{-8} \ \mu M^{-1} \text{ s}$$

The apparent Michaelis constant is given by:

$$K_m^* = (slope) \times (rate_{max})$$

$$= (5.01 \times 10^8 \text{ s}) \times (4.98 \times 10^{-8} \ \mu M^{-1} \text{ s})$$

$$K_m^* = 24.9 \ \mu M$$

Finally, the K_I value is given by:

$$K_I = \frac{[I]}{\dfrac{K_m^*}{K_m} - 1} = \frac{200 \ \mu M}{\left(\dfrac{24.9 \ \mu M}{6.49 \ \mu M}\right) - 1}$$

$$= 70.4 \ \mu M$$

P36.23   Reciprocal plots provide a relatively straightforward way to determine if an enzyme demonstrates Michaelis–Menten kinetics and to determine the corresponding kinetic parameters. However, the slope determined from these plots can require significant extrapolation to regions corresponding to low substrate concentrations. An alternative to the reciprocal plot is the Eadie–Hofstee plot, where the reaction rate is plotted versus the rate divided by the substrate concentration and the data are fit to a straight line.

a. Beginning with the general expression for the reaction rate given by the Michaelis–Menten mechanism:

$$R_0 = \frac{R_{max}[S]_0}{[S]_0 + K_m}$$

Rearrange this equation to construct the following expression, which is the basis for the Eadie–Hofstee plot:

$$R_0 = R_{max} - K_m\left(\frac{R_0}{[S]_0}\right)$$

b. Using an Eadie–Hofstee plot, determine R_{max} and K_m for hydrolysis of sugar by the enzyme invertase using the following data:

[Sucrose]$_0$ (M)	Rate (M s^{-1})
0.029	0.182
0.059	0.266
0.088	0.310
0.117	0.330
0.175	0.362
0.234	0.361

(a) $$R_0 = \frac{R_{max}[S]_0}{[S]_0 + K_m}$$

$$R_0([S]_0 + K_m) = R_{max}[S]_0$$
$$R_0[S]_0 = R_{max}[S]_0 - R_0 K_m$$

$$R_0 = R_{max} - K_m\left(\frac{R_0}{[S]_0}\right)$$

(b) Using the data provided, a plot of $rate_0$ versus $rate_0/[S]_0$ is constructed:

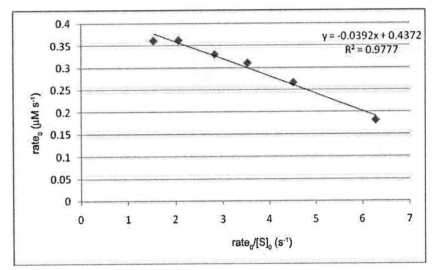

Best fit to a straight line yields a y intercept of 0.437 M s^{-1}, which is equal to $Rate_{max}$. The slope of the line is equal to -0.0392 M, which is equal to $-K_m$ such that $K_m = 0.0392 \text{ M}$.

P36.24   Determine the predicted rate law expression for the following radical-chain reaction:

$$A_2 \xrightarrow{k_1} 2A\cdot$$

$$A\cdot \xrightarrow{k_2} B\cdot + C$$

$$A\cdot + B\cdot \xrightarrow{k_3} P$$

$$A\cdot + P \xrightarrow{k_4} B\cdot$$

The differential rate of P formation is

$$\frac{d[P]}{dt} = k_3[A\cdot][B\cdot] - k_4[A\cdot][P]$$

The rate expressions for $A\cdot$ and $B\cdot$ are

$$\frac{d[A\cdot]}{dt} = 2k_1[A_2] - k_2[A\cdot] - k_3[A\cdot][B\cdot] - k_4[A\cdot][P]$$

$$\frac{d[B\cdot]}{dt} = k_2[A\cdot] - k_3[A\cdot][B\cdot] + k_4[A\cdot][P]$$

Applying the steady-state approximation for [B],

$$k_3[A\cdot][B\cdot] - k_4[A\cdot][P] = k_2[A\cdot]$$

$$[B\cdot] = \frac{k_2 + k_4[P]}{k_3}$$

Substituting this result into the differential rate expression for P yields:

$$\frac{d[P]}{dt} = k_3[A\cdot]\left(\frac{k_2 + k_4[P]}{k_3}\right) - k_4[A\cdot][P]$$

$$= k_2[A\cdot]$$

Now, adding the steady-state expressions for [Ag] and [Bg] yields:

$$0 = 2k_2[A_2] - 2k_3[A\cdot][B\cdot]$$

$$[A\cdot] = \frac{k_2[A_2]}{k_3[B\cdot]} = \frac{k_2[A_2]}{k_2 + k_4[P]}$$

Substituting this expression into the differential rate expression for P yields the final result:

$$\frac{d[P]}{dt} = \frac{k_2 k_1[A_2]}{k_2 + k_4[P]}$$

P36.28   The adsorption of ethyl chloride on a sample of charcoal at 0 °C measured at several different pressures is as follows:

$P_{C_2H_5Cl}$ (Torr)	V_{ads} (mL)
20.	3.0
50.	3.8
100.	4.3
200.	4.7
300.	4.8

Using the Langmuir isotherm, determine the fractional coverage at each pressure and V_M.

The Langmuir equation can be written as

$$\frac{P}{V} = \frac{P}{V_m} + \frac{1}{KV_m}$$

And the fractional coverage, θ, is simply the ratio of adsorbed volume to the volume of maximum adsorption (V_m):

$$\theta = \frac{V}{V_m}$$

Therefore, V_m is required to determine the fractional coverage versus pressure. The plot of $\frac{P}{V}$ vs. P should yield a straight line with slope equal to the inverse of V_m. This plot is as follows:

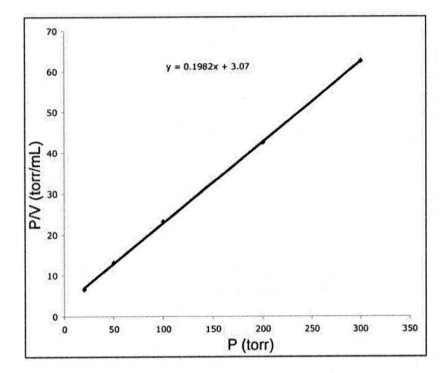

The equation for the best-fit line is:

$$\frac{P}{V} = 0.198 \text{ mL}^{-1}(P) + 3.07 \text{ torr mL}^{-1}$$

Thus, the V_m value is

$$V_m = \frac{1}{slope}$$
$$V_m = 5.04 \text{ mL}$$

With V_m, θ can be determined, resulting in the following:

P (atm)	θ
20.	0.595
50.	0.754
100.	0.853
200.	0.932
300.	0.952

P36.31 Many surface reactions require the adsorption of two or more different gases. For the case of two gases, assuming that the adsorption of a gas simply limits the number of surface sites available for adsorption, derive expressions for the fractional coverage of each gas.

If we assume that the only effect of the gases is to limit the number of sites available for adsorption, then the Langmuir model yields the two following relationships:

$$k_{d1}\theta_1 = k_{a1}P_1(1 - \theta_1 - \theta_2)$$
$$k_{d2}\theta_2 = k_{a2}P_2(1 - \theta_1 - \theta_2)$$

where k_a, k_d, and θ are the rate of adsorption, desorption, and fractional coverage, respectively. In addition, the subscript denotes either gas 1 or 2 in the two gas mixture. The two equations can be solved to yield θ_1 and θ_2 as follows:

$$\theta_1 = \frac{K_1P_1}{1 + K_1P_1 + K_2P_2}$$

$$\theta_2 = \frac{K_2P_2}{1 + K_1P_1 + K_2P_2}$$

P36.33 Another type of autocatalytic reaction is referred to as cubic autocatalytic and corresponds to the following elementary process:

$$A + 2B \rightarrow 3B$$

Write the rate law expression for this elementary process. What would you expect the corresponding differential rate expression in terms of ξ (the coefficient of reaction advancement) to be?

$$R = k[A][B]^2$$
$$\frac{d\xi}{dt} = k\xi^2(1 - \xi)$$

P36.38 If $\tau_f = 1 \times 10^{-10}$ s and $k_{ic} = 5 \times 10^8$ s^{-1}, what is Φ_f? Assume that the rate constants for intersystem crossing and quenching are sufficiently small that these processes can be neglected.

Φ_f is related to k_f and τ_f by the expression

$$\Phi_f = k_f\tau_f$$

and k_f is related to τ_f by

$$\frac{1}{\tau_f} = k_f + k_{ic} \quad \text{(assuming } k_{isc}^s, k_q \text{ are small)}$$

$$\tau_f = 1 \times 10^{-10} \text{ s} \quad \text{and} \quad k_{ic} = 5 \times 10^8 \text{ s}^{-1}$$

Thus

$$\frac{1}{1 \times 10^{-10} \text{ s}} = k_f + 5 \times 10^8 \text{ s}^{-1}$$

and

$$k_f = 1 \times 10^{10} \text{ s}^{-1} - 5 \times 10^8 \text{ s}^{-1}$$
$$k_f = 9.5 \times 10^9 \text{ s}^{-1}$$

And finally,

$$\Phi_f = k_f\tau_f$$
$$= 9.5 \times 10^9 \text{ s}^{-1} \cdot 1 \times 10^{-10}$$
$$\Phi_f = 0.95$$

P36.43 A central issue in the design of aircraft is improving the lift of aircraft wings. To assist in the design of more efficient wings, wind-tunnel tests are performed in which the pressures at various parts of the wing are measured generally using only a few localized pressure sensors. Recently, pressure-sensitive paints have been developed to provide a more detailed view of wing pressure. In these paints, a luminescent molecule is dispersed into an oxygen-permeable paint and the aircraft wing is painted. The wing is placed into an airfoil, and luminescence from the paint is measured. The variation in O_2 pressure is measured by monitoring the luminescence intensity, with lower intensity demonstrating areas of higher O_2 pressure due to quenching.

a. The use of platinum octaethylporphyrin (PtOEP) as an oxygen sensor in pressure-sensitive paints was described by Gouterman and coworkers [*Review of Scientific Instruments* 61 (1990): 3340]. In this work, the following relationship between luminescence intensity and pressure was derived: $\dfrac{I_0}{I} = A + B\left(\dfrac{P}{P_0}\right)$, where I_0 is the fluorescence intensity at ambient pressure P_0, and I is the fluorescence intensity at an arbitrary pressure P. Determine coefficients A and B in the preceding expression using the Stern–Volmer equation:

$k_{total} = \dfrac{1}{\tau_f} = k_f + k_q[Q]$. In this equation τ_l is the luminescence lifetime, k_f is the luminescent rate constant, and k_q is the quenching rate constant. In addition, the luminescent intensity ratio is equal to the ratio of luminescence quantum yields at ambient pressure, Φ_0, and an arbitrary pressure, Φ: $\dfrac{\Phi_0}{\Phi} = \dfrac{I_0}{I}$.

b. Using the following calibration data of the intensity ratio versus pressure observed for PtOEP, determine A and B:

I_0/I	P/P_0	I_0/I	P/P_0
1.0	1.0	0.65	0.46
0.9	0.86	0.61	0.40
0.87	0.80	0.55	0.34
0.83	0.75	0.50	0.28
0.77	0.65	0.46	0.20
0.70	0.53	0.35	0.10

c. At an ambient pressure of 1 atm, $I_0 = 50,000$ (arbitrary units) and 40,000 at the front and back of the wing. The wind tunnel is turned on to a speed of Mach 0.36 and the measured luminescence intensity is 65,000 and 45,000 at the respective locations. What is the pressure differential between the front and back of the wing?

(a) Starting with the version of the Stern–Volmer equation provided in the problem:

$$k_{tot} = k_l + k_q[Q]$$

The luminescence quantum yield can be expressed in terms of k_l and k_{total} as:

$$\Phi = \frac{k_l}{k_{total}}$$

Therefore:

$$\frac{\Phi_0}{\Phi} = \frac{k_{total}}{k_{total_0}}$$

$$= \frac{k_l + k_q P}{k_l + k_q P_0}$$

$$= \frac{k_l}{k_l + k_q P_0} + \frac{k_q P_0}{k_l + k_q P_0}\left(\frac{P}{P_0}\right)$$

$$= A + B\left(\frac{P}{P_0}\right)$$

(b)   The plot of (I_0/I) versus (P/P_0) is as follows:

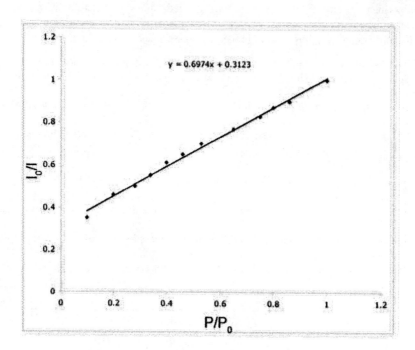

Best fit by a straight line to the data yields the following equation:

$$\frac{I_0}{I} = 0.697\left(\frac{P}{P_0}\right) + 0.312$$

Thus, $A = 0.312$ and $B = 0.697$.

(c) $$\left[\left(\frac{I_0}{I}\right)_{front} - \left(\frac{I_0}{I}\right)_{back} =\right] = 0.697\left(\frac{P_{front} - P_{back}}{P_0}\right)$$

$$\left(\frac{50,000}{65,000} - \frac{40,000}{45,000}\right) = 0.697\left(\frac{P_{front} - P_{back}}{1\ atm}\right)$$

$$-0.120 = 0.697\left(\frac{P_{front} - P_{back}}{1\ atm}\right)$$

$$-0.172\ atm = P_{front} - P_{back}$$

P36.45   The pyrene/coumarin FRET pair $(r_0 = 39\ \text{Å})$ is used to study the fluctuations in enzyme structure during the course of a reaction. Computational studies suggest that the pair will be separated by $35\ \text{Å}$ in one conformation, and $46\ \text{Å}$ in a second configuration. What is the expected difference in FRET efficiency between these two conformational states?

$$Eff = \frac{r_0^{\ 6}}{\left(r_0^{\ 6} + r^6\right)}$$

Using the above expression to calculate the difference in FRET efficiency:

$$Eff(35\text{Å}) - Eff(46\ \text{Å}) = \frac{(39\ \text{Å})^6}{((39\ \text{Å})^6 + (35\ \text{Å})^6)} - \frac{(39\ \text{Å})^6}{((39\ \text{Å})^6 + (46\ \text{Å})^6)}$$

$$= 0.386 \approx 0.39$$

P36.49 In Marcus theory for electron transfer, the reorganization energy is partitioned into solvent and solute contributions. Modeling the solvent as a dielectric continuum, the solvent reorganization energy is given by:

$$\lambda_{sol} = \frac{(\Delta e)^2}{4\pi\varepsilon_0}\left(\frac{1}{d_1} + \frac{1}{d_2} - \frac{1}{r}\right)\left(\frac{1}{n^2} - \frac{1}{\varepsilon}\right)$$

where Δe is the amount of charge transferred, d_1 and d_2 are the ionic diameters of ionic products, r is the separation distance of the reactants, n^2 is the square of the index of refraction of the surrounding medium, and ε is the dielectric constant of the medium. In addition, $(4\pi\varepsilon_0)^{-1} = 8.99 \times 10^9$ J m C^{-2}.

   a. For an electron transfer in water ($n = 1.33$ and $\varepsilon = 80$.) for which the ionic diameters of both species are 6 Å and the separation distance is 15 Å, what is the expected solvent reorganization energy?

   b. Redo the above calculation for the same reaction occurring in a protein. The dielectric constant of a protein is dependent on sequence, structure, and the amount of included water; however, a dielectric constant of 4 is generally assumed consistent with a hydrophobic environment.

   (a) For the transfer of one electron ($\Delta e = 1$) and $\varepsilon = 80$. we obtain:

$$\lambda_{sol} = (1.602 \times 10^{-19}\ C)^2 \times (8.99 \times 10^9\ J\ m\ C^{-2})$$

$$\times \left(\frac{1}{(6 \times 10^{-10}\ m)} + \frac{1}{(6 \times 10^{-10}\ m)} - \frac{1}{(15 \times 10^{-10}\ m)}\right) \times \left(\frac{1}{(1.33)^2} - \frac{1}{(80)}\right)$$

$$= 3.4 \times 10^{-19}\ J$$

   (b) With the transfer of one electron ($\Delta e = 1$) and $\varepsilon = 4$ we obtain:

$$\lambda_{sol} = (1.602 \times 10^{-19}\ C)^2 \times (8.99 \times 10^9\ J\ m\ C^{-2})$$

$$\times \left(\frac{1}{(6 \times 10^{-10}\ m)} + \frac{1}{(6 \times 10^{-10}\ m)} - \frac{1}{(15 \times 10^{-10}\ m)}\right) \times \left(\frac{1}{(1.5)^2} - \frac{1}{(4)}\right)$$

$$= 1.2 \times 10^{-19}\ J$$

NOTES

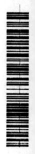